T0342509

Onshore and Offshore Wind Energy

Onshore and Offshore Wind Energy

Evolution, Grid Integration, and Impact

Second Edition

Vasilis Fthenakis
Columbia University and Brookhaven National Laboratory
NY, USA

Subhamoy Bhattacharya
University of Surrey
England, UK

Paul A. Lynn
Formerly Imperial College
London, UK

Registered Offices
John Wiley & Sons, Inc., 111 River Street, Hoboken, NJ 07030, USA
John Wiley & Sons Ltd., The Atrium, Southern Gate, Chichester, West Sussex, PO19 8SQ, UK

For details of our global editorial offices, customer services, and more information about Wiley products visit us at www.wiley.com.

Wiley also publishes its books in a variety of electronic formats and by print-on-demand. Some content that appears in standard print versions of this book may not be available in other formats.

Library of Congress Cataloging-in-Publication Data:

Names: Fthenakis, Vasilis M., 1951- author. | Bhattacharya, Subhamoy, author. | Lynn, Paul A., author.
Title: Onshore and offshore wind energy : evolution, grid integration, and impact / Vasilis Fthenakis Columbia University NY, USA, Subhamoy Bhattacharya, University of Surrey, England, UK, Paul A. Lynn, Butcombe, North Somerset, UK.
Description: Second edition. | Hoboken, NJ, USA : Wiley, 2025. | Includes index.
Identifiers: LCCN 2024025371 (print) | LCCN 2024025372 (ebook) | ISBN 9781119854470 (hardback) | ISBN 9781119854487 (adobe pdf) | ISBN 9781119854494 (epub)
Subjects: LCSH: Wind power. | Offshore electric power plants. | Wind energy conversion systems.
Classification: LCC TJ820 .L95 2025 (print) | LCC TJ820 (ebook) | DDC 621.31/2136–dc23/eng/20240802
LC record available at https://lccn.loc.gov/2024025371
LC ebook record available at https://lccn.loc.gov/2024025372

Cover Design: Wiley
Cover Images: © Sharomka/Shutterstock, © zhengzaishuru/Shutterstock

Set in 9/11pt TimesNewRomanMTStd by Straive, Chennai, India
Printed and bound by CPI Group (UK) Ltd, Croydon, CR0 4YY

C9781119854470_031024

Contents

About the Authors

Vasilis Fthenakis

Vasilis Fthenakis obtained his Diploma in Chemistry from the University of Athens, Greece; his MS in Chemical Engineering from Columbia University; and his PhD in Fluid Dynamics and Atmospheric Science from New York University. He is the founder and director of the Center for Life Cycle Analysis (CLCA) and an adjunct Professor at Columbia University, New York, USA. He is also a distinguished scientist emeritus at Brookhaven National Laboratory (BNL), where he conducted energy-environmental research for 36 years. Dr. Fthenakis is the coauthor and editor of six books and about 400 scientific publications on topics at the interface of energy and the environment.

In 2008, he co-authored "Grand Plan for Solar Energy," a leading study published in Scientific American in 11 languages, showing the technical, economic, and geographical feasibility of solar with other renewables to satisfy 69% of the electricity needs of the USA by 2050.

Since 2002, to safeguard the environmental friendliness of photovoltaics, he has been defining and promoting a proactive, long-term environmental strategy, including recycling of photovoltaics at the end of their useful lives. His work produced a patented technology for CdTe recycling and models for optimizing the collection of end-of-life photovoltaics. He started the International Energy Agency (IEA) PV-EH&S Task (Task 12) on LCA and Recycling and led it as the U.S. Operating Agent from 2007 to 2012.

He is a Fellow of the American Institute of Chemical Engineers (AIChE) "in recognition and appreciation of superior attainments, valuable contributions, and service to Chemical Engineering," a Fellow of the Institute of Electrical and Electronic Engineers (IEEE) for "outstanding contributions to photovoltaics technology," and recipient of a number of awards including the 2006 US DOE Certificate of Appreciation "for superior technical, management and communications skills exhibited in photovoltaic environmental research," the 2018 IEEE William Cherry Award for "pioneering research at the interface of energy and the environment that catalyzed photovoltaic technology advancement and deployment world-wide," and the 2022 Karl Boer Medal of Merit for "distinguished contributions to quest for sustainable energy." Currently, he is leading nationally and internationally research on solar desalination, solar hydrogen, energy systems modeling, life-cycle analysis, and photovoltaic recycling.

Subhamoy Bhattacharya

Subhamoy Bhattacharya, a Chartered Engineer (CEng) and Fellow of ICE (Institution of Civil Engineers, London), holds the Chaired Professor in Geomechanics at the University of Surrey since 2012, where he leads the Geotechnical Research group and directs the SAGE (Surrey Advanced Geotechnical Engineering) laboratory. He is also the Co-founder and Chief Scientific Officer of Renew Risk™ (based at Lloyd's of London), a university spin-off company specializing in risk analytics for renewable energy/energy transition, which build the world's first commercially available catastrophe models for offshore wind farms.

He is an expert in geotechnical and earthquake engineering, where his main interests are offshore foundations and earthquake engineering. He commenced his undergraduate education at the Indian Institute of Technology (IIT-Kharagpur) in geology and geophysics, and after one year of completion switched to civil engineering degree program at Indian Institute of Engineering Science and Technology (IIEST, Shibpur) and graduated with a first-class degree in civil engineering. Following his undergraduate education, he worked at Consulting Engineering Services (Jacobs) designing buildings and bridges before moving to Cambridge for doctoral studies.

He earned his PhD from the University of Cambridge (as a Cambridge Nehru Commonwealth Scholar) in 2003, investigating the failure of pile-supported structures (bridges and buildings) during seismic liquefaction and proposing a new theory of pile failure during earthquakes. Following his PhD, he worked at Offshore Geotechnical Consultancy Fugro designing foundations for offshore oil and gas structures, including the anchors for FPSO's (Floating Production Storage and Offloading). In 2005, he transited to academia, first at the University of Oxford, where he was a Departmental Lecturer in Engineering Science and JRF (Junior Research Fellow) at Somerville College, and then at the University of Bristol, where he was a Senior Lecturer in Soil Dynamics before taking up the Chair at Surrey in 2012. He regularly lectures in the UK and abroad and acts as a consultant in civil, earthquake, geotechnical, and offshore engineering, including legal and arbitration cases.

Paul A. Lynn

Paul A. Lynn obtained his BSc (Eng) and PhD degrees from Imperial College London, UK. After five years in the electrical/electronics industry, mainly as a radar engineer with the Marconi Company, he lectured at Imperial College and the University of Bristol, latterly as a Reader in Electronic Engineering with a special interest in digital signal processing. A strong and growing interest in renewable energy then led him to accept an invitation to become the founding managing editor of Wiley's journal "Progress in Photovoltaics," a role he held for 14 years. He also wrote a trilogy of Wiley books on solar, wind, and wave/tidal energy. In retirement, he designed and built a 22-ft solar-powered catamaran that carried him and his wife on the first-ever solar voyage along the entire River Thames from Gloucestershire to London, and in recent years he has continued as an author, with numerous books, papers, and articles.

Foreword

Wind energy is one of only two renewable energy sources that can potentially power the world for all purposes multiple times over. The other is solar energy. The growth of wind has been remarkable, as has its decline in cost. Transforming the world's energy infrastructure to address global warming, air pollution, and energy security problems requires the continued growth of this clean, renewable energy resource. To that end, this book is *an essential and outstanding* tool for professionals, technicians, policymakers, students, and interested lay readers who are in the wind energy industry, work on wind-related topics, need to make decisions regarding wind energy, or want to learn about it. This book contains not only case studies but also appendices, questions and answers, and problems. An instructor's manual with solutions to the problems is also available.

The book methodically marches through the important aspects of wind energy. First up, in Chapter 1, is a discussion of the difference between windmills and wind turbines and of wind turbine development and installation to date. Chapter 2 details the power in the wind, wind speed probability distributions, and wind resources. Chapter 3 covers everything you want to know about wind turbine design and efficiency. Chapter 4 focuses on offshore wind systems, including their foundations and transmission systems. Chapter 5 examines the engineering of an offshore wind farm, including the spacing needed between wind turbines. Chapter 6 moves on to discuss wind turbine operation and maintenance. Integrating wind into the grid and dealing with wind's variability are the subjects of Chapter 7. Lastly, Chapter 8 is on the cost, material needs, and impacts on the environment and wildlife of wind turbines.

Through this book, readers will gain the essential knowledge needed to help accelerate the growth of wind energy but also to bust myths about it often amplified by its detractors. One such myth is that wind turbines cause substantial bird deaths. Another is that large-scale wind development will cause blackouts due to wind's variability. A third is that wind farms take up more land than coal generators and mines.

Readers will also have access to important data and equations necessary to analyze the efficiency of wind turbines, to integrate wind farms into the electrical grid, and to ensure that wind contributes to a reliable, clean, renewable energy future. Those who finish this book will have gained a wealth of knowledge, arming them with the information needed to speak accurately and authoritatively about all aspects of wind energy. It is time to start your learning experience with this *fantastic book*. There is no time to waste!

By Mark Z. Jacobson
Prof. of Civil and Environmental Engineering, Stanford University

Preface

The wind energy industry, offshore as well as onshore, is growing at a remarkable pace. To many of us, it symbolizes a desire to harness one of nature's most widespread sources of renewable energy, exploiting a "fuel" that is eternal and carbon-free. This book presents a concise account of large turbines and utility-scale wind energy aimed at a wide readership including professionals, policymakers, and employees in the energy sector needing an appreciation of the basic principles underlying wind energy or a quick update. Its style and level will also appeal to undergraduate and postgraduate students, as well as the large and growing number of thoughtful people who are interested in onshore and offshore wind farms and the contribution they are making to electricity generation in the 21st century. The first edition of this book, which was published in 2012, was one of the first books to emphasize today's exciting developments in offshore wind, and it was designed as an appetizer rather than a formal textbook, with copious color photographs to illustrate the industry's progress as it moves, apparently inexorably, toward 1000 GW of global installed capacity.

The current second edition describes fundamental technical aspects of offshore wind with new chapters on "Fundamentals of Offshore Systems" and on "Offshore Wind Engineering". In addition, a new chapter on wind turbine "Operation and Maintenance" and greatly expanded and updated chapters on "Grid Integration" and "Wind Energy Growth and Sustainability" holistically cover the techno-economic and environmental-social aspects of wind energy for the high growth envisioned in climate change mitigation scenarios.

The effective harnessing of wind power involves many aspects of engineering science, from rotor aerodynamics to electrical generators, control systems, foundation structures, resilience, reliability, and grid networks. We have tried to introduce this huge field in a way that explains the essential theoretical background and indicates the main engineering challenges. The overall tone is deliberately accessible rather than overly technical, but several key topics, especially those pertaining to the emerging offshore wind turbine technologies, are covered in sufficient technical detail for students of wind energy.

We hope the new edition will serve as an essential primary resource for entrants to the wind energy industry needing an up-to-date appreciation of the subject. It also offers a unique treatise on the sustainability of emerging transformative technologies, making it valuable to system analysts and energy policy strategists. Last but not least, we have included end-of-chapter questions and problems to support instructors and the ever-increasing number of college and university students taking courses in renewable energy technologies.

Vasilis Fthenakis
Columbia University, New York, NY, USA
Subhamoy Bhattacharya
University of Surrey, Surrey
England, UK
Paul A. Lynn
Imperial College (Retired), London, UK
January 2024

Acknowledgment to the Second Edition

Following Paul Lynn's lead in including color photographs in the first edition, illustrating wind turbine beauty and extraordinary promise, we have added plenty more in this new edition.

We also added a lot of material on off-shore wind and problems and solutions in each chapter of this new edition.

Many of the materials presented in this book are based on the research work of our past and current students, which includes undergraduate, master's, and PhD students. A special mention goes to Dr. Samet Ozturk, Zuoran Zhang, Ulvi Rahmanli, Sayan Bhattacharyya, Dr. Aleem Mohammad, and Dr. Georgios Nikitas.

This second edition is dedicated to our wives, Christina and Paromita, for their love and encouragement. *Onshore and Offshore Wind Energy* defines a pragmatic renewable energy solution to the risks that climate change presents to our children and the generations that follow, and we thank our children, Antonia, Menelaos, and Ishan, for being our constant source of motivation for a book that is more about their generation than it is about ours.

This is a new area, and the technology development is very fast. Offshore turbines may not only be sited in deeper waters and further offshore but also in seismic and typhoon zones. Much of the information presented is expected to be outdated in the next few years, and the book could need a new edition. There can also be errors and omissions in the book, and we would like to know them. Please email us at Subhamoy.Bhattacharya@gmail.com and vmf5@columbia.edu; your comments will be duly acknowledged in the next edition.

It has been a pleasure for the two of us and Paul Lynn to work together on this new edition, a transatlantic link appreciated by the three of us.

Vasilis Fthenakis
Subhamoy Bhattacharya

Acknowledgment to the First Edition

I am grateful to a number of wind energy companies and organizations for permission to use their excellent color photographs. In alphabetical order, including the names of individuals who have offered generous help and advice, they are:

- Canadian Wind Energy Association, Ottawa, Canada (Lejla Latifovic)
- DONG Energy A/S, 2820 Gentofte, Denmark (Kathrine Westermann)
- Ecotricity Ltd., Stroud, England (Mike Cheshire)
- ENERCON GmbH, 26581 Aurich, Germany (Anne-Kathrin Gilberg)
- French Wind Energy Association, 75008 Paris, France (Benoit Seveno)
- London Array Ltd., London SW1H 0RG, England (Joanne Haddon)
- REpower Systems AG, 22297 Hamburg, Germany (Caroline Zimmermann)
- Vestas Wind Systems A/S, 8940 Randers SV, Denmark (Michael Holm)

The publishers acknowledge the use of the above photographs, which are reproduced with the permission of the copyright holders.

The book also includes 80 color illustrations by David Thompson, who has interpreted my sometimes rough and ready sketches with great skill. Dave worked closely with me on my previous book, *Electricity from Sunlight*, and it has been a pleasure to repeat the collaboration.

The author of a short but wide-ranging book on wind energy inevitably draws on many sources for information and inspiration. In my case, various books, articles, and websites have helped clarify the subject's scientific basis, technological development, and current worldwide status, and I have tried to cite them adequately in the chapter reference lists. A special mention should be made of two books that have proved invaluable for clear explanations of difficult concepts that I have attempted to summarize:

- Wind Energy Explained: Theory, Design and Application by J.F. Manwell, J.G. McGowan and A.L. Rogers (Wiley 2009)
- Wind Power in Power Systems, edited by T. Ackermann (Wiley 2005)

I freely acknowledge the debt I owe the authors and recommend the books to anyone wishing to take their understanding of wind energy to a higher level.

Paul A. Lynn

About the Companion Website

This book is accompanied by a companion website.

www.wiley.com/go/fthenakis/windenergy2e

This website includes solution manual.

1 Introduction

1.1 Wind energy and Planet Earth

Half a century ago, it would have taken a brave person to predict today's extraordinary renaissance of machines powered by the wind. Traditional windmills for milling grain and pumping water had been largely consigned to technological history, overtaken by electric motors fed from centralized power plants burning fossil fuels. But by a curious twist of history, large numbers of wind turbines, installed both onshore and offshore, are today injecting energy into electricity grids for the benefit of us all and helping usher in a new age of renewable energy.

The background to this development is, of course, the massive redirection of energy policy that most experts and politicians now agree is essential if Planet Earth is to survive the 21st century in reasonable shape. For the last few hundred years, humans have been using up fossil fuels that nature took around 400 million years to form and store underground. A huge effort is now underway to develop and install energy systems that make use of natural energy flows in the environment including wind and sunlight, with a major contribution from large wind turbines. This is not simply a matter of fuel reserves, for it is becoming clearer by the day that, even if those reserves were unlimited, we could not continue to burn them with impunity. Today's scientific consensus assures us that the resulting carbon dioxide emissions would lead to a major environmental crisis. So the danger is now seen as a double-edged sword: on the one side, fossil fuel depletion; and on the other, the increasing inability of the natural world to absorb emissions caused by the burning of what fuel remains, leading to accelerated climate change.

Back in the 1970s, there was very little public discussion about energy sources, including electricity. In the industrialized world, we had become used to the idea that electricity is generated in large, centralized power plants, preferably out of sight as well as mind, and distributed to factories, offices, and homes by a grid network with far-reaching tentacles. Few people had any idea how the electricity they took for granted was produced, or that the burning of coal, oil, and gas was building up global environmental problems. Those who were aware tended to assume that the advent of nuclear power would prove a panacea; a few

Onshore and Offshore Wind Energy: Evolution, Grid Integration, and Impact, Second Edition.
Vasilis Fthenakis, Subhamoy Bhattacharya, and Paul A. Lynn.
© 2025 John Wiley & Sons Ltd. Published 2025 by John Wiley & Sons Ltd.
Companion website: www.wiley.com/go/fthenakis/windenergy2e

even claimed that nuclear electricity would be so cheap that it would not be worth metering! It was all very reassuring and convenient—but, as we now realize, dangerously complacent.

Yet, even in those years, a few brave voices suggested that all was not well. In his famous book *Small is Beautiful*,[1] first published in 1973, E.F. Schumacher poured scorn on the idea that the problems of production in the industrialized world had been solved. Modern society, he claimed, does not experience itself as part of nature, but as an outside force seeking to dominate and conquer it. And it is the illusion of unlimited powers deriving from the undoubted successes of much of modern technology that is the root cause of our present difficulties. We are failing to distinguish between the capital and income components of the Earth's resources. We use up capital, including oil and gas reserves, as if they were a steady and sustainable income. But they are once-and-only capital. It is like selling the family silver and going on a binge.

Schumacher's message, once ignored or derided by the majority, is now seen as mainstream. For the good of Planet Earth and future generations, we have started to distinguish between capital and income and to invest heavily in renewable technologies—including wind energy—that produce electricity free of carbon emissions. The message was powerfully reinforced by former US Vice President Al Gore, whose inspirational video in 2006 *An Inconvenient Truth*[2] has been watched by many millions of people around the world.

The fossil fuels laid down by solar energy over hundreds of millions of years must surely be regarded as capital, but the winds that blow over the world's land surfaces and oceans day by day, year by year, and century by century are effectively free income to be used or ignored as we wish. Nothing is "wasted" or exhausted if we don't use it because it is there anyway. The challenge for the future is to harness such renewable energy effectively, designing and creating efficient and hopefully inspiring machines to serve humankind without disabling the planet.

This is a good moment to consider the meaning of renewable energy a little more carefully. It implies energy that is sustainable in the sense of being available in the long term without significantly depleting the Earth's capital resources or causing environmental damage that cannot readily be repaired by nature itself. In his excellent book *A Solar Manifesto*[3], German politician Hermann Scheer considered Planet Earth in its totality as an energy conversion system. He noted how, in its early stages, human society was itself the most efficient energy converter, using food to produce muscle power and later enhancing this with simple mechanical tools. Subsequent stages—releasing relatively large amounts of energy by burning wood and focusing energy where it is needed by building sailing ships for transport and windmills to grind grain and pump water—were still essentially renewable activities in the above sense (Figure 1.1).

What really changed things was the 19th-century development of the steam engine for factory production and steam navigation. Here, almost at a stroke, the heat energy locked in coal was converted into powerful and highly concentrated motion. The industrial society was born. And ever since, we have continued burning coal, oil, and gas in ways which pay no attention to the natural rhythms of the Earth and its ability to absorb wastes and by-products, or to keep providing energy capital. Our approach has become the opposite of renewable, and it is high time to change priorities (Figure 1.2).

It would, however, be unfair to pretend that renewable energy is an easy answer. For a start, it is diffuse and intermittent. Often, it is unpredictable. And although the "fuel" is free and the waste products are minimal, up-front investment costs tend to be large. There are

Figure 1.1 Bring in the new: a scene in Portugal (Source: with permission of Repower).

certainly major challenges to be faced and overcome as we move toward a new energy mix for the 21st century.

Our story now moves on to modern wind energy, which is already one of the most mature of the renewable technologies, and still advancing rapidly. But before getting involved in the details, we should consider the gift of a global wind resource that is helping wean us away from our addiction to fossil fuels (Figure 1.3).

1.2 Winds of the world

The winds of the world are produced by the Sun's uneven heating of the Earth's atmosphere and may be thought of as a form of solar energy. Variations in atmospheric pressure caused by differential heating propel air from high-pressure to low-pressure regions, generating winds that are also greatly affected by the Earth's rotation and surface geography.[4] On a large scale, they may be broadly divided into latitudinal and longitudinal patterns.

The most consistent latitudinal wind patterns are found over the great oceans of the world, well away from large land masses and mountain ranges. For many centuries, the captains of sailing ships depended on reliable *trade winds* to speed them on their way, trying to avoid the *horse latitudes* at around 30° north and south and the equatorial *doldrums* that threatened to becalm them for days on end. It is hardly surprising that wind meteorology exercised some famous minds throughout the great age of sail. Edmond Halley (1656–1742), an English astronomer best known for computing the orbit of *Halley's comet*, published his ideas on the formation of trade winds in 1686, following an astronomical expedition to the island of

Figure 1.2 New horizons: a Danish offshore wind farm (Source: with permission of Orsted).

Figure 1.3 The renaissance of wind energy (Source: with permission of VESTAS).

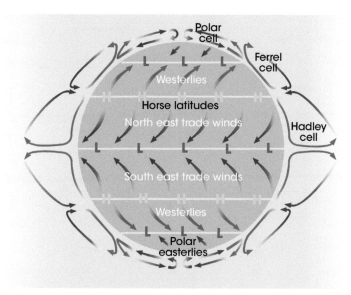

Figure 1.4 Atmospheric cells and latitudinal wind belts.

St. Helena in the South Atlantic. The atmospheric mechanism proposed by George Hadley (1685–1768), a lawyer who dabbled productively in meteorology, attempted to include the effects of the Earth's rotation—a theory that was subsequently corrected and refined by American meteorologist William Ferrel (1817–1891).

The contributions of Hadley and Ferrel to our understanding of latitudinal wind patterns are acknowledged in the names given to atmospheric "cells" shown in Figure 1.4, which illustrates major wind belts encircling the planet. Essentially, these are generated by the steady reduction in solar radiation from the equator to the poles. The associated winds, rather than flowing northward or southward as we might expect, deflect to the east or west in line with the *Coriolis effect*, named after French engineer Gaspard Coriolis (1792–1843), who showed that a mass (in this case, of air) moving in a rotating system (the Earth) experiences a force acting perpendicular to both the direction of motion and the axis of rotation.[4]

The *Hadley cells*—closed loops of air circulation—begin near the equator as warm air is lifted and carried toward the poles. At around 30° latitude, north and south, they descend as cool air and return to complete the loop, producing the *northeast* and *southeast trade winds* that have had such a major historical impact on ships sailing between Europe and the Americas. A similar mechanism produces *polar cells* in the Arctic and Antarctic regions, giving rise to *polar easterlies*. If you live in northwest Europe, you will know all about freezing winter winds from Siberia!

The *Ferrel cells* of the mid-latitudes, sandwiched between the Hadley and polar cells, are less well-defined and far less stable. Meandering high-level *jet streams* tend to form at their boundaries with the Hadley cells, generating localized passing weather systems. This makes the coastal wind patterns of countries such as Denmark, Germany, and Britain famously variable. So although the prevailing winds are *westerlies*, they are often displaced by flows from other points of the compass, especially during the winter months.

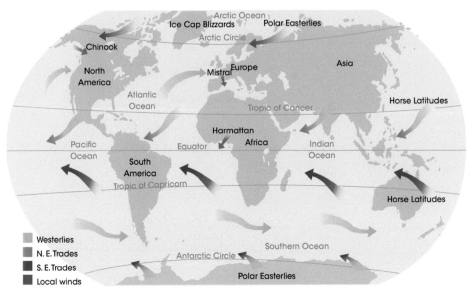

Figure 1.5 Wind patterns of the world.

Temperature and pressure gradients caused by the Sun also drive longitudinal "cells" that produce their own wind patterns. For example, there is a vast loop of winds over the Pacific Ocean known as the *Walker Circulation*, named after Sir Gilbert Walker, who, in the early 20th century, tried to predict Indian monsoon winds. The loop is caused by differences in surface temperature between the eastern and western Pacific and normally produces easterly trade winds, exerting major influences on the climates of southeast Asia and the western coasts of the Americas.

In general, we see that the complexity of the world's major wind patterns, illustrated in Figure 1.5, is the result of air rising over warmer areas of oceans and continents and subsiding over cooler ones. The effects occur at all scales over the Earth's surface, from the vastnesses of the Atlantic, Pacific, and Southern Oceans and the Sahara Desert down to mid-scale phenomena that generate famous winds such as the *Mistral* in France, the *Chinook* in North America, and the *Harmattan* in West Africa.

When we come to consider the wind regimes of smaller regions or countries, the patterns of Figure 1.5 are increasingly modified by local geography. Although it certainly helps to keep the big picture in mind, conditions are often affected by hills and mountains, valleys, forests, and variations in terrain—as well as by the time of day and season of the year.[4, 5] Coastal areas often experience "sea breezes" that carry air ashore during the day, followed by opposite "land breezes" at night, powered by the fluctuating temperatures caused by the different thermal capacities of the land and the sea.

Some of the above points are illustrated in Figure 1.6, which shows a typical wind distribution during the summer months over the UK and Ireland, two of the windiest nations in Europe. The main flow is from the Atlantic, corresponding to the prevailing westerlies, but a complex low-pressure feature (a depression) over Scotland, driven by a high-altitude Jetstream, produces localized winds that tend to circle anticlockwise. On a day such as this,

Figure 1.6 A wind pattern over the UK and Ireland.

not all UK and Irish wind turbines face the same way! In winter, the main flow quite often swings easterly and comes from the Arctic via continental Europe. On top of these broad seasonal effects, winds in a particular location can be greatly affected by local geography.

What does this highly complex story tell us about generating large amounts of electricity from the wind? There are several points that bear on turbine design and installation. Highly variable wind patterns such as those of northwest Europe demand turbines that can align themselves easily with the flow. If turbines are placed in clusters, the shadowing effect of each on its neighbors may be quite serious when the wind blows from certain directions—a point needing careful consideration during the planning phase. Generally speaking, offshore wind farms are less problematic than onshore locations, for if installed well out to sea they are more likely to find themselves in the consistent company of prevailing winds, undisturbed by land features (Figure 1.7).

The unpredictability of wind means that commercial developers of large turbines go to considerable lengths to assess a site's actual potential,[5] monitoring and recording variations in speed and direction over a year or more before proceeding with installation. The winds of the world remain wild and free—and it will always be a challenge to harness their power as efficiently and economically as possible.

1.3 From windmills to wind turbines

Windmills have a long and venerable history. Some of the earliest practical machines, developed in Persia around the 10th century, were based on rectangular sails rotating

Figure 1.7 Making the most of prevailing winds: a wind farm off the shores of Denmark (Source: with permission of Orsted).

about vertical axis. Windmill technology subsequently spread through the Middle East into southern Europe, and by the late 12th century, windmills were being built in England, Holland, and Germany, where horizontal axis machines were always preferred.[4]

Although "windmill" implies a machine devoted to the milling or grinding of wheat and other grains, another application proved extremely valuable over the centuries—pumping water to drain low-lying land. The Dutch became world-renowned for using windmills to help reclaim large areas of land from the North Sea, and English farmers drained land in East Anglia. Many other nations contributed to developing highly effective windmills for grinding grain, pumping water, and other mechanical tasks. Around a quarter of a million windmills were installed over the centuries in Western Europe, and although many experts put their heyday in the years 1750–1850, there were still tens of thousands in operation at the start of the 20th century.

We see that windmills were a major source of mechanical energy in Europe before the Industrial Revolution gathered pace. Replacing or supplementing the muscle power of humans and animals, they were an important part of the economic, social, and cultural landscape. The flour miller was a key member of many communities; stories of millers and their families abounded in local folklore; and windmills entered the canon of European literature—never more famously than in Cervantes' *Don Quixote* (1605/15), whose romantic but delusional hero attacked a set of Spanish windmills believing them to be ferocious giants. Don Quixote's imaginary enemies survive to this day in the central Spanish region of La Mancha, carefully restored and eagerly sought by tourists (see Figure 1.8).

The highly variable winds of northwest Europe, already mentioned in the previous section, demanded windmills that could easily be turned to face the wind. In early *post mills*, the

Figure 1.8 The Spanish windmills made famous in Cervantes' *Don Quixote* (Source: Lourdes Cardenal/Wikimedia Commons/CC BY SA 3.0).

complete timber structure turned on a vertical post; in later *tower mills*, also known as *smock mills*, most of the building remained stationary, and only the top section, or cap, rotated. Tower mills could be built higher and heavier than post mills, allowing them to support larger sails. As the years went by, increasingly sophisticated features were incorporated: secondary rotors, known as *fantails*, to turn the cap into the wind automatically; sails (or blades) with a degree of twist to increase efficiency; and in some advanced designs, speed governors (Figure 1.9).

It was also essential to protect against violent winds because, as we shall explain later, the power intercepted by a windmill rotor is proportional to the cube of the wind speed. A doubling of speed increases the power eight times; a trebling, 27 times. This could spell disaster. Various approaches were used to prevent damage including reducing the area of sailcloth covering lattice blades, designing blades with tiltable wooden shutters that could spill the wind, and a variety of mechanisms, manual and automatic, to turn the rotor blades away from the oncoming blast. By the end of the 19th century, European windmills had achieved a high level of technical sophistication.

Various numbers of main blades were tried, including 6, 8, and 12, but over the years a consensus developed that four-bladed designs represented the best compromise between construction and maintenance costs, rotor weight, and performance. By the 18th century, it was well understood that the speed of the blade tips should ideally be kept proportional to the wind speed and that the mechanical power developed was proportional to the area "swept" by the rotor. Taken together, these factors mean that a large windmill is necessarily a low-speed machine developing a high torque on its main shaft, but smaller windmills operate at relatively high speed and low torque (Figure 1.10).

(a)   (b)

Figure 1.9 A Dutch windmill dating from 1757 and an English tower windmill of 1790 incorporating a fantail (Source: (a) Massimo Catarinella/Wikimedia Commons/CC BY 3.0. (b) Oliver Dixon/Wikimedia Commons/CC BY SA 2.0).

(a)   (b)

Figure 1.10 An eight-bladed English windmill and a multibladed American windmill (Source: (a) Richard Croft/Wikimedia Commons/CC BY SA 2.0. (b) Ben Franske/Wikimedia Commons/CC BY SA 4.0).

In the mid-19th century, a range of smaller machines known as *American windmills* were developed as pumps for watering livestock, irrigating land, and supplying water for steam engines on the American railroads. With multibladed steel rotors mounted on lattice towers, they featured self-regulating mechanisms to turn them away from damaging high winds and could be left unattended. Hundreds of thousands were installed by the 1930s. Similar designs are still manufactured around the world today, mainly used in remote locations lacking a convenient electricity supply.

It was inevitable that the Industrial Revolution would sooner or later spell the demise of the traditional tower windmill. Steam engines, and subsequently internal combustion engines, could supply highly concentrated power on demand at any time of day or night, unaffected by the vagaries of the wind. Electric motors offered highly convenient power for grinding grain and pumping water. So, by the mid-20th century, it was rare to find a traditional windmill spinning its sails delightfully in the breeze.

Yet even as the tower windmill continued its remorseless decline, developments in electrical technology began to inspire imaginative engineers with the idea of using the wind to generate electricity.[4] One of the earliest and most remarkable designs for driving an electrical generator with a windmill saw the light of day in 1888, when Charles Brush erected a pioneering machine in the backyard of his home in Cleveland, Ohio (see Figure 1.11). It is clear that Mr. Brush did not do things by halves. A multibladed rotor 17 m in diameter was kept facing the wind by a large tail vane. A 12-kW dynamo, driven at 50 times rotor speed, charged a battery bank that fed various motors and hundreds of incandescent lights in his home over the next 15 years. This was one of the first machines that could properly be described as a *wind turbine*.

The following decades produced many new designs for electricity generation. Danish inventor Poul La Cour built over 100 machines in the period 1891–1918, with power ratings up to 35 kW. Large numbers of smaller machines, in many ways the natural successors to the American wind pump, were used in the USA and elsewhere to charge batteries and provide modest amounts of electricity for farmsteads and other remote locations. As the 20th century progressed, there was a trend toward using blades with true airfoil shapes based on ideas gained from the fast-developing aircraft industry. By the late 1930s, the 1-MW power landmark had been exceeded by a single machine—the famous two-bladed Smith–Putnam turbine installed in Vermont, USA.[4]

However, one of the most influential machines in the story of modern wind power was the somewhat later 200-kW turbine built in 1957 in Gedser, a windswept coastal region of Denmark (see Figure 1.12). Designed and built by Johannes Juul, the Gedser turbine used automatic *stall control* to limit rotor power output and speed in high winds and incorporated emergency aerodynamic tip brakes, deployed by centrifugal force to prevent damage in extreme conditions. Its inbuilt safety features kept it running reliably for 11 years, and it was refurbished in 1975 at NASA's request to provide valuable data for the USA's developing wind energy program. This remarkable machine presaged many aspects of modern turbine design and helped cement Denmark's reputation for innovation in wind energy, which continues to this day.

In spite of the technical advances of the Smith–Putnam and especially the Gedser turbines, the 30 years following World War II were a lean period for international wind energy. It was generally assumed that cheap fossil fuels, plus the advent of nuclear power, would "keep the lights on" into the indefinite future, and the idea that wind turbines could make a significant

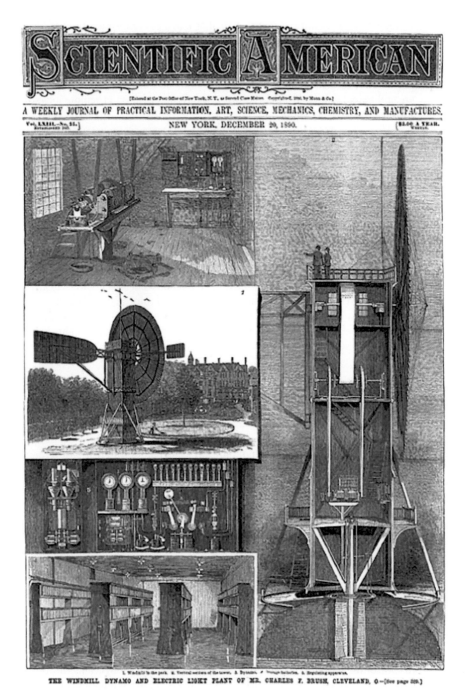

Figure 1.11 The Brush Wind Turbine of 1888, with associated equipment (Source: Scientific American/Wikimedia Commons/Public domain).

Figure 1.12 The 200-kW Gedser wind turbine in an old photograph of 1957 (Source: Energimuseet/Wikimedia Commons/CC BY 4.0) and on a commemorative Danish postage stamp issued in 2007.

contribution to global electricity generation was little more than a pipe dream for enthusiasts. Yet by the mid-1970s, the emerging environmental movement was starting to challenge official complacency about energy supplies, and the first "oil shock" emphasized the determination of oil-producing nations in the Middle East to exert greater control over the price and availability of their "black gold."

At this point, the US administration under President Jimmy Carter started serious support for renewable energy technologies, including wind. A number of large turbine projects were financed and, perhaps more importantly, a new legal framework was introduced requiring electric utilities to allow turbine connection to the grid. By the early 1980s, the state of California—true to its reputation as an enthusiast for technical adventures—accommodated thousands of wind turbines with a combined power rating of more than a gigawatt, and installations in the windswept Altamont Pass achieved international status as harbingers of a new age of wind. However, all was not well: although imported Danish machines generally performed satisfactorily, other designs proved unreliable; when government support was subsequently withdrawn by the unsympathetic Reagan administration, the initial "wind rush" ground to a halt. International wind power's center of gravity now moved back across the Atlantic to Europe, and especially to Denmark and Germany.

Many turbine designs and configurations have been tried and tested over the years, including one-bladed and two-bladed rotors mounted both upwind and downwind of the support tower, and a variety of vertical axis machines. However, the vast majority of today's large turbines are horizontal axis machines with a three-bladed rotor upwind of the tower, supported by a housing, or *nacelle*, containing a gearbox and electrical generator. The turbines previously shown in Figures 1.1–1.3, and 1.7 all have this configuration, and a typical layout of the main components is illustrated in Figure 1.13. The rotor is kept facing the wind (or deliberately turned away from it) by a *yaw motor*, and a brake is provided to prevent rotation,

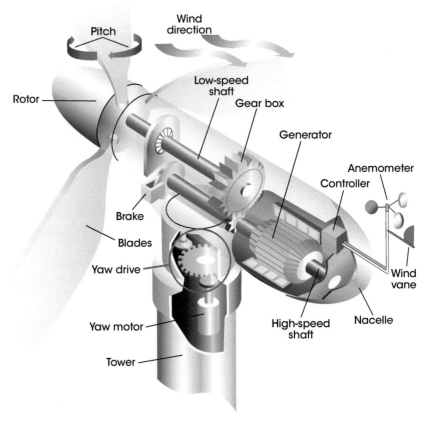

Figure 1.13 Main components of a typical modern high-power wind turbine.

for example during maintenance. Wind speed is measured by an anemometer, and the main blades are swiveled to vary their pitch, controlling or limiting the amount of power captured as the wind varies.

So far we have concentrated on the development of large wind turbines—the main focus of this book—but smaller machines suitable for individual homes, farms, and leisure applications have a parallel history. Although their total impact on electricity generation is very modest compared with today's large wind farms, they are highly valued by individuals and organizations wishing to generate relatively small amounts of electricity in remote locations and are well-described elsewhere.[5]

We now move on from this brief history of wind energy to consider how the enthusiasm of its pioneers, often criticized or dismissed by conventional voices, has developed into a worldwide industry with high hopes for the future (Figure 1.14).

Figure 1.14 Sky-high: working on the nacelle of a large wind turbine (Source: with permission of Repower).

1.4 Power, energy, and performance

Many people are confused about how much electricity wind turbines actually generate and what contribution they make to reduce carbon dioxide emissions. The basic issue is wind energy's intermittency. Professional engineers and scientists are quite used to dealing with it, understanding that turbine output and performance can only be sensibly discussed in terms of statistical averages over significant time scales. But the general public, fully aware that turbines stop turning when the wind dies, wonder what is going on and need assurance that today's wind technology is what it claims to be. Their uncertainty is played upon by climate-change skeptics, many with vested interests in opposing renewable energy.

In many ways, this seems strange. After all, solar panels are not generally criticized because the Sun disappears at night; nor water reservoirs when they run dry during droughts; nor conventional power plants because they often operate below full output—and occasionally have to shut down. Most of our inventions and machines are only used intermittently. Perhaps, wind turbines' misfortune is to be so visible: no wind, no power, no movement!

We will tackle such issues by quantifying turbine performance from a general point of view and relating it to the electricity consumption of households. This will help set the scene for a discussion of wind energy's international status and future potential. However, before we start, it is extremely important to realize that large electricity grids are not concerned with the intermittency of individual turbines, or even of wind farms, unless they are very large. What matters is the *total* generation by all wind farms, which are normally spread over a wide area and experience different wind conditions. Although total wind generation is certainly variable, it is not "on–off" like that of an individual turbine, and intermittency is a minor concern to the grid system as a whole. Of course, this is not to say it is trivial from the point of view of turbine design and annual energy production. The owners and operators of turbines are certainly interested in how much electricity their machines produce!

For over a century, advances in wind engineering have resulted in turbines of increased size and power rating. As already noted, Charles Brush's windmill of 1888 (see Figure 1.11) drove a 12-kW dynamo; the Gedser turbine of 1957 (Figure 1.12) was rated at 200 kW; and although the Smith–Putnam machine[4] of the 1940s managed to exceed the 1-MW landmark, it was so far before its time that another half-century passed before megawatt turbines became commonplace. Now things have moved on again, and we are quite used to the idea of individual turbines rated at 2 or 5 MW, with some 10-MW machines already installed offshore and even larger designs glinting in the eyes of turbine designers.

How do such power ratings relate to wind energy's practical contribution to national and global electricity supplies? A key point is also an obvious one: a turbine only generates its full-rated power when the wind reaches, or exceeds, a certain speed. Much of the time it produces considerably less power. Given an onshore site with a good wind regime, the *average* power output of a large modern turbine is around 30% of its rated maximum. For example, a turbine rated at 2 MW typically produces an average of about 0.6 MW, measured day and night over a complete year. The precise percentage, referred to as the *capacity factor* or *load factor*, depends on the technical efficiency of the turbine and the quality of the site, which may be affected by features such as hills, forests, or rough terrain that impede or disturb the airflow. Generally speaking, offshore turbines do better than onshore ones, reaching capacity factors well above 40% in some cases.

Turbine manufacturers and electric utilities rarely mention capacity factors, probably regarding them as too technical for the general public. Instead, they try to relate turbine performance to personal experience by comparing the amount of electricity produced with household demand. For example, in Germany, the Netherlands, and the UK, a 1-MW onshore turbine is often stated to meet the needs of around 600 households. No doubt this is a reasonable way of explaining things to consumers, but it is obvious that a wind turbine cannot supply households on an hour-by-hour or day-by-day basis because the wind does not blow to order. Supply and demand are often out of step, and this can lead to confusion. It would be more accurate to say "over an average year" the 1-MW turbine generates electricity equivalent to the annual consumption of about 600 households. But this is rather a mouthful, and its subtleties would probably be lost on most people (Figure 1.15).

In any case, there are several reasons to treat such estimates with caution:

- The annual electricity production of a particular type of turbine (and therefore the number of "households equivalent") is site-dependent and fluctuates to some extent from year to year due to the variable nature of wind.
- Small but significant power losses occur during transmission, especially when turbines are placed far from consumers.
- Consumption of electricity by households tends to increase year-by-year as living standards rise.
- Consumption patterns within a country often vary considerably from region to region and between city and rural communities (so which "households" are being used in the calculation?).
- It is also worth noting that average household electricity consumption varies greatly from country to country. For example, the average USA figure is roughly twice that of Western Europe.

Figure 1.15 Delivering power, energy, and performance: a 5-MW offshore wind turbine (Source: with permission of Repower).

We see that estimates of "households equivalent" are necessarily approximate. This is not to say they are wrong, simply that they rely on certain assumptions and are subject to statistical variation. Used sensibly, they give an easily understood indication of turbine performance.

In the discussion that follows, it is important to remember that power is a *rate* of energy production or consumption and therefore has dimensions of (energy/time). Conversely, energy has dimensions of (power × time). To take a familiar example, if an electric heater is rated at 1000 W or 1 kW, this is the amount of *power* it consumes when switched on. If left on for one hour, the *energy* used is 1 kWh (generally referred to as 1 "unit" of electricity). A 2-kW heater switched on for half an hour also uses 1 kWh of energy—power is being consumed at twice the rate but for only half the time. As householders, it is the energy we pay for, expressed in kWh or units of electricity. When assessing the contribution made by wind power to electricity generation, the key quantity is annual energy production.

We may describe the annual energy production of a turbine in terms of its rated power and capacity factor. For a turbine rated at P_r MW operating at a capacity factor C_f, the average power output, measured over a complete year, is given by:

$$P_{av} = (P_r \times C_f)\text{MW} \tag{1.1}$$

Since there are 8760 hours in a year, the turbine's annual energy production E_a is

$$\begin{aligned} E_a &= (8760 \times P_{av})\text{MWh} \\ &= (8760 \times P_r \times C_f)\text{MWh} \end{aligned} \tag{1.2}$$

For example, a machine rated at 2 MW, operating at a capacity factor of 30%, is expected to generate about $8760 \times 2 \times 0.3 = 5256$ MWh per year.

Alternatively, the capacity factor may be estimated if we know the turbine's rated power and the annual amount of energy it produces:

$$C_f = E_a/(8760 \times P_r) \tag{1.3}$$

As we move up the power scale from individual turbines to wind farms, and then on to national and global electricity production, the numbers increase dramatically, and it is often more convenient to work with gigawatts or even terawatts. The various units for measuring power are related as follows:

1000 W = 1kW (kilowatt)	1000kW = 1MW (megawatt)
1000MW = 1GW (gigawatt)	1000GW = 1TW (terawatt)

There is a corresponding set of units for energy, measured in kilowatt-hours (kWh), megawatt-hours (MWh), gigawatt-hours (GWh), and terawatt-hours (TWh).

Figure 1.16 shows some typical power and energy figures for electrical consumption and generation. Power is expressed either as a rated (peak) value or as an average measured over a complete year. Energy is shown as an annual total. The various items are:

- *Household.* The average power consumption of Western European households, measured night and day over a complete year, is about 0.5 kW. Since there are 8760 hours in a year, this corresponds to an annual energy consumption of about $0.5 \times 8760 = 4380$ kWh $= 4.4$ MWh. (Peak power consumption depends on how many appliances are switched on simultaneously and is not normally of great interest—provided the household's fuses are not tripped!)

- *2-MW onshore turbine.* The peak (rated) power is 2 MW and assuming a 30% capacity factor the average power is 0.6 MW, producing annual energy of $8760 \times 0.6 = 5256$ MWh $= 5.256$ GWh. This is equivalent to the annual electricity requirements of about $5256/4.4 = 1200$ households.

- *5-MW offshore turbine.* With a peak power of 5 MW and a 40% capacity factor, the average power is 2 MW giving an annual energy of $8760 \times 2 = 17,520$ MWh $= 17.52$ GWh, equivalent to about 4000 households.

- *1-GW conventional power plant.* Large modern power plants (fossil fuel or nuclear) are typically rated between 1 and 2GW. This is their peak power and also, in principle, their average power assuming continuous operation at maximum output. However, in practice, their capacity factors are less than 100% and we will use 90% as a typical figure. So in a full year, the output of a 1-GW plant is about 8760×0.9 GWh $= 7.9$ TWh, equivalent to the needs of about $7,900,000/4.4 = 1,800,000$ Western European households. This puts into perspective the challenge of substituting wind power for conventional power plants: about 450 very large offshore turbines, rated at 5 MW, are needed to produce annual electricity equivalent to a 1 GW conventional power plant.

Consumption		Power		Annual energy (E_a)	Capacity factor (C_f)
		rated (P_r)	av $^{ge}(P_{av})$		
	Household		0.5 kW	4400 kWh = 4.4 MWh	
Generation					
	2-MW onshore turbine	2 MW	0.6 MW	5300 MWh = 5.3 GWh	30%
	5-MW offshore turbine	5 MW	2 MW	17000 MWh =17 GWh	40%
	Power plant	1 GW	0.9 GW	7900 GWh = 7.9 TWh	90%

Figure 1.16 Electricity consumption and generation.

Note that we have given values in Figure 1.16 to two significant figures. Generally speaking, the accuracy of wind energy calculations, and of statistical data relating to household electricity usage, does not justify more (unfortunately values are sometimes given to four or five significant figures, implying unwarranted accuracy).

Of course, the wind does not blow to order, and there is often a mismatch between electrical supply and demand. This is why it is so important to distinguish between a turbine's peak power and its average power over a complete year. Intermittency and variability are features of all renewable energy technologies that harness natural energy flows in the environment—wind, solar, tide, and wave. However, the variability of renewable energy is drastically decreased when different sources of renewable energy are coupled together. One of the major challenges of wind engineering is to integrate variable generation successfully into grid networks that supply a wide range of industrial and domestic consumers who hardly notice when the wind blows!

Table 1.1 Performance data for six wind farms.

Wind farm	Onshore or offshore	Turbines No.	MW	Rated power MW	Annual energy GWh	Households equivalent Total	per MW	C_f (%)
(a) Horse Hollow (Texas USA)	ON	291 130	1.5 2.3	735	1690	180,000	245	26
(b) Rudong (China)	ON	100	1.5	150	333	150,000	1000	25
(c) Maranchon (Spain)	ON	104	2.0	208	500	140,000	670	27
(d) Marienkoog (Germany)	ON (coastal)	7	3.6	25	78	17,500	700	36
(e) Burbo Bank (UK)	OFF	25	3.6	90	315	80,000	890	40
(f) Horns Rev 2 (Denmark)	OFF	91	2.3	209	800	200,000	960	44

Source: Adapted from GWEC.

Table 1.1 puts the discussion into context with technical data from six wind farms. The top four are onshore installations in the USA, China, Spain, and Germany; the bottom two lie off the coasts of the UK and Denmark. The selection is intended to illustrate an interesting range of sites, turbine numbers and sizes, capacity factors, and "households equivalent." It is certainly not intended as a comparison between different countries and their expertise in wind energy.

We first note the large range of turbine numbers, sizes, and total rated power (between 25 and 735 MW). When discussing wind farms, or national wind power totals, the total rated power is also widely referred to as the installed capacity. Estimated (or measured) annual energy totals range between 78 and 1690 GWh, and the estimated number of "households equivalent" is between 17,500 and 200,000. The capacity factors are particularly interesting: 25–27% for three onshore sites, but an impressive 36% for the German installation consisting of a few large modern turbines on a flat site next to the coast. The two offshore farms achieved 40% and 44%, underlining the benefits of placing turbines out to sea.

The estimates of the number of households supplied for each megawatt of installed capacity ("households per MW") show large variations, partly due to differences in turbines and wind regimes from one site to another, but mainly to differences in average household consumption between countries. Most striking is the contrast between the wind farms in Texas, USA (245 households/MW), and in Rudong, China (1000 households/MW), reflecting the far greater electricity consumption in the USA. Once again, we see that "households equivalent" needs careful interpretation.

Wind energy's potential for reducing carbon dioxide emissions is often mentioned in the press and on various websites. But this, too, turns out to be a tricky issue. Certainly, electricity generated by the wind offsets or replaces electricity produced by other means. But which means? In a country such as Norway with its abundant supplies of hydropower, does new wind turbine capacity replace hydroelectric generation, with its very low attendant carbon emissions? In India and China, currently burning a great deal of coal in conventional power plants, do more wind turbines mean less coal burning? Many countries produce electricity from a range of fossil fuels, nuclear power, and renewables. The environmental benefits of wind energy must clearly depend on the energy strategy and current "energy mix" of the country concerned. No wonder, the claims made for carbon dioxide reductions by wind and other renewable technologies differ widely. It is a lot simpler to stay with comparisons based on electricity generation!

1.5 Coming up-to-date

It is now time to outline the international status of wind energy, the challenges it currently faces, and the issues that will affect its future. We focus initially on the worldwide growth of the industry, paying particular attention to countries that are using large turbines to spearhead the surge in installed capacity, both onshore and offshore.[6] To make sense of the figures, we must first be clear about how data on wind power and energy is presented.

A country's wind power capacity is generally expressed in terms of installed megawatts (MW), equal to the total rated power of all its wind turbines. This figure is often supported by estimates of annual energy production measured in gigawatt-hours per year (GWh/yr). For example, in 2023 Germany had a cumulative installed capacity of about 59 GW, yielding about 140,000 GWh/yr. Recalling that a Western European household typically uses 4.5 MWh/yr, this translates into the needs of about $140 \times 10^{12}/4.5 \times 10^6 = 31$ million households. It is certainly an impressive figure although we must remember that householders are not the only consumers of electricity. Much is needed by industry, commerce, and public services, and in modern developed economies household consumption typically accounts for around 30% of the total.

Naturally, the numbers are even bigger for global wind. As of 2023, cumulative global onshore wind capacity stood at about 842 GW onshore plus 64 GW offshore, yielding some 2100 TWh/yr. Using Equation (1.3), these figures can be used to estimate the average capacity factor of the world's 2023 stock of wind turbines:

$$C_f = E_a/(8760 \times P_r) = 2100 \times 10^{12}/(8760 \times 906 \times 10^9) = 0.26 = 26\% \tag{1.4}$$

This covers machines installed over many years in many countries with a wide range of wind regimes. The continuing trend toward larger and more efficient turbines, probably with an increasing proportion installed offshore, means that the "global" capacity factor is expected to nudge steadily upward in the years ahead.

The meteoric increase in the world's cumulative installed capacity[6] is illustrated in Figure 1.17. Back in 1995, the figure was about 5 GW, and it kept increasing by around 30% per annum during 1995–2010 and by about 20% during 2010–2023.

But we must not get carried away—at least, not yet. Electricity generated by the wind still provides only about 7% of global demand. True, there are big differences between countries: Denmark, the leader, currently gets 55% of its electricity from the wind; Germany gets 32%, Ireland 35%, Greece 19%, USA 10%, China 7.5%, and India 4.4%. Some other countries with large populations have virtually no wind turbines—yet. Table 1.2 shows the "top 10" in order of cumulative installed capacity as of 2023, showing China in the lead

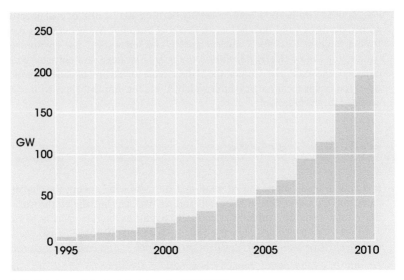

Figure 1.17 Global cumulative installed capacity, 1995–2010.

with almost 40% of the total. All countries on the list have certainly been very ambitious in promoting and installing wind turbines over the past few years. Emerging markets in Latin America—especially Brazil—and Asia—especially India—are now taking off.

But ambition is certainly not confined to the major players.

Denmark's installed capacity of 1250 W per head of population is remarkable. It is equivalent to every Danish man, woman, and child having their own personal wind turbine rated at 1.2 kW—or indeed rather more, because small machines almost always operate with lower capacity factors than large ones. A turbine to generate this much power in strong winds would need a rotor diameter of about 1.8 m, not the easiest thing for every citizen to mount on a domestic roof or tall pole in the garden! However, Denmark's vision has resulted so far in generating 55% of its total electricity using wind turbines, both onshore and offshore.

One of today's hottest topics in wind energy is the development of offshore systems (Figure 1.18); this book devotes Chapters 4 and 5 to offshore technology and project development.

Table 1.2 The "top ten" countries for cumulative installed capacity at the start of 2023, expressed as (a) MW total and (b) % of the total.

	Onshore	Total (GW)	% of Total
1	China	333,998	40
2	USA	144,184	17
3	Germany	58,951	7
4	India	41,930	5
5	Spain	29,793	4
6	Brazil	25,632	3
7	France	20,653	2
8	Canada	15,261	2
9	UK	14,575	2
10	Sweden	14,393	2

Source: Adapted from GWEC.

Figure 1.18 Wind energy moves offshore (Source: with permission of Repower).

Until quite recently, the extra challenges and costs of anchoring turbines to the seabed and bringing their power ashore seemed daunting. But two major issues have changed the consensus view. First, as wind energy grows, new onshore sites in leading wind energy countries inevitably become scarcer—sometimes for technical reasons including a lack of good wind regimes, but also because of public resistance to the erection of turbines near homes or in locations with high landscape value.

The second issue is far more positive. Developments in turbine design and technology are leading to highly reliable and efficient machines rated at 8 MW or more, and their deployment offshore looks increasingly attractive. Better wind regimes, coupled with the improved technical efficiency of very large turbines, help offset the undoubtedly higher costs of installation and maintenance in a marine environment. Offshore wind capacity currently accounts for only about 7% of the global total; 10 years ago, it was only 2% and is poised to rise in the coming years. The countries most active in this area are China, UK, Germany, Denmark, and the Netherlands (Figure 1.19).

So far, we have focused on the growth of installed capacity, national and global, onshore and offshore. Recent experience and future projections are full of optimism. But as global wind energy moves inexorably onward, what are the major technical interests and concerns of the engineers and other professionals behind this inspiring modern industry? We may group them under several headings:

- ◼ *Turbine and system design and development.* The average rated capacity of large new turbines is presently about 8 MW and continues to rise, giving increased technical efficiencies and economies of scale. Among the components and systems undergoing

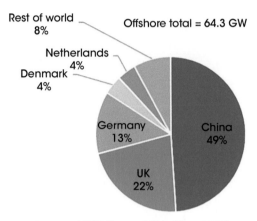

Figure 1.19 Offshore total installation as of 2023 (Source: Adapted from GWEC).

constant development are blade airfoils to suit a variety of wind regimes, very long turbine blades for high-power machines, drive trains including gearboxes, electrical generators including low-speed designs for direct coupling to rotors, and advanced electronic systems for turbine monitoring and control.

- *Offshore challenges.* The marine environment poses special challenges for wind turbines including the design, manufacture, and placement of towers and seabed anchorages, effective protection against corrosion and storm conditions at sea, and transport and installation of very large rotors and nacelles.

- *Grid integration and expansion.* For over a hundred years, electricity grids in developed countries have been designed to distribute the electricity generated by large, centralized power plants. The advent of wind and other renewables injects an increasing percentage of smaller-scale, decentralized generation, often in quite different locations. Grid networks must be adapted and expanded. Transmission systems must cope with fluctuating generation, a growing challenge as wind's share of total generation increases. Offshore, the economics of wind power depend greatly on the provision of efficient cable connections and hubs, including shared international distribution—for example, in countries bordering the North Sea. For a comprehensive picture of grid integration challenges and solutions, see Chapter 7.

- *Planning and public acceptance.* Planning procedures for wind power developments vary greatly between countries. If complex and time consuming, they can act as serious inhibitors to growth. Closely connected is the delicate issue of public acceptance and the degree to which governments allow their citizens to influence planning decisions. There are also important issues of public education and information, community involvement, and possible compensation for loss of property values.

- *Environmental issues.* The environmental credentials of wind energy are beyond doubt, but there are also concerns. The visual impact of large land-based turbines and wind farms is probably the most serious. Others include turbine noise, bird and bat fatalities, and, in the case of offshore installations, effects on marine conservation and fishing. We will discuss this in detail in Chapter 8 (Figure 1.20).

Figure 1.20 A bright future for wind power (Source: with permission of Repower).

We are witnessing the development of a great new industry with major implications for technological research and development, "green" employment opportunities, and reduction of carbon emissions while at the same time enhancing energy security. From a historical perspective, the story of practical "windmills" that began over a thousand years ago in the Middle East is embarked on a remarkable new chapter.

Appendix 1.A: Energy units and conversions

Thermal (Primary) Energy	Electricity
Joule (J); kilojoule (kJ), 1 kJ = 10^3 J; Megajoule (MJ), 1 MJ = 10^6 J; Exajoule (EJ), 1 EJ = 10^{16} J	Wh; kWh; MWh; Gwh, 1 GWh = 10^9 W; TWh, 1 TWh = 10^{12} W
Btu; QBtu(Quad), 1 QBtu = 10^{15} Btu	1 kWh = 1000 W × 3600 s = 3.6 MW s = 3.6 $MJ_{electricity}$ (MJ_e)
1 kBtu = 0.293 kWh	1 kWh = 3.41 kBTU

1 MJ_e = ~3 MJ (Conversion of electricity to primary energy assuming an average grid conversion factor of 0.33).

CO_2 Emissions per fuel type

Fuel type	Kg of CO_2 per unit of consumption
Grid electricity	40–60 per kWh (depending on grid mix)
Natural gas	3142 per metric ton
Diesel fuel[1]	2.68 per liter (L)
Petrol[2]	2.31 per liter
Coal	2419 per metric ton

[1]One liter of diesel weighs 0.83 kg. Diesel consists for 86% of carbon, or 0.72 kg of carbon per liter. According to the stoichiometry of the combustion reaction $C + O_2 = CO_2$ 12 kg of C produce 3.6 times more (44 kg) of CO_2. Thus, combustion of 1 l of diesel would produce approximately 2.6 kg of CO_2.
[2]One liter of petrol weighs 0.75 kg. Petrol consists for 87% of carbon, or 0.65 kg of carbon per liter of petrol. In order to combust this carbon to CO_2, 1.74 kg of oxygen is needed. The sum is then 2.39 kg of CO_2/liter of petrol.

CO_2 emissions in transportation

Vehicle type	Miles/gal (mpg)	km/l	kg CO_2/km
Medium hybrid car[1]	50	18	0.11
Medium gasoline car[1]	28	12.5	0.20
Diesel car[2]	47	20	0.12
Trailer truck, diesel engine		3	2.68
Rail			0.06 per person
Air, short haul (500 km)			0.18 per person
Air, long haul			0.11 per person
Shipping			0.01 per ton

[1] A car rated at 20 km per L (47 mpg) uses 5 l/100 km; this corresponds to 5 l × 2.6 kg/l/(per 100 km), thus 0.13 kg CO_2/km.
[2] An average consumption of 5 l/100 km then corresponds to 5 l × 2.39 g/l/(per 100 km) = 0.12 kg CO_2/km.

Self-assessment questions

Q1.1 Which countries are the largest developers of wind turbines?

Q1.2 Name three countries which have the largest deployment of inland wind turbine (a) total and (b) per capital?

Q1.3 What are the major challenges with offshore wind turbine deployment?

Q1.4 What is the size and power capacity evolution of wind turbines?

Q1.5 During the last decade, wind turbine global cumulative capacity has been increasing at an average annual rate of about 30%. What is the equivalent percentage for your nation (or country of residence) and what factors (climatic, economic, and political) seem responsible for any difference?

Q1.6 What is the difference between primary energy and electricity and what losses are accounted for in converting the first to the latter?

Q1.7 What are the (a) the total annual electricity demand and (b) total annual energy demand for your nation (or country of residence) expressed in kWh, GWh, and TWh for electricity and in MJ, EJ, and quads for primary energy?

Q1.8 What is a fundamental issue with wind energy being the sole energy source in a country?

Q1.9 Why taller wind turbines can produce more electricity than shorter ones in the same location?

Q1.10 Do you expect the same turbine to produce more electricity in offshore or inland locations and why?

Problems

1.1 The US annual electricity demand is approximately 4000 TWh. What is the corresponding primary energy for this demand if the energy mixture consists of 37% natural gas, 25% coal, 20% nuclear, 10% hydro, and 8% wind and solar? How would the primary energy demand change if the renewable energy contribution to the mixture increases to 70% (according to the Solar Grand Plan) in a 10,000 TWh/yr 2050 scenario. Assume the following primary to electrical energy conversion factors: natural gas 0.4, coal 0.28, nuclear 0.33, hydro 1.0, wind and solar 0.1.

1.2 What is the height required for a flow of metric ton of water per second to produce 1 kWh of electricity in a hydrodynamic plant with efficiency of 100%?

1.3 Derive the conversion of Joule to units of mass, length, and time in the metric systems using the kinetic energy equation.

1.4 How many tons of CO_2 are produced annually by an internal engine car and by a hybrid car? (Use the table in the Appendix of Chapter 1.)

1.5 Burning a metric ton of coal with 67% carbon content generates about 2.4 tons of CO_2 and burning a litter of petroleum generates about 2.7 kg of CO_2.

(a) If a total of 20 million tons of coal are burned everyday world wide, calculate the weight of carbon dioxide emitted into the atmosphere from burning coal and associated ppm increase if we don't consider the absorption of CO_2 on land and the sea.

(b) Calculate the same for the total combustion of petroleum assuming a global consumptions of 70 million barrels per day (a barrel of petroleum contains 42 gallons and a gallon is 3.78 l).

(c) Calculate the same for an equivalent consumption of natural gas.

1.6 The USA with a population of 300 million accounts for about 18% of the 550 quad (quadrillion Btu) global energy demand.

(a) What would be the increase in the global demand if all the 7.4 billion people on the globe use the same per capita energy as the US population.

(b) what would be the corresponding CO_2 emissions assuming that the energy source mix doesn't change.

Answers to questions

Q1.1 Denmark and China.

Q1.2 China, USA, Germany, and Denmark.

Q1.3 Installation, especially under adverse weather conditions, also maintenance and repairs for the same reason.

Q1.4 From 1-MW 60-m tall wind turbines, we are now at 10-MW 100-m tall wind turbines.

Q1.5 Search for country-specific data.

Q1.6   Primary is the energy embedded in a fuel, and there are losses during combustion and other processing that converts primary energy to electricity.

Q1.7   Search for country-specific data.

Q1.8   Intermittency.

Q1.9   Because the wind velocity profile is parabolic with velocity increasing with height.

Q1.10 It depends on the location. Generally, a WT placed offshore is expected to perform better than one on land, as winds are typically stronger offshore as friction from ground reduces wind speed.

References

1. E.F. Schumacher. *Small Is Beautiful*, 1st edn 1973, republished by Vintage. Random House: London (1993).
2. A.L. Gore. *An Inconvenient Truth*. Bloomsbury Publishing: London (2006).
3. H. Scheer. *A Solar Manifesto*. James & James: London (2005).
4. J.F. Manwell *et al. Wind Energy Explained: Theory, Design and Application*, 2nd edition. John Wiley & Sons Ltd: Chichester (2009).
5. P. Gipe. *Wind Energy Basics: A Guide to Home and Community-scale Wind Energy Systems*, 2nd edition. Chelsea Green Publishing: Vermont (2009).
6. Global Wind Energy Council, *Global Wind Report* (2023). https://gwec.net/globalwindreport2023/.

2 Capturing the wind

2.1 Wind speed and power

A wind turbine must capture as much of the wind's power as possible, convert it efficiently into electricity, and protect itself against damage in violent weather. Performance depends crucially on the conditions at a particular site including the wind's average speed and variability and the occurrence of extreme events.

We start by considering how much power impinges on a turbine rotor placed in a steady airstream. A well-known equation of fluid mechanics gives this as

$$P = 0.5\rho A U^3 \tag{2.1}$$

where ρ is the density of the air, A is the area of the intercepted airstream (equal to the area "swept" by the rotor), and U is the speed of the wind. In standard conditions (sea level, temperature 15°C), the density of air is 1.225 kg/m^3. So the amount of power intercepted by each square meter of rotor is

$$P = 0.612\, U^3 \text{ watts} \tag{2.2}$$

For example, if the wind speed is 6 m/s (a moderate breeze), the power intercepted per square meter is $0.612 \times 6^3 = 132$ W; but if the speed rises to 24 m/s (a severe gale), the power becomes $0.612 \times 24^3 = 8460$ W. This massive increase is due to the *cubic* relationship between speed and power given by Equation (2.2).

We have used the word "intercepted" rather than "captured" because the above figures relate to the power in the wind, not the amount actually extracted by a turbine rotor. As we shall see later, there are fundamental limitations to rotor efficiency and large modern turbines typically capture up to about 50% of the wind power presented to them.

The density of air varies with both elevation and temperature. Cold air at sea level is considerably denser than warm air at a high upland site. A turbine produces more power in the winter than in the summer, in winds of the same speed. Differences of 10% or more can

Onshore and Offshore Wind Energy: Evolution, Grid Integration, and Impact, Second Edition.
Vasilis Fthenakis, Subhamoy Bhattacharya, and Paul A. Lynn.
© 2025 John Wiley & Sons Ltd. Published 2025 by John Wiley & Sons Ltd.
Companion website: www.wiley.com/go/fthenakis/windenergy2e

occur between different sites or at the same site between summer and winter, so elevations and temperatures should be considered when predicting energy yields.

It is hardly an exaggeration to say that the cubic relationship between wind speed and power is the single most important factor affecting wind turbine design and performance, for several reasons:

- A wind turbine has to operate over a huge range of wind power levels. For example, it may be designed to cut in and start generating when the wind speed reaches 4 m/s and cut out to prevent damage at 24 m/s. This 6:1 range of wind speeds corresponds to a 216:1 range of intercepted power that the turbine must convert efficiently into electricity.

- The cubic relationship also means that short periods of strong wind contribute disproportionately to a turbine's annual energy output. In the above example, a single hour of severe gale possesses as much energy as 216 hours—nine days and nights—of moderate breeze.

- The suitability of a particular site, onshore or offshore, is often discussed in terms of its average or mean wind speed. Although this is a key indicator, speed variability is also important. The cubic relationship means that "gusty" sites generally produce more annual electricity than "steady" ones with the same average wind speed.

- Wind turbines need reliable safety systems to prevent damage in extreme winds. If a machine that is designed to cut out at 24 m/s fails to do so and a 35 m/s storm develops, the intercepted power increases threefold and may threaten disaster.

The above points are illustrated and amplified by Figure 2.1 showing the all-important cubic relationship between wind power and speed. Power is given in kilowatts per square meter of intercepted area (kW/m^2). Speed is indicated in meters per second (m/s), and also in knots (nautical miles per hour), which are often used to measure wind strengths at sea. The relationship between various units of speed is

$$1 \text{ m/s} = 1.95 \text{ knots} = 3.60 \text{ kph} = 2.24 \text{ mph} \tag{2.3}$$

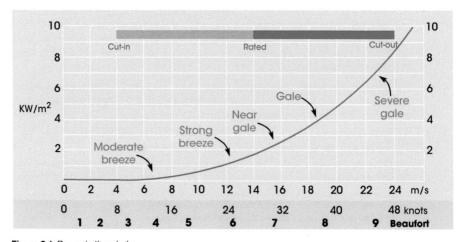

Figure 2.1 Power in the wind.

For convenience, it is helpful to remember that the speed in knots is very close to twice its value in meters per second.

Also shown is the *Beaufort Scale* used by sailors and meteorologists to describe wind strengths at sea. We have included Beaufort numbers 1–9 and indicated some of the corresponding descriptive terms against the curve. For example, Beaufort Force 4, with an average wind speed of 6.7 m/s, is described as a "moderate breeze." Beaufort Force 9, with an average speed of 22.6 m/s, is a "severe gale." Not shown are Force 10 ("storm"), 11 ("violent storm"), and 12 ("hurricane"), which are dangerous territory for wind turbines.

The horizontal bar at the top of the figure indicates typical cut-in and cut-out speeds of a large wind turbine, together with the speed at which it reaches its rated (peak) output power. Between cut-in and rated speeds, we expect the power generated to follow an approximately cubic curve; between rated speed and cut-out, it stays close to the rated value. Turbine manufacturers provide tables, or graphs, showing the electrical power produced by their machines at different wind speeds. A typical *power curve* for a 2-MW turbine is shown in Figure 2.2. Its precise form depends on the machine's efficiency at capturing winds of different speeds and on the systems—aerodynamic and electrical—that are designed to protect it from damage as it approaches cut-out.

We have already emphasized that short periods of strong wind, well above average speed, contribute disproportionately to a turbine's annual energy output. To quantify this vital point, we now consider a site with an average wind speed of 8 m/s measured over a complete year at the intended height of the turbine rotor. Such an annual average is normally very productive for a large wind turbine. However, referring back to Figures 2.1 and 2.2, we see that an 8 m/s wind, a "moderate breeze" on the Beaufort scale, would not be much use if it remained constant. It is the wind's variability that is so valuable.

Suppose variability is assessed using an anemometer and data logger to record the average wind speed hour by hour over a complete year. There are 8760 hours in a year, so we obtain 8760 readings. The data is now separated into intervals or "bins" 1 m/s wide and the number of occurrences (N) in each bin is noted. We now construct a histogram showing the distribution of hourly wind speeds over the year, giving the result in Figure 2.3(a). The "skewed" shape, tailing off gradually at high speeds, is typical of sites in temperate climates

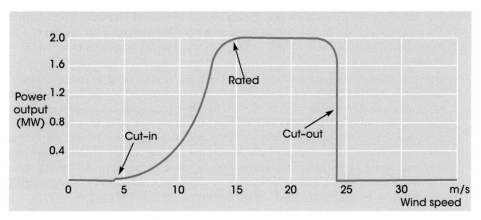

Figure 2.2 Typical power curve for a 2-MW wind turbine.

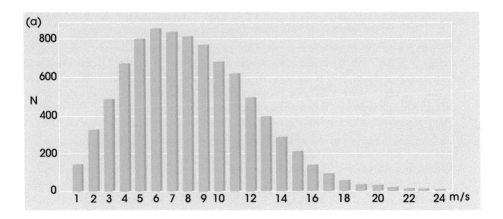

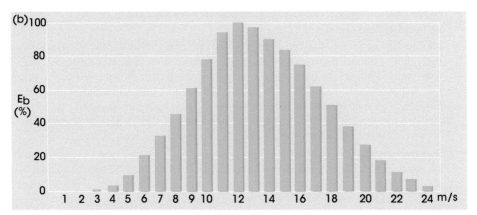

Figure 2.3 (a) A distribution of hourly average wind speeds over a complete year and (b) the relative amount of energy at each speed.

with variable winds, for example, near the coastlines of Western Europe and North America. The *mode* (most likely speed) is about 6 m/s, which is lower than the average value of 8 m/s. Most of the time the speed is between 2 and 15 m/s, but there are occasional gales and periods of calm.

If we now multiply the number of hours (N) in each bin by the cube of the speed, we obtain a measure of the bin's energy, giving the histogram shown in part (b) of the figure. Its vertical scale, normalized to a peak value of 100%, is not important here, but the shift of the peak toward higher wind speeds, compared with part (a), is highly significant. It shows that the most productive wind speeds, in terms of their annual energy contributions, are between about 8 and 20 m/s, with a peak at 12 m/s. Almost none of the total energy is contained below 4 m/s and above 24 m/s, so little is lost by choosing these as the cut-in and cut-out speeds. It is worth noting that histograms based on hourly average wind speeds ignore variability within the hour—which, once again, tends to increase the available energy.

We first mentioned the issue of intermittency at the start of Section 1.4, noting that stationary turbines tend to make people skeptical about wind energy. We now see that the main culprit (probably an unfair description!) is the cubic relationship between wind speed and power. For although a light breeze is quite noticeable when it blows in our faces, especially on a winter's day, to a wind turbine it is insignificant. Unfortunately, the public often fails to realize the enormous difference between the energy content of a light breeze and a really good blow—and why it is perfectly reasonable for turbines to spend some of their time in a state of suspended animation.

We have chosen a fairly high average wind speed of 8 m/s for the above illustration, but of course there are wide variations among sites (Figure 2.4). The cubic relationship between wind speed and power means that relatively small differences in average speed have a big impact on the available energy. A hilltop or offshore site offering 10 m/s would be a lot better than 8 m/s, assuming similar variability. At the other end of the scale, average speeds below 5 or 6 m/s rarely justify the installation of large turbines. So, in practice, most of today's machines operate within a fairly narrow range of average wind speeds.

This brings us to a very important sitting issue: how high should a turbine be mounted above the ground or sea surface? In general, the higher the better because wind strength increases with height, sometimes dramatically. In addition, short-term turbulence that produces unwelcome stresses on turbine blades and towers tends to decrease. So, there are big advantages in raising the hub height of a turbine. Over the years, many disappointing performances have been caused by ignoring this advice. The increase of average wind speed with height is referred to as *wind shear* and is often approximated by a simple *power law* of the form:

$$U/U_0 = (h/h_0)^\alpha \tag{2.4}$$

where U is the speed at height h, U_0 is the speed at some reference height h_0 (speed measurements are often made 10 or 20 m above the surface), and α is a *power law exponent* that depends on the surface roughness and the atmospheric stability.

Clearly, the rougher the surface, the more it retards the flow of wind and the greater the benefit of raising a turbine's hub height. The exponent α is greater for rough surfaces than for smooth ones and greater under stable than unstable atmospheric conditions. For example, snow-covered flat land and calm seas have α values around 0.1; arable crops around 0.2; and landscapes with trees, hedges, and buildings around 0.3. It is important to realize, however, that all are very site dependent and influenced by atmospheric conditions, ground undulations, hills, and other local features—and, offshore, by the state of the sea. In many

(a)   (b)

Figure 2.4 Different surfaces, different wind shear: scenes from England and Germany (Source: (a) with permission of Orsted. (b) With permission of Repower).

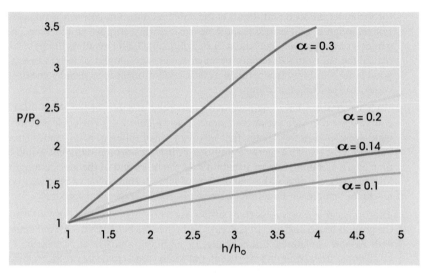

Figure 2.5 Variation of wind power with height, depending on surface roughness and atmospheric stability.

cases, the amount of wind shear varies with wind direction because of changes in the surface or landscape. Yet, in spite of the difficulty of assigning precise values of α to particular sites, the message is clear; raise the height of turbines, especially over rough surfaces, as much as possible consistent with the economics of providing high towers and possible concerns about visual intrusion.

Equation (2.4) describes the variation of wind speed with height, but of course it is the variation of wind *power* with height that affects turbine output. Using the by-now familiar cubic relationship between speed and power, we may write:

$$P/P_0 = (h/h_0)^{3\alpha} \tag{2.5}$$

where P and P_o are the power at heights h and h_0, respectively.

Figure 2.5 illustrates this result for the three α values mentioned earlier, for increases in height up to five times. For example, on an extensive flat site covered with arable crops having $\alpha = 0.2$, doubling the tower height may increase the available wind power by about 50%, trebling it by about 90%. Also shown is a curve for $\alpha = 0.14$, which is relevant to smooth grassland including large areas of the Great Plains in the USA. As we shall see in our discussion of wind mapping in Section 2.2.3, this is widely used in North America as a default value to estimate increases in wind power and speed with height. And in general, we conclude that wind shear is an important consideration when choosing sites for wind turbines, and that raising hub heights is highly beneficial to performance.

2.2   Wild wind

2.2.1   Introduction

The winds of the world are to varying degrees unpredictable, intermittent, fickle in speed and direction, and occasionally extreme. All this increases the challenge and fascination of designing effective wind energy systems.

We have already seen that wind's variability has a big impact on the annual energy intercepted by a turbine rotor. Periods of above-average speed are highly productive due to the cubic relationship between wind speed and power, and gusty winds tend to be more productive than consistent ones. But it is important to realize that variability occurs over different timescales, with important implications for turbine design, installation, and economics:

- *Long term (interannual)*. The wind resource at a particular site varies from year to year. For example, a coastal site in Western Europe may experience a series of strong "autumn gales" 1 year, but not the next. So, a single year's wind speed measurements (such as we used to construct histograms in Figure 2.5), although widely used to assess a site's potential, may not give an accurate long-term picture. Interannual variations up to 5% in average wind speed are quite common, with larger effects on energy production. An added complication is the climate change due to global warming and its impact on wind patterns. These are important issues for turbine installers and energy planners wanting to predict the performance of wind farms over 20 years or more.

- *Annual (seasonal)*. Most locations experience substantial variations in wind speed over the course of a year. As an example, Figure 2.6 represents average monthly values for a UK site where the autumn and winter months tend to be the most windy and summer months the calmest. Such annual variations are important when it comes to assessing wind energy generation in relation to seasonal electricity demand, or for planning turbine maintenance. The dots represent values for a single year; the vertical bars show the range of values recorded over a 10-year period, indicating substantial interannual as well as seasonal variation.

- *Daily (diurnal)*. In many cases, there are systematic daily variations in wind strength due to differential heating by the Sun as the Earth spins on its axis. Typically, winds are stronger during daylight hours than at night, with the largest variations occurring in the spring and summer months. Diurnal effects depend greatly on local geography, proximity to the sea and altitude (Figure 2.7).

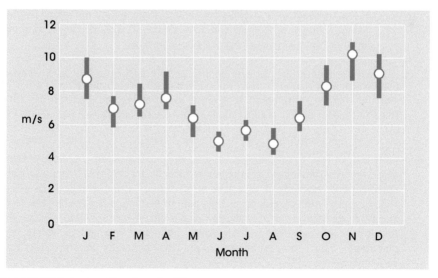

Figure 2.6 Wind speed variations throughout the year.

Figure 2.7 High uplands, wild winds (Source: With permission of Repower).

- *Short term.* Random fluctuations in wind speed and direction over periods of less than about 10 minutes are generally classed as *turbulence*. A discrete, and possibly violent, event within a turbulent wind stream is called a *gust*—or, when accompanied by rain or sleet and especially at sea, a *squall*. Turbulence, gusts, and squalls are worrying for turbine designers and operators since they place heavy dynamic loads on blades and structures, make big demands on turbine control systems, and may affect the quality of electricity delivered to the grid. Figure 2.8 shows a typical plot of wind speed over a 2-min period, with several strong gusts occurring around a high average speed of 14 m/s. Here the wind speed can fluctuate by 50% (and its power by about 140%) in less than 20 s—a big challenge for a large turbine.

The rapid development of wind technology over the past half-century has encouraged scientists and engineers to look ever more closely at the characteristics of "wild wind," developing and adapting statistical models to describe its variability and predict turbine performance. It is now time for us, too, to turn our attention to some key ideas in probability and statistics.

2.2.2 Wind statistics

When the world-famous Danish physicist Niels Bohr (1885–1962) famously remarked that "prediction is very difficult, especially of the future," it seems unlikely he was thinking of wind turbines—even the pioneering ones built by his compatriot Poul La Cour between 1891 and 1918 (see Section 1.3) of which he was presumably aware. But the statistical techniques that Bohr knew well have certainly proved important in wind engineering, especially for describing the fluctuating nature of the wind resource and for predicting its effect on turbine performance.

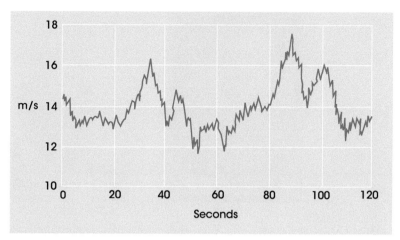

Figure 2.8 Turbulence and gusts.

We will focus here on some important practical questions:

- If we know the *average* wind speed at a particular site, can we make an accurate prediction of the *speed distribution* over the course of a full year and use it to estimate the annual energy yield of a wind turbine?
- What can wind statistics tell us about the proportion of time that a turbine is likely to be stationary and the chance that extreme events may damage or even destroy it?

We have already gone some way toward answering the first of these questions. In Section 2.1, we considered a site with an average annual wind speed of 8 m/s and imagined taking 8760 hourly readings to determine the speed distribution over a complete year. The data was separated into "bins" 1 m/s wide, and the number of occurrences in each bin noted. We then constructed histograms showing the distribution and the relative amount of energy contained within each "bin" (Figure 2.3). It is clear that this approach involves time-consuming measurements, and many years' data may be needed to give an accurate long-term picture. Measurements at the intended hub height of a large modern turbine—say between 70 and 120 m above the surface—are likely to be cumbersome and expensive. So the question naturally arises: if we know the average wind speed for a particular site, typically measured at a height of 10 or 20 m, can we extrapolate to the required hub height and describe the wind regime by a theoretical distribution rather than a comprehensive set of practical measurements?

Before answering this question, we should say a few words about probabilities. The histogram in Figure 2.3(a) allows us to estimate the probability of finding the wind speed in a particular "bin." For example, if the bin centered on 10 m/s contains 680 scores (out of a total of 8760) and bearing in mind that it is 1 m/s wide, the probability of the speed falling in the range 9.5–10.5 m/s may be estimated as 680/8760 = 0.0776. If we use bins half as wide, we expect half as many scores in each—and so on. So the value 0.0776 is in effect a *probability density per m/s*, rather than a simple probability. And we may regard the histogram as an approximation to an underlying continuous function representing the probability of finding

(a)   (b)

Figure 2.9 Lord Rayleigh and Waloddi Weibull (Source: (a) The Royal Society/Wikimedia Commons/Public domain. (b) Robert B. Abernethy/https://www.quanterion.com/wp-content/uploads/2014/11/ChapterOne .pdf/last accessed February 14, 2024).

the wind speed within a narrow range of values. Such a distribution is known as a *probability density function (pdf)*.

Not surprisingly, meteorologists have long been interested in describing wind speed distributions using pdfs. Two of the most valuable functions are named after the English physicist Lord Rayleigh (1842–1919) and the eminent Swedish engineer Waloddi Weibull (1887–1979) (Figure 2.9).

Rayleigh, a Cambridge professor and Nobel Prize winner with a polymath's interest in the physical sciences, developed a pdf to describe the scattering of radio waves. The *Rayleigh distribution*, which has since found widespread application in areas including wind engineering, is defined by a single parameter and is normally written in the form:

$$p(x) = (x/\sigma^2)\exp(-x^2/2\sigma^2) \tag{2.6}$$

where p is the probability of value x and σ is the mode of the distribution. Applying it to wind statistics, we may replace x by the wind speed U and denote the mean (average) wind speed by U_a. The mode, mean, and standard deviation (SD) of the Rayleigh distribution are then related as follows:

$$\sigma = U_a\sqrt{2/\pi} = 0.798\,U_a \tag{2.7}$$
$$SD = U_a\sqrt{(4-\pi)/\pi} = 0.523\,U_a \tag{2.8}$$

and the distribution may be restated in terms of wind speed:

$$P(U) = (\pi/2)\left(U/U_a^2\right)\exp\left\{(-\pi/4)\left(U^2/U_a^2\right)\right\} \tag{2.9}$$

Figure 2.10 shows its form for average speeds of 4, 6, 8, and 10 m/s. Note that in each case, the modal (peak) value occurs at a lower speed than the average value, due to the skewed shape of the curves.

How successful is the Rayleigh distribution at modeling annual wind speed distributions? For sites in temperate climates with variable wind regimes, including many in Western

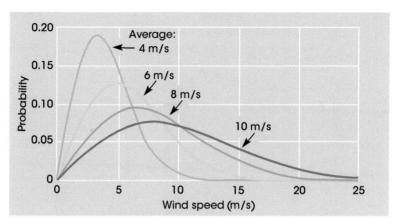

Figure 2.10 Rayleigh distributions for four average wind speeds.

Europe and continental North America, it generally gives an accurate description of the ratio between mode and average wind speeds and the tail off at high speeds. In cases where the only available information about a site is its average wind speed, the Rayleigh model is a very useful approximation.

The alternative *Weibull distribution*, championed by its Swedish namesake in a famous paper published by the American Society of Mechanical Engineers in 1951,[1] has found applications in fields as diverse as reliability engineering, weather forecasting, and radar—not to mention insurance and infant mortality! It has an extra parameter that may be independently adjusted, varying the ratio between mean and mode, the amount of "skew," and the way it approaches extreme values. Applied to wind statistics it covers a wider variety of wind regimes than the Rayleigh distribution and is often preferred when more data is available than just the average wind speed.[2] Its pdf takes the form:

$$p(x) = (kx^{k-1}/A^k)\exp(-x^k/A^k) \qquad (2.10)$$

where A is a *scale factor* and k is a *form factor* allowing the shape of the curve to be adjusted. Note that if $k = 2$ Equation (2.10) reduces to the same form as Equation (2.9), showing that the Rayleigh distribution is a special case of the Weibull or, alternatively, that the Weibull is a generalization of the Rayleigh. Again replacing x by the wind speed U, the Weibull pdf becomes:

$$P(U) = (kU^{k-1}/A^k)\exp(-x^k/A^k) \qquad (2.11)$$

Figure 2.11 shows three examples of Weibull pdfs for an average wind speed of 6 m/s, with k set to 1.5, 2.0 (Rayleigh) and 2.5. Note how, as k decreases, the peak moves to the left and the curve spreads out, denoting a wind regime with a wider speed range. In practice it is found that k values between 1.5 and 2.0 cover the great majority of cases, with a tendency for inland sites to have lower values than coastal and offshore ones.[2] Sites with relatively constant wind speeds, for example, those swept by consistent trade winds, are likely to display higher values. The relationships between mode, mean, and standard deviation are more complex than for a Rayleigh distribution, involving *gamma functions*, and are described elsewhere.[3]

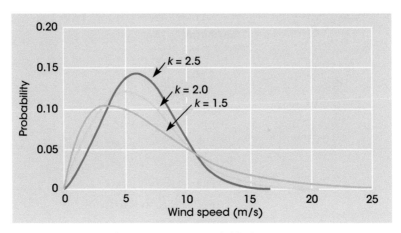

Figure 2.11 Weibull distributions for an average wind speed of 6 m/s.

We have labeled the abscissa "wind speed" and the ordinate "probability" but, as already noted, such curves are actually *density* functions. The probability of finding the wind speed at some precise value—say exactly 10 m/s—is vanishingly small; we must instead talk about the probability of it occupying a small range around 10 m/s. To illustrate this important point, the Rayleigh distribution for an average wind speed of 8 m/s is drawn again in Figure 2.12(a). Its value at a wind speed of 10 m/s may be calculated using Equation (2.9):

$$p(10) = (1.57 \times 10/64) \exp(-0.786 \times 10/64) = 0.0718 \qquad (2.12)$$

To find the probability of the speed falling within a small range of around 10 m/s, we multiply this figure by the "bin width." For example, if we choose a bin width of 1.0 m/s, denoting a speed range between 9.5 and 10.5 m/s, the probability is $0.0718 \times 1.0 = 0.0718$, equal to the small rectangular area in the figure. In other words, there is a 7.18% chance, over the year as a whole, of the speed falling within this range. More generally, the probability that the speed falls between any two values is found as the *area* under the curve between them—equivalent to integrating the function between the two limits. Note also that the total area under a pdf is unity because the wind speed must always be *somewhere* between zero and infinity.

Having found a suitable pdf, we can estimate the probability that the wind speed is below, or above, certain values. If we set these as a turbine's cut-in and cut-out speeds, we can predict the proportion of time the turbine is expected to be stationary. Suppose, for example, the speeds in this case are 5 m/s for cut-in and 20 m/s for cut-out, also indicated in Figure 2.12(a). The area under the curve between 0 and 5 m/s represents the probability that the wind blows at less than 5 m/s; the much smaller area above 20 m/s shows that it exceeds 20 m/s. And although we could find these areas by integrating the pdf between the relevant limits, it is more convenient to use the *cumulative distribution function* (*cdf*) shown in part (b) of the figure.

A cdf represents a "running integration" of the corresponding pdf. Starting at the left-hand end of the pdf, it progressively sums the area under the curve. Intuitively, it is "the area covered so far." And since the total area under a pdf is always unity, a cdf must eventually reach a value of 1.0. The cdf of a Rayleigh distribution is given by

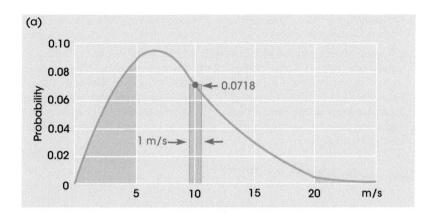

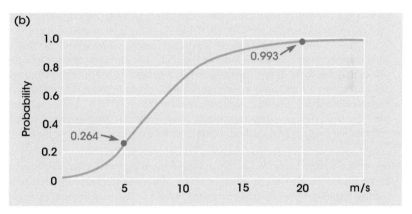

Figure 2.12 (a) Rayleigh pdf for an average wind speed of 8 m/s and (b) the cumulative distribution function.

$$d(x) = 1 - \exp(-x^2/2\sigma^2) \tag{2.13}$$

and is plotted in part (b) of the figure for the same average wind speed of 8 m/s. Its value at 5 m/s is 0.264, so there is a 26.4% probability that the wind speed is below cut-off and the turbine is stationary. In other words, over the course of a full year, we expect the turbine to be idle about 26% of the time due to low wind speeds (disappointing at first sight—until we recall the cubic relationship between wind speed and power!). The value of the cdf at 20 m/s is 0.993, so there is a 0.7% chance of the wind speed exceeding 20 m/s, causing the turbine to cut out for its own protection.

So far, we have been considering annual statistics—the distribution of wind speeds over a whole year, and the proportion of time spent above, below, or close to a particular value—for which the Rayleigh and Weibull pdfs are most appropriate. We now turn to two additional and important aspects of wind variability that may be characterized using statistical models: short-term fluctuations of wind speed in a turbulent air stream and the chance occurrence of extreme events.

In the previous section, we discussed random fluctuations in wind speed and illustrated a typical two-minute history of gusts in a turbulent air stream in Figure 2.8. Gusts and squalls place heavy dynamic loads on turbine blades and structures and test the effectiveness of control systems, so it may be useful to represent them by a statistical model.

When a random variable such as short-term wind speed displays variations above and below its average (mean) value, and when the variations are due to multiple independent factors, the *Gaussian* distribution is the classic way of describing them. Devised by the German mathematician Carl Friedrich Gauss (1777–1855), (Figure 2.14) who originally used it to describe random errors in astronomical observations, it finds widespread application throughout science and technology—so much so that it is also known as the *normal* distribution. Its pdf is usually written as

$$p(x) = (1/\sigma\sqrt{2\pi})\exp(-(x-\mu)^2/2\sigma^2) \tag{2.14}$$

where μ is the mean (average) value, σ is the standard deviation, and σ^2 is the variance. Gaussian curves are symmetrical and have a "bell" shape, implying that fluctuations of a given size above and below the mean are equally probable. Figure 2.13 shows three typical examples: a red curve for an average wind speed of 10 m/s with a high degree of speed variability or turbulence ($\sigma^2 = 4.0$); an orange curve for the same average speed but less variability ($\sigma^2 = 0.8$); and a green curve for a lower average speed of 6 m/s and intermediate variability ($\sigma^2 = 2.0$). Note that this form of pdf shows the relative probability of different wind speeds occurring in a turbulent air stream over comparatively short time scales, but it tells us nothing about *temporal structure*, for example, how rapidly the speed can change between one moment and the next, or how often gusts occur. Such questions are generally tackled by *spectral* or *correlation analysis* of wind speed data, a more advanced topic covered elsewhere.[3]

Our final topic in this section concerns "extreme events" that could seriously damage, or even destroy, a wind turbine. We have previously noted that a wild and turbulent air stream will, from time to time, produce wind speeds far above the average value. A machine designed to withstand severe gales cannot necessarily survive a "once in a lifetime" gust superimposed on

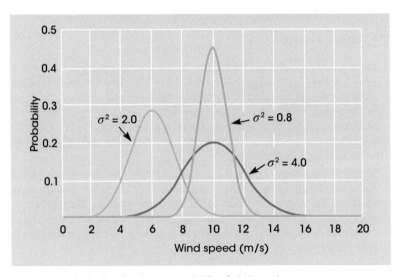

Figure 2.13 Gaussian distributions for short-term variability of wind speed.

a storm-force blast. Turbine designers need to sleep at night, feeling confident that a machine intended for a working life of, say, 25 years, is extremely unlikely to face such a threat.

Once again, it is all a question of probability and statistics. We can never be certain that a damaging gust will not occur next year, or in the next decade. But it is reassuring to know that the *return time* of potentially catastrophic events—the average interval between recurrences—is considerably greater than the design lifetime of wind turbines. Similar concerns arise in other fields. Ships occasionally meet freak waves; coastal defenses are sometimes overwhelmed; and insurance companies face huge claims for earthquake damage. But it is not only natural forces and disasters that have encouraged statisticians to develop what is known as *extreme value theory*; one of its pioneers spent a working lifetime investigating the strength of cotton thread.

Leonard Tippett (1902–1985), a physics graduate of Imperial College London who worked for a cotton research institute in Manchester, England, realized that the strength of yarn for weaving depended on its weakest fibers and that occasional failure was, in his terms, an "extreme event." The statistical techniques he pioneered were taken up and formalized by Emil Gumbel (1891–1966), a Heidelberg mathematics professor who was forced from his post by the Nazis and eventually became an American citizen(Figure 2.14). We should perhaps not be surprised that the *Gumbel distribution* has since found widespread application to extreme events including the freak winds that can threaten wind turbines.

The type of problem addressed by Tippett and Gumbel may be summarized as follows. We cannot generally build up the history and statistics of extreme events that occur very infrequently by waiting for them to happen, because it is likely to involve a very long wait! For example, a freak gust sufficient to destroy a turbine may only occur at a particular site once a century on average. Many centuries of measurements would be needed to build up a reliable database. So, the question arises: is it possible to record extreme events over a much shorter period and extrapolate the data using a statistical model?

We should be clear about the very small probabilities involved here. For example, if we are interested in the strongest gust over, say, a 10-second interval that is expected to occur once in a hundred years at a particular location, we are talking about a probability of occurrence in *any* 10-second interval of $1/(100 \times 8760 \times 360) = 3.2 \times 10^{-9}$. This simply underlines the fact that very extreme events are extraordinarily unlikely!

Figure 2.14 Carl Friedrich Gauss and Emil Gumbel (Source: (a) Christian Albrecht Jensen/Wikimedia Commons/Public domain. (b) Courtesy of Tuncel M. Yegulalp).

To illustrate the use of the *Gumbel distribution*, suppose that the speed of the strongest gust at a particular site has been recorded each year for ten successive years. We find the mean value m and standard deviation σ of the 10-point dataset and calculate two parameters:

$$\beta = (\sigma\sqrt{6})/\pi = 0.780\,\sigma \tag{2.15}$$

$$\mu = m - 0.577\,\beta \tag{2.16}$$

The probability density of annual "extreme events" (i.e., peak gusts) occurring at other wind speeds U_e may now be found using the Gumbel distribution:

$$p(U_e) = (1/\beta)\exp(-\gamma)\exp\{-\exp(-\gamma)\} \tag{2.17}$$

where $\gamma = (U_e - \mu)/\beta$ $\tag{2.18}$

We shall find the corresponding cdf especially useful in this case:

$$d(U_e) = \exp\{-\exp(-\gamma)\} \tag{2.19}$$

As previously noted, a cdf computes "the area covered so far" under the corresponding pdf and gives the probability of finding the wind speed *below* a certain value. However, we are concerned here with extreme winds *above* a certain value, in other words with the complementary function:

$$1 - d(U_e) = 1 - \exp\{-\exp(-\gamma)\} \tag{2.20}$$

Suppose the strongest gust at an exposed hilltop site in Scotland has been recorded each year for 10 years. The 10 values have a mean of 38 m/s and standard deviation of 6.5 m/s, giving $\beta = 5.07$ m/s and $\mu = 35.1$ m/s. Using Equation (2.20), we may construct the curve in Figure 2.15, showing the probability that the highest gust recorded in any 1 year will exceed a certain value. For example, the probability of it exceeding 40 m/s is about 0.3, or 30%.

However, we wish to predict the chance of a dangerous gust over a century, not just a single year. For a return time (average time between occurrences) of 100 years, the probability in any one year must be 0.01. Figure 2.15 shows that this corresponds to a speed of 59 m/s. We conclude that once in every hundred years, a gust in excess of 59 m/s can be expected. This is, of course, far above the average annual peak of 38 m/s and massively greater than the mean wind speed of this or any other site. Turbine designers and would-be installers on Scottish hilltops take note!

We have introduced a number of statistical concepts in this section, and it may be helpful to end by reviewing the four types of probability density function used to characterize the variability of wind:

- *The Rayleigh distribution* is valuable for modeling the annual distribution of wind speeds at a particular site when the only available information is the average (mean) wind speed.
- *The Weibull distribution* is generally preferred to the Rayleigh distribution when more comprehensive wind speed data is available since it gives more flexibility. It includes the Rayleigh as a special case.

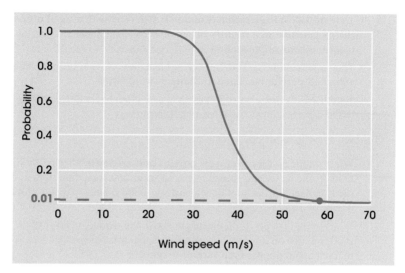

Figure 2.15 The probability of peak wind gusts, based on a Gumbel distribution.

- *The Gaussian (normal) distribution* may be used to model short-term fluctuating wind speeds in a turbulent air stream.
- *The Gumbel distribution* is valuable for predicting the maximum strength of gust to be expected over a long-time scale (e.g., a century), based on annual records over a much shorter time scale (e.g., a decade).

As Niels Bohr said, prediction, especially of the future, is very difficult. But in wind engineering, it is certainly important to make the effort!

2.2.3 Mapping and forecasting

The extraordinary growth of the global wind industry over the past 30 years has coincided with big efforts to map the world's wind resources. As far back as 1987, a wind atlas was produced in the USA showing national and state wind data. By 2005, the National Renewable Energy Laboratory (NREL) located in Colorado had developed advanced digital modeling and mapping techniques to produce revised maps with greatly enhanced resolution.[4] A high-quality *European Wind Atlas* was published in 1989 under the auspices of the European Community.[5] Since then the wind mapping techniques pioneered in America and Europe have been adopted in well over 100 countries. Using data from the European Wind Atlas, modified to take account of local features and conditions, it is now possible to predict the wind regime at a particular site with the aid of a *Wind Atlas Analysis and Application Program (WAsP)*.

Back in Section 1.2, we considered the world's major wind patterns, illustrating them in Figures 1.4 and 1.5. But this was big scale; we will now home in on more detailed national and local wind atlas data, followed by some comments on wind forecasting.

The uses and limitations of modern, high-resolution, wind atlases may be summarized as follows:

- They provide energy planners and policy makers with a rapid overview of the wind energy potential of a country or region, its geographical distribution, and proximity to local centers of population.
- They offer commercial developers and installers of wind turbines indications of likely areas and sites for individual turbines or wind farms.
- Together with specialist software, they may often be used to calculate the wind potential of a particular site with reasonable accuracy.

However, as we have previously emphasized:

- Wind resources vary from year to year and are much affected by hills and undulations in the landscape, impediments to wind flow including trees and buildings, and the roughness of the land or sea surface. Wind atlases may not provide the required degree of detail.
- Wind strength depends on the height above the ground or sea surface. Wind atlas data is often based on low-level measurements that are extrapolated to turbine hub heights making assumptions about surface roughness.

Although wind atlases have improved greatly over the years and have undoubted value for overall planning and prediction, their use should be tempered with caution, especially when attempting to estimate wind resources at sites with complex geography and surface conditions. With these caveats in mind, we now consider three examples of wind atlas data.

Figure 2.16, based on a portion of the European Wind Atlas, gives an overview of wind resources around the coasts of Western Europe. The various countries are denoted by their initial letters (N = Norway, Sw = Sweden, D = Denmark, G = Germany, Ne = the Netherlands, E = England, and so on). Color coding is used to denote average wind speeds 50 m above the surface, and the values in the table refer to flat coastal sites, appropriate for many large onshore turbines. Noteworthy features include:

- The good-to-excellent wind resources along the coastline stretching from Norway right down to northwest France; also in the UK, Ireland, and a small region of northwest Spain exposed to the Atlantic Ocean. All these coasts have prevailing westerly winds (see also Figure 1.5).
- A high wind resource on the Mediterranean coast of France, related to the famous *Mistral* wind that channels down the valley of the river Rhone (see also Figure 1.5), and the north-westerly *Tramontane* that accelerates between the Pyrenees mountains and the Massif Central.
- A small region of strong winds in northeast Spain, just to the south of the Pyrenean chain dividing Spain from France (and a correspondingly calm area on the other side of the mountains).

Although useful as a summary, a figure on this scale hides many important details. For example, the whole of Scotland is depicted as a turbine installer's dream; yet, there are big differences in wind resource between Scottish mountains and valleys, inland and coastal areas, and the east and west coasts. Such mapping gives a general indication of wind energy potential and is probably most valuable to energy planners.

We have concentrated on wind speeds at European coastal sites because of their interest in wind energy. However, the big differences in wind potential caused by surface roughness

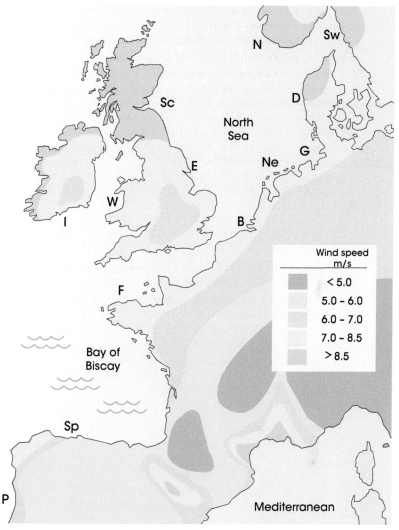

Figure 2.16 Large-scale wind resources of Western Europe. The wind speeds in the table are annual averages at a height of 50 m above flat coastal sites.

and local geography are acknowledged in the European Wind Atlas by listing average wind speeds for five different site categories. In order of increasing wind potential, these are

- *Sheltered terrain.*
- *Open plain.*
- *At a sea coast.*
- *Open sea.*
- *Hills and ridges.*

The speeds given in Figure 2.16 relate to the third of these categories. Values for sheltered terrain are considerably lower; for hills and ridges, considerably higher. Of course, all such estimates are approximate and depend on the characteristics of a particular site.

In the USA, wind potential is assessed rather differently by defining seven *wind power classes*. Each class covers a range of average power densities in W/m^2 and equivalent wind speeds at heights of 10, 30, and 50 m above the surface. Increases in power and speed with height are generally based on the so-called *1/7 power law*. Referring to our discussion of the effects of surface roughness in Section 2.1, you may recall that a *power law exponent α* was used to predict the increase of wind speed and power with height above the surface. Smooth surfaces have α values as low as 0.1 and rough surfaces as high as 0.3 or 0.4 (see Figure 2.5). A value of 0.14, or 1/7, is widely accepted for extensive grassland sites including large areas of the Great Plains in the USA, so the *1/7 power law* is often used as a default in the absence of more specific information. It predicts the following increase in wind power with height (see also Equation (2.5)):

$$P/P_0 = (h/h_0)^{3\alpha} = (h/h_0)^{0.42} \tag{2.21}$$

If the height increases from 10 to 30 m, the power is multiplied by 1.586 (say 1.6); from 10 to 50 m, by 1.966 (say 2.0). These factors are often used for tabulations of wind power classes.[3, 6]

Figure 2.17 shows a wind map for northern California, based on information in the NREL atlas.[4] Its area resolution is about 10 times that of Figure 2.16, with correspondingly greater

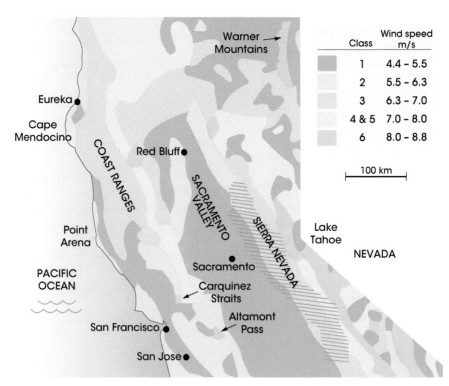

Figure 2.17 Wind resources in northern California.

detail. We have used color coding to indicate the various wind power classes (combining the relatively close-spaced classes 4 and 5). Average wind speeds are given for a height of 50 m taking local topology and surface conditions into account. In coastal ranges and mountain regions, the wind classes generally refer to sites on hilltops or ridges. We see that classes 1 and 2 typify much low-lying land including the Sacramento Valley, with class 6 attained in parts of the Sierra Nevada and Warner Mountains. Sites suitable for wind turbines are normally rated class 3 or above.

One apparent anomaly is the small, isolated, class 6 area about 100 km east of San Francisco. The Altamont Pass through the Diablo Range is an internationally famous cradle of the American wind energy industry, and its wind farm has one of the densest populations of wind turbines in the world. Initially developed around 1980 in the heady days of the "Californian wind rush," it still supports several thousand turbines, mostly fairly small by modern standards. Another limited class 6 area is in the Carquinez Straits, a tidal gap between coastal hills where the Sacramento River flows toward the Pacific Ocean.

As turbines installed by electric utilities become larger and more powerful, their rotor diameters demand greater hub heights. Wind atlases generally extrapolate wind data to heights of 50 m or perhaps 80 m, but we are now moving into the era of 100–150 m hub heights for multi-megawatt machines. The table in Figure 2.18 shows what happens to the speeds and power densities of the various wind power classes when extrapolated from 50 to 125 m, again assuming a 1/7 power law ($\alpha = 0.14$): speeds increase by about 14% and power densities by about 50%. Power class 6 now reaches up to 10 m/s average wind speed. Once again, the advantages of "moving skywards" are emphasized.

Actually, very few onshore turbines operate in average wind speeds of 10 m/s and above, but offshore it is a different story. Turbines tend to be larger and higher and the wind stronger and less turbulent. So, this seems a good moment to leave the safety of the land and venture into the wild North Sea, where many of Europe's largest turbines are being installed. Sandwiched between the coasts of Scotland and England on one side and Norway, Denmark, Germany, the Netherlands, and Belgium on the other, the North Sea's prevailing westerly winds offer an impressive wind resource, and its limited depth makes turbine installation relatively economical.

Wind power class	Height = 50 m		Height = 125 m	
	speed m/s	power W/m²	speed m/s	power W/m²
1	4.4 – 5.5	100 – 200	5.0 – 6.3	150 – 290
2	5.5 – 6.3	200 – 300	6.3 – 7.2	290 – 440
3	6.3 – 7.0	300 – 400	7.2 – 8.0	440 – 590
4	7.0 – 7.5	400 – 500	8.0 – 8.6	590 – 730
5	7.5 – 8.0	500 – 600	8.6 – 9.1	730 – 880
6	8.0 – 8.8	600 – 800	9.1 – 10.0	880 –1200
7	8.8 – 11.9	800 – 2000	10.0 – 13.6	1200 – 3000

Figure 2.18 Wind speeds and power densities at heights of 50 and 125 m.

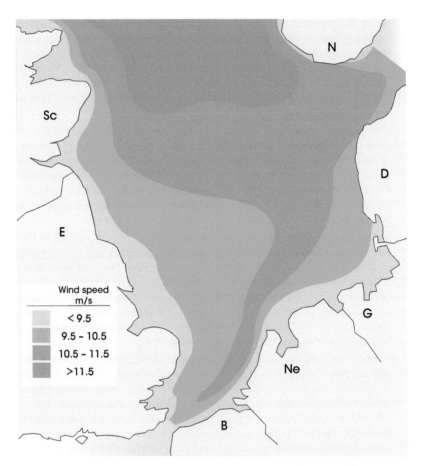

Figure 2.19 Estimated average wind speeds at 125 m above mean sea level in the North Sea.

Figure 2.19 draws on information from the European Wind Atlas, but extrapolated to a height of 125 m above mean sea level using the 1/7 power law ($\alpha = 0.14$). It may seem surprising to use this value of α once again, but it is generally reckoned about right for moderate-to-rough sea states—although we must bear in mind that wind shear above the sea depends on many factors including wave heights and distances from windward shores.

Countries bordering the North Sea are indicated by their initial letters, as shown in Figure 2.16, and average wind speeds are arranged in four color bands. We have now moved decisively up the power scale, especially between Scotland and Norway, and areas in the top two-color bands would be rated as wind power class 7 under the American system. As we might expect—and unlike the two land maps already discussed—the North Sea has huge areas of consistent wind resource undisturbed by land irregularities. Prevailing westerlies blow especially hard against exposed coasts of Norway and Denmark, whereas the English landmass provides some shelter. The figure nicely illustrates the North Sea's attractions for wind engineering, in spite of the challenges and costs of working in a harsh marine environment.

In this brief account of wind mapping, we have concentrated on three large-scale areas that are familiar to most people. Of course, much of the value of modern wind atlases stems from their high resolution, allowing planners and installers to home in on much smaller areas. You may wish to consult the NREL,[4] European,[5] or a national atlas to explore the wind resource in a location near to you. Also, as we shall see in the next section, it is often possible to predict the wind resource at a particular site with reasonable accuracy using specialized software.

We conclude this section with some comments on wind forecasting. Actually, the distinction between mapping and forecasting is quite subtle. After all, a map based on historical wind data is a type of long-term forecast that assumes past information can be used to predict the future. However, we generally reserve the word "forecasting" for relatively short-term predictions covering hours or days. The weather forecasts we see on our TV screens fall in this category.

What is the importance of short-term forecasting for the wind energy industry? There are powerful reasons for wanting to know what the wind is going to do in the next few hours, or days, especially for the operators of electricity grids. For example, if an Atlantic weather front with associated high winds is approaching the North Sea, the combined output of coastal and offshore wind turbines will increase dramatically. This may impact heavily on the electricity grids to which the turbines are connected.

A guiding principle of renewable energy, including wind energy, is "use it or lose it." Unlike conventional electricity generators that burn fossil fuels, wind turbines have very low operating costs because their "fuel" is free. Wind power should ideally be accepted by a grid whenever it is available. However, a second guiding principle applies to electricity grids: they cannot store electricity, so a balance must be maintained at all times between generation and consumption. To achieve this in a cost-effective way, the various power plants supplying a grid are brought in and out of operation, or *scheduled*, according to their operating costs. A sudden increase in wind power must be offset by reductions from other generators. But this is not as simple as it sounds because different types of generators vary greatly in their flexibility and speed of control. This makes accurate wind forecasting over periods of a few hours especially important for systems with a substantial proportion of wind generation.

A further challenge relates to the operation of modern deregulated electricity markets, in which electricity is traded like other commodities. Although market details vary a lot among countries, the basic idea is that the producer's contract to supply electricity to the grid at certain prices and times in the future—typically up to 24 or 48 hours ahead. Clearly, this places great emphasis on the value of accurate wind forecasting over a few days, for wind farm owners as producers and for grid operators and electric utilities as purchasers.

Wind forecasting is therefore important from both technical and operational perspectives. A great deal of effort has recently been put into improving its meteorological and analytical background,[6] a trend that will surely accelerate as more and more wind energy is integrated into grid networks.

2.3 Predicting turbine output

We now come to an extremely important question: how much electrical energy can a turbine be expected to produce over the course of an average year? The answer is the key to its economic justification—the need to offset capital, installation, and running costs against the value of the electricity generated.

We have already emphasized the crucial importance of the wind regime for predicting the performance of a wind turbine. The cubic relationship between wind speed and power means that quite small changes in average wind speed have a big impact on turbine output. It is therefore vital to assess a site's wind regime as accurately as possible before deciding whether to proceed with installation. A minimum requirement is a reliable estimate of the mean wind speed at the proposed hub height.

One way of assessing the wind potential at a particular site, known as the *wind atlas method*, may be used when it is too difficult or time consuming to make practical wind speed measurements. The method takes data from a high-definition wind atlas, modifies it to allow for the site's special features, and produces a site-specific prediction. In Europe, it is widely implemented using the European Wind Atlas[5] together with WAsP software,[7] illustrated by Figure 2.20. The atlas was originally produced from meteorological data collected at some 220 European stations over many years, together with information about sheltering by obstacles, surface roughness, and surface geography (known as *orography*) at each location. The meteorological data was processed by a series of three theoretical models to remove site-specific influences, illustrated by the upper line in the figure. The resulting atlas summarizes regional and local wind resources unhindered by purely local conditions. To estimate the resource at a new site—one being considered for a turbine or wind farm—the process is reversed. Data from the wind atlas is now modified by the WAsP program using information about the site's orography, roughness, and obstacles, producing the required prediction. This is represented by the lower line in the figure.

Clearly, the accuracy of the method depends on the validity of the theoretical models that generate the wind atlas in the first place and on the quality and quantity of information about the new site fed into the WAsP program. It is most successful for predicting wind conditions at uncomplicated sites, but less likely to yield accurate results in highly structured landscapes, including mountain territory and sites with multiple, complex, obstructions.

Of course, the best way of defining the wind resource is by careful onsite measurements, assuming the necessary time and money are available. Back in Section 2.1, we considered using an anemometer and data logger to record average wind speed hour by hour over a complete year. The data was then separated into "bins" 1 m/s wide to produce a speed distribution—see Figure 2.3(a). In our discussion of wind statistics in Section 2.2.2, we explained how such a distribution may be fitted to a Rayleigh or Weibull function. This is normally the starting point for a prediction of annual turbine output.

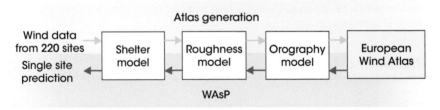

Figure 2.20 Predicting a site's wind regime by the wind atlas method.

Suppose that wind speed measurements have been made over a complete year, typically at a relatively low height, and extrapolated to the hub height of the intended turbine. One year's data is unlikely to be entirely representative of the wind regime at the site, so we fit a Rayleigh or Weibull function to "iron out" chance fluctuations and provide a smoother version of the distribution. This is shown in part (a) of Figure 2.21, which gives the number of hours spent in each speed "bin".

We now turn to the turbine's *power curve* shown in part (b), which shows the manufacturer's estimate of power output at each value of wind speed. This turbine is rated at 3 MW with a cut-in speed at 4 m/s and cut-out at 25 m/s and reaches its rated power at about 14 m/s. We multiply each power value by the corresponding number of hours in part (a) to give the electrical energy in each speed bin. For example, the 10 m/s bin is occupied for about 580 hours in a year and generates a turbine output of about 1.8 MW. The bin energy, referred to as the *class yield*, is therefore $580 \times 1.8 = 1040$ MWh. The class yields for all wind speeds form the histogram in part (c) of the figure.

The total annual energy is found by summing the individual class yields to give the *cumulative yield* in part (d). It amounts to 9.2 GWh, allowing us to predict the *capacity factor* of this particular turbine at this particular site. First discussed in Section 1.4 and defined by Equation (1.3), the capacity factor is the ratio between the annual energy actually produced (E_a) and the maximum that would be produced if the turbine worked throughout the year at its full rated output. In this example, we have

$$C_f = E_a/(8760 \times P_r) = 9.2 \, \text{GWh}/(8760 \times 3 \, \text{MW}) = 0.35 = 35\% \qquad (2.22)$$

which is fairly typical for a large, efficient, modern turbine operating at a good onshore site with a high average wind speed.

Figure 2.21 illustrates several important points that we have discussed previously:

- Although the turbine is quite often stationary due to low wind speeds (in this example, for about 1400 hours out of the annual total of 8760 hours, or 15% of the time), little is lost because wind speeds below 4 m/s contain very little energy.
- Wind speeds above cut-out are certainly very energetic, but since they only occur for a very small proportion of the time (in this example, well under 1%), again very little is lost.
- Class yields are highest between about 9 and 15 m/s, well above the mean and mode wind speeds, due to the cubic relationship between speed and power.
- The turbine's power curve approximates a cubic between cut-in and rated speeds, but its precise form depends on the efficiency (aerodynamic, mechanical, and electrical) with which it converts different wind speeds into electrical output.

Hopefully, the calculations work out favorably, the decision to purchase the turbine is made, and it is installed without mishap. Large wind turbines are invariably equipped with an anemometer for measuring wind speed on the nacelle, very close to hub height (see Figure 2.22). So now comes the opportunity to check theory against practice by actual recordings of wind speed and electrical output over a complete year or more. Time for turbine designers, manufacturers, and wind farm operators to cross their fingers!

We have now covered broad issues of turbine power and energy production, relating them to the statistical properties of the wind—the free "fuel" that drives a modern industry. In

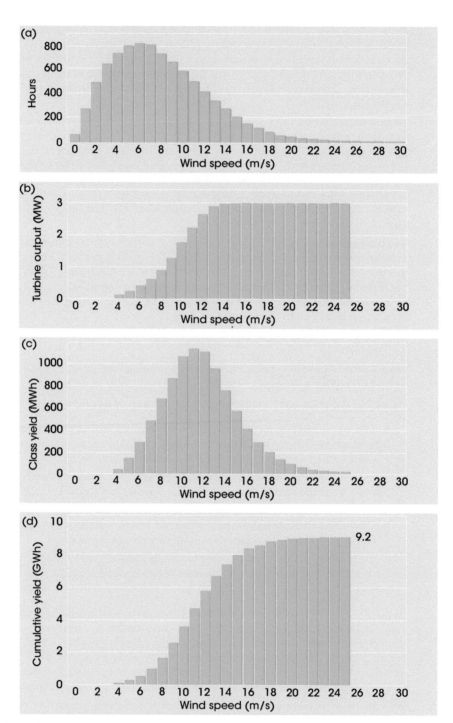

Figure 2.21 Predicting turbine output over the course of a year.

Figure 2.22 Checking the anemometer on a high nacelle (Source: With permission of Repower).

the next chapter, we turn our attention to more detailed, but equally fascinating, matters: the principles underpinning the design, manufacture, and installation of today's megawatt turbines.

Self-assessment questions

Q2.1 A wind turbine operates between 5 and 25 m/s; calculate this range in mph.

Q2.2 Why do wind turbines have a cut-out wind speed?

Q2.3 How does wind speed change with height from the ground?

Q2.4 How does wind turbine power output change with height from the ground?

Q2.5 How is wind velocity affected by the ground roughness?

Q2.6 What are the similarities and differences between the Raleigh and the Weibull distributions of wind speed?

Q2.7 When is a Gumbel distribution more applicable than Gaussian distribution?

Q2.8 How can wind speed data at 10 m be used for estimating the performance of a 60 m tall wind turbine?

Q2.9 What is a wind capacity ratio and how is it calculated?

Q2.10 Give ranges of wind turbine capacity ratios, from average to maximum in land and offshore installations.

Problems

2.1 Calculate the energy (in Joules) of $2\,\text{m}^3$ of air moving at $5\,\text{m/s}$. How long could this energy be used to illuminate a 60-W light bulb if it could be converted to electricity with 50% efficiency?

2.2 What is the velocity of the wind that provides five times the power provided by wind with a velocity of 8 m/s?

2.3 What fraction of the total daily energy produced by a wind turbine is produced during the daylight hours if the wind velocity is 10 m/s during 14 hours of day and 4 m/s during 10 hours of night?

2.4 Use the figure below to estimate the relative power provided by a wind turbine at an altitude of 600 m compared with one at the ground level in March.

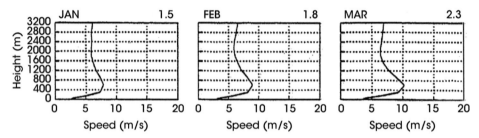

Figure: Typical variation of wind speed with altitude .(Source: https://www.nrel.gov/docs/legosti/old/7809.pdf)

2.5 Annual energy production estimate

Based on an average wind speed of 10 m/s, estimate the annual energy production from a horizontal axis wind turbine with a 15 m diameter. Assume that the wind turbine is operating under standard atmospheric conditions ($\rho = 1.22\ \text{kg/m}^3$).

2.6 Wind speed variation with height
   (a) Determine the wind speed at a height of 50 m over flat terrain with a few trees, if the wind speed measured at a height of 10 m is 4 m/s. For your estimate, use two different wind speed estimation methods.
   (b) Using the same methods as part a, determine the wind speed at 50 m if the trees were all removed from the terrain.

2.7 Weibull distribution calculations

Assuming the Weibull parameters to be $c = 5$ m/s, and $k = 1.6$.
   (a) What is the average velocity at this site?
   (b) Estimate the number of hours per year that the wind speed will be between 6 and 8 m/s during the year.
   (c) Estimate the number of hours per year that the wind speed is above 15 m/s.

2.8 Annual power estimation problem—Betz-type machine

Estimate the annual energy production of a wind turbine with a 10 m diameter operating at standard atmospheric conditions ($\rho = 1.225\ \text{kg/m}^3$) under 6 m/s average wind speeds. Assume that the site wind speed probability density is given by the Rayleigh density distribution.

2.9 Actual data analysis and power prediction

Based on the spreadsheet (TX_WindSpeed.xls), obtained from NOAA Local Climatological Dataset, which contains 3 months of data (mph) for Alpine Casparis Municipal Airport, TX, determine:

(a) The average wind speed for the 3 months
(b) The standard deviation
(c) A histogram of the velocity data (via the method of bins—suggested bin width of 2 mph)
(d) From the histogram data develop a velocity–duration curve
(e) From above develop a power–duration curve for a given 20 kW turbine at the TX site.
For the wind turbine, assume:

$$P = 0\,\text{kW}\ 0 < U \leq 6\,(\text{mph})$$
$$P = U^3/400\,\text{kW}\ 6 < U \leq 30\,(\text{mph})$$
$$P = 20\,\text{kW}\ 30 < U \leq 40\,(\text{mph})$$
$$P = 0\,\text{kW}\ 40 < U\,(\text{mph})$$

(f) From the power duration curve, determine the energy that would be produced during these 3 months in kWh.

Answers to questions

Q2.1 11.25–56.25 mph.

Q2.2 To protect it from damage caused by very strong winds.

Q2.3 It increases according to power law.

Q2.4 By the cube of wind velocity.

Q2.5 Higher ground roughness reduces low-elevation wind velocity.

Q2.6 Raleigh is a special case om the Weibull distribution applicable when the only information is the mean wind speed. The Weibull distribution is preferred whem more detailed wind speed data are available.

Q2.7 The Gaussian distribution is used to describing short-term fluctuations of wind speed, whreas the Gumbel distribution is more applicable to prediciting maximum strength of wind gusts over a long period of time. Note that this is an oversimplified answer and you are expected to elaborate in more detail about thes differences of these statistical wind speed representations.

Q2.8 Calculate the wind speed at 100 m by using a power law with an exponent determined by the stability of the atmosphere.

Q2.9 Ratio of hrs during a year that wind turbines produce power at rated capacity over the total hours of a year (i.e., 8760 hours).

Q2.10   Capacity ratios depend on location and season; typicall higher off shore that onshore and higher in the winter than in the summer, in the same location, as stronger winds over longer times are more frequent in the winter in most locations. Capacity ratios typically range from 0.2 to 0.45.

References

1. W. Weibull. A statistical distribution function of wide applicability. *Transactions of the American Society of Mechanical Engineers*, 18, 293–297 (1951).
2. S. Heier. *Grid Integration of Wind Energy Conversion Systems*, 2nd edition. John Wiley & Sons Ltd: Chichester (2006).
3. J.F. Manwell *et al*. *Wind Energy Explained: Theory, Design and Application*, 2nd edition. John Wiley & Sons Ltd: Chichester (2009).
4. NREL Wind Energy Resource Atlas of the United States. http://rredc.nrel.gov/wind/pubs/atlas/maps.html. (Accessed on May 4, 2024)
5. European Wind Atlas. www.windatlas.dk/europe., 1989 (Accessed on May 4, 2024)
6. T. Ackermann (ed.). *Wind Power in Power Systems*. 2nd edition, John Wiley & Sons Ltd: Chichester (2012).
7. Wind Atlas Analysis and Application Program (WAsP). www.wasp.dk.

3 Wind turbines

3.1 Turbine types and sizes

Wind turbines of many shapes and sizes have been invented, constructed, and marketed over the past century. The brief history given in Section 1.3 outlined some key developments:

- The steady decline of the traditional windmill for grinding corn and pumping water.
- An increasing interest in using the wind to generate electricity, especially in Denmark in the first half of the 20th century.
- The development of global wind turbine capacity which began in the late 1970s, accelerated as the new millennium approached, and now delivers substantial amounts of clean, renewable power to the world's electricity grids.

In previous chapters, we considered some general aspects of turbine performance including annual electricity production in a given wind regime, and how this relates to average household consumption. There are a few more scene-setting aspects to be covered before we get down to detail and explore the various components of large modern turbines—rotors and blades, gearboxes, electrical generators, power electronics, and control systems—and explain the principles that underpin their design and operation.

In this book, we concentrate on large *horizontal-axis wind turbines* (*HAWTs*) rated between about 1 and 12 MW that dominate the current international surge in wind energy capacity. The vast majority have three-bladed rotors mounted upwind of their towers, and every major manufacturer in the world today promotes this basic configuration as the best compromise between cost, technical performance, and long-term reliability (you may refer back to Figure 1.13). Alternative designs are proposed and championed from time to time, and some large vertical-axis machines are currently under development. We shall consider them a little later. But our basic standpoint from now on is the near-universal one: tomorrow's global wind turbine industry, like today's, will be overwhelmingly based on HAWT technology.

Why do almost all large HAWTs employ three blades rather than one, two, four, or even more? In fact, a few one-bladed designs, and many two-bladed ones, have seen the light of

Onshore and Offshore Wind Energy: Evolution, Grid Integration, and Impact, Second Edition.
Vasilis Fthenakis, Subhamoy Bhattacharya, and Paul A. Lynn.
© 2025 John Wiley & Sons Ltd. Published 2025 by John Wiley & Sons Ltd.
Companion website: www.wiley.com/go/fthenakis/windenergy2e

Figure 3.1 Established favorite: the three-bladed HAWT rotor (*Source:* With permission of Repower).

day in the last 30 years. But they are now fading from view. It is also interesting that traditional windmills, following centuries of development, nearly always had four sails—and their designers presumably knew a thing or two. But, as the wind industry has embarked on its multi-gigawatt era, the three-bladed model has moved inexorably into pole position, favored for its technical qualities and cost-effectiveness. The high cost of individual blades means there is no virtue in using more of them than necessary to achieve high performance. And public acceptability, which depends largely on visual impact and attractiveness, definitely favors three blades over one or two (Figure 3.1).

We have previously noted that there is a fundamental limit to the percentage of energy in an airstream that can be captured by any wind turbine, regardless of the number of blades—a point we will develop a little later. Today's large three-bladed rotors typically capture up to about 50% of the wind's energy (provided the wind speed is above cut-in and below cut-out) for conversion into electricity and, as we shall see in the next section, basic physics assures us that little more can be achieved. We may therefore use this figure with some confidence to estimate the approximate size of the turbine needed to achieve a given power rating, regardless of its detailed design and technology. And, of course, size is a key indicator of weight and cost and an important criterion for public acceptance.

As explained in Section 2.1, the amount of power intercepted by a rotor of area A m^2 facing a steady airstream of speed U m/s is given by:

$$P = 0.612\,AU^3 \text{ watts} \tag{3.1}$$

A HAWT rotor with a diameter d meters has a swept area $\pi d^2/4$, so in this case:

$$P = 0.612\big(\pi d^2/4\big)U^3 = 0.481 d^2 U^3 \text{ watts} \tag{3.2}$$

If we assume that the rotor is able to capture 50% of the power in the airstream and that this mechanical power is converted into electrical power with a typical efficiency of 90%, the overall efficiency of the turbine is $0.50 \times 0.9 = 0.45$, or 45%, and the electrical power it generates is given by:

$$P_e = (0.45)(0.481)d^2U^3 = 0.216d^2U^3 \text{ watts} \tag{3.3}$$

Most large modern turbines are designed to reach their rated power outputs at wind speeds between 12 and 14 m/s. Taking 13 m/s as a typical value, and rearranging it to make the d the subject of the equation, we arrive at the following relationship between rotor diameter and rated power:

$$d = \left(P_e/0.216U^3\right)^{0.5} = 0.0459\, P_e^{0.5} \text{ metres} \tag{3.4}$$

We can now predict the rotor diameters needed to produce various power ratings. For example:

P_e (MW)	2	3	5	7	10
d (meters)	65	80	101	121	145

These results are illustrated in Figure 3.2. The left-hand side is colored grass-green to denote that 2-MW and 3-MW machines are often sited onshore. The right-hand side is sea-blue because giant turbines rated at 5 MW and above are more likely to ply their trade offshore. Tower heights are typically about one rotor diameter, giving a blade clearance of about one rotor radius above the ground or sea surface, although this depends on the local wind regime and the importance of visual intrusion.

Of course, we can hardly expect our calculations to apply precisely in all cases. Not all turbines reach their rated power at a wind speed of 13 m/s. Small ones are often less efficient and operate onshore in less favorable wind regimes, so need somewhat bigger rotors to achieve a given power rating; the largest and latest machines tend to be more efficient and are sited offshore on higher towers. But overall, our analysis gives a good indication of the rotor diameters of today's large HAWTs.[1] Satisfyingly, we have based it on just a few simple assumptions: the amount of power in the wind, the efficiency of wind capture by a three-bladed rotor;, and the efficiency of conversion from mechanical to electrical power.

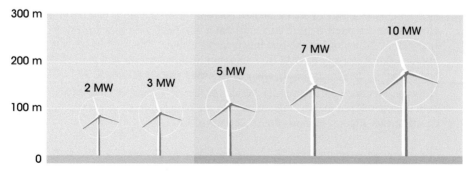

Figure 3.2 Typical dimensions of large HAWTs.

Now seems a good moment to make a general point: the most important indication of a turbine's power-generating capacity is the area swept by its rotor blades. This applies to a wide range of designs, vertical as well as horizontal axis, assuming their rotors and electrical generators have comparable efficiencies. In other words, if you wish to make a quick estimate of a turbine's power rating, first find out *the area swept by its rotor*. Over the years, there have been many examples of manufacturers of new machines—especially small ones aimed at individuals and local groups[2]—making outlandish claims (in extreme cases, the advertised electrical output has exceeded the amount of power in the wind at the stated speed!). Fortunately, the purchasers of megawatt turbines are generally well-informed and unlikely to be taken in by false promises. But every now and again, a new design is trumpeted with exaggerated claims, so caution is certainly in order. A rotor's swept area will always be a key indicator of power capacity—so much so that some experts would prefer to see machines rated on this basis.

This leads us to a brief discussion of vertical-axis wind turbines (VAWTs), which are also limited in their power output by the swept areas of their rotors. Although we shall not devote much time and space to them here, it must be admitted that turbine rotors spinning about vertical axes have some inherent advantages:

- A VAWT rotor does not notice which direction the wind is coming from. It always "faces the wind" so, unlike a HAWT, does not require a yaw mechanism.
- The gearbox and electrical generator may be placed beneath the rotor, keeping much of the weight close to the base of the machine.
- A large rotor may be accommodated relatively close to the ground or sea surface, avoiding the need for a high tower and nacelle.

However, there are compensating disadvantages:

- A VAWT cannot "turn away from the wind" to save itself from damage.
- A VAWT blade creates downstream turbulence that interferes with other blade(s). Unlike a HAWT, the blades of a VAWT experience wide variations in wind strength and direction as they rotate, impacting aerodynamic efficiency and threatening serious vibration.

Over the years, a multitude of designs have been proposed, many prototypes built, and a few machines (especially small ones) were offered in the marketplace.[2] VAWTs continue to have protagonists who, of course, like to concentrate on the advantages mentioned above. However, we are surely right to conclude that in this field, as in so many other branches of engineering, "the proof of the pudding is in the eating." The harsh truth is that the total installed global capacity of VAWTs remains miniscule compared with their HAWT cousins, and no major company presently offers large VAWT machines for sale.

However, it is never wise to say never. We should therefore mention a few VAWT designs that have been developed and manufactured over the years, some of which may impact future developments. Figure 3.3 shows four well-known examples:

a) *Cup anemometer*. Hardly a "turbine" in the full sense of the word, this small vertical-axis machine is widely used to measure wind speed at meteorological stations and on tall buildings and structures – including the nacelles of wind turbines

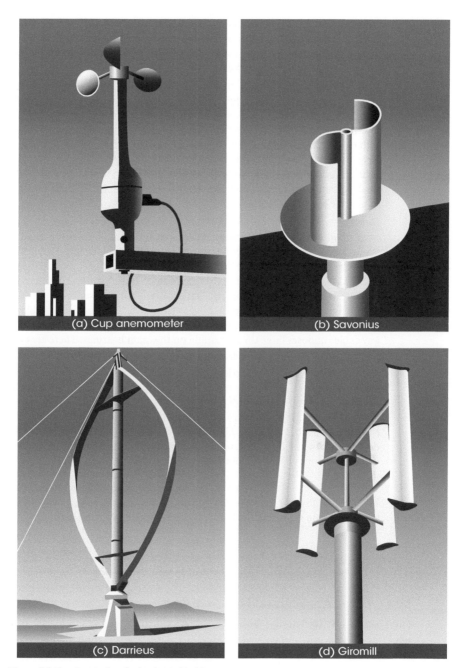

Figure 3.3 Four types of vertical-axis wind turbines.

(see also Figure 2.22). The wind is "caught" by the open ends of the cups, but "slips past" the closed ends, ensuring that the torque on the shaft is always in the same direction. The device is not required to deliver significant amounts of power and runs more or less freely in the wind.

b) *Savonius turbine.* Invented by the Finnish engineer Sigurd Savonius in 1922, this form of turbine works rather like a two-cup anemometer. The wind catches each "scoop" in turn, but it slips past its rear surface. In the figure, the turbine is shown with an open top to reveal its characteristic "S" shape. Although inherently inefficient compared with modern blade turbines, Savonius machines have often been used to provide small amounts of mechanical power, for example, in pumping applications. Amateur enthusiasts have fashioned them out of oil drums. But the most widespread application of small Savonius machines today is on vehicle roofs, coupled directly to extractor fans to provide air conditioning.

c) *Darrieus turbine.* This type of VAWT is claimed to be the best-researched and developed of all vertical-axis machines. Prototypes have been built up to rated powers of about 0.5 MW, with many smaller designs marketed to individual homeowners and small organizations. Patented by the French aeronautical engineer Georges Darrieus in 1927, it is popularly known as the "eggbeater." Based on curved blades with airfoil sections, it is more efficient than the Savonius but tends to suffer from self-generated turbulence, fluctuating torque, and blade vibration. Some of these problems may be reduced by shaping the blades in the form of a helix. In larger machines, the need for guy cables imposes large thrust loading on the main bearings. There are currently no commercial Darrieus machines in competition with large HAWTs.

d) *Giromill turbine.* Georges Darrieus' patent of 1927 also covered turbines with straight blades supported from the main axis. Although potentially more robust than the "eggbeater," the giromill is subject to many of the same problems. A more sophisticated version, known as the *cycloturbine*, orients the blades at optimum angles to the wind as they rotate—but at increased complexity and cost. Cycloturbines have never been developed beyond the prototype stage although there is some chance that large commercial versions will be deployed successfully offshore.

From now onward, we will return to our favorite topic—large grid-connected HAWTs with three-bladed rotors that promise so much for the future of global electricity supplies. In the next section, we begin exploring the basic principles underlying their operation.

3.2   Aerodynamics

3.2.1   Rotor efficiency and the Betz limit

The rotor of a HAWT intercepts the wind and aims to transform as much of its kinetic energy as possible into useful mechanical energy. Although it is tempting to imagine that an ideal rotor could do this perfectly, a few moments' thought reveals a serious problem. To extract all the kinetic energy from a moving stream of air, it must be brought to a complete standstill, yet this is impossible because the air must pass through the rotor's swept area and continue its journey on the far side. It cannot "pile up" in front of the rotor. So, how much kinetic energy can actually be extracted?

This fundamental question was tackled by Albert Betz (1885–1968), a German naval engineer and professor at the University of Göttingen whose interests turned toward wind energy at the end of the First World War, quite possibly stimulated by the pioneering work being done on wind turbines in neighboring Denmark. The rotor problem had strong links with a naval one he had worked on earlier—the efficiency of marine propellers—and the theoretical limitation to rotor efficiency he discovered is known as the *Betz Limit*.

The Betz Limit states that a rotor in a steady airstream cannot convert more than 16/27, or 59%, of the wind's kinetic energy into mechanical energy. Rather intriguingly, this figure is based on the fundamentals of fluid dynamics and is independent of the precise type or design of the rotor. Betz's insight produced one of those moments when the elegance of theoretical physics meets practical engineering, issuing a warning to designers not to attempt the impossible. And, we need hardly be surprised that the practical difficulties of approaching a theoretical limit mean that large modern HAWT rotors achieve maximum aerodynamic efficiencies of about 50%, a figure we have used in the previous section.

How was the Betz Limit derived? We start by considering the *stream tube* shown in Figure 3.4. Air enters the tube with a speed U_1 and, after passing through the rotor, exits with a lower speed U_2. The only flow is across the ends of the tube, and the speed reduction means that the tube's cross-sectional area must increase toward the downstream side. This is accompanied by a change in pressure across the rotor disc. There are some important assumptions:

- The air is incompressible and the flow homogeneous.
- There is no energy loss due to friction.
- The change in air pressure is uniform over the rotor disc (equivalent to specifying a rotor with an infinite number of blades).

A full analysis based on classical linear momentum theory can be used to calculate the rotor's efficiency, thrust, and power output.[3] From the point of view of efficiency—our primary interest in this section—we may condense the argument considerably. First, it is clear that the rotor causes the reduction of wind velocity from U_1 to U_2 shown in Figure 3.4, and this occurs at the rotor plane. The average of these two velocities is therefore a good estimate of

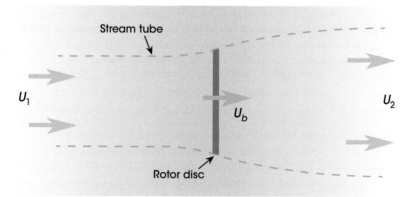

Figure 3.4 A rotor and its stream tube.

the velocity U_b actually "seen" by the rotor blades:

$$U_b = (U_1 + U_2)/2 \tag{3.5}$$

We now introduce an *axial induction factor (a)* defined as the fractional decrease in wind velocity between the input to the stream tube and the rotor plane:

$$a = (U_1 - U_b)/U_1 = (U_1 - U_2)/2U_1 \tag{3.6}$$

giving:

$$U_b = (1 - a)U_1 \tag{3.7}$$

If $a = 0$, the wind passes through the rotor unhindered, and no energy is captured. A value $a = 0.5$ implies that the velocity at the exit from the stream tube is zero, which is not allowed. So, the Betz theory only holds over the range $0 \leq a < 0.5$.

The rotor efficiency, equal to the fraction of the incoming wind power extracted by the rotor, is also referred to as the *power coefficient (Cp)* and is related to the axial induction factor as follows:[3]

$$C_p = 4a(1 - a)^2 \tag{3.8}$$

What value of the axial induction factor produces the greatest value of the power coefficient? Differentiating Equation (3.8) and equating to zero, we obtain:

$$(3a - 1)(a - 1) = 0, \text{giving } a = 1/3 \text{ or } a = 1. \tag{3.9}$$

The first of these values represents the required maximum. Substituting $a = 1/3$ in Equation (3.8), we obtain Betz's famous result for the maximum value of power coefficient:

$$C_{pmax} = 4/3(2/3)^2 = 16/27 = 59\% \tag{3.10}$$

Another interesting aspect is the change in the stream tube cross section from the upstream to downstream side of the rotor. If $a = 1/3$, maximizing the rotor power, the initial velocity U_1 is reduced by 1/3 at the rotor and by 2/3 at the exit from the stream tube. Since the air is assumed incompressible, any velocity changes must be offset by changes in cross section. Therefore, the stream tube must have an input cross section equal to 2/3 of the rotor disc (swept area), expanding to twice the disc area downstream.

You may feel that these results, widely quoted and respected, rest on slightly shaky foundations. It is hard for nonexperts to assess the validity of the aerodynamics involved, but we know that assumptions such as homogeneous flow through the rotor and absence of friction cannot truly apply in practice. This implies that the 59% is rather optimistic. But it has been accepted as the absolute upper bound on the efficiency of a wind turbine rotor for nearly a century, and it seems clear that nobody ever expects to exceed it!

There is one more significant caveat, and this concerns the nature of the disturbed airstream, or *wake*, downstream of the rotor. Figure 3.4 shows air leaving the stream tube parallel to the axis, at a reduced speed but without any rotation. But as wind passes through a HAWT and

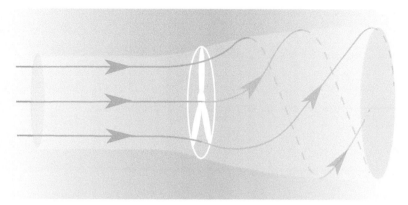

Figure 3.5 Vortex formation in a turbine wake.

exerts a torque on the rotor, there must be an opposite reaction on the airstream. In other words, there is some vortex formation in the wake, illustrated by Figure 3.5. This imparts unwanted rotational kinetic energy to the airstream and reduces the rotor's power coefficient. The greater the torque on the rotor, the stronger the vortex, so machines running at low speed with high torque are most affected. Fortunately, vortex formation rarely subtracts more than a few percent from the power coefficient of a well-designed rotor.

Albert Betz set the scene for rotor designers, placing an absolute limit on the height of their ambitions. In the following sections, we shall see how they have attempted to climb ever closer to the summit.

3.2.2   Lift and drag

The Betz Limit discussed in the previous section estimates the maximum efficiency of an ideal rotor in a steady airstream as 59%. This famous result pays no attention to the actual design of the rotor, leaving such small details to the designers of blades and hubs! Our own story must pick up at this point since we wish to clarify the principles underlying today's large HAWT rotors.

Two basic forces dictate the interaction between the wind and the blades of a rotor, determining its aerodynamic performance. The first is *lift*, a beneficial force on which all modern bladed turbines depend. And the second is *drag*, normally a detrimental force to be reduced as far as possible—except in the case of machines such as the cup anemometer and Savonius turbine (illustrated in Figure 3.3), which actually press it into service. A clear understanding of lift and drag is essential for appreciating the fundamental aerodynamic principles on which all wind turbines depend.

We start by considering the forces acting on a thin, flat plate placed in a steady wind. Figure 3.6(a) shows an edge view, with the plate aligned exactly with the wind direction. Assuming the plate is very thin and smooth, there is negligible disturbance to the airstream, and no forces are generated.

In part (b) the plate is inclined at an angle to the oncoming wind, which tries to push it away with a drag force, and also to move it at right angles with a lift force. The net force on the plate is the resultant (vector sum) of these two forces. Part (c) shows the plate positioned at

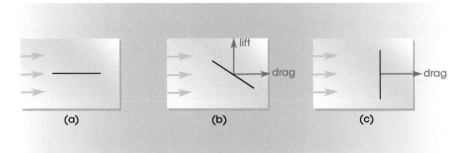

Figure 3.6 Forces on a thin plate in a steady wind.

right angles to the wind. There is no lift, but the drag force is large. We see that the relative amounts of lift and drag depend greatly on the inclination of the plate. To summarize:

- Lift forces are produced perpendicular to the oncoming airstream.
- Drag forces are produced parallel to the oncoming airstream.

Wind turbines may be broadly divided into two classes. *Drag machines* aim to present a large surface area to the oncoming wind, like the plate in part (c) of the figure, maximizing drag. The earliest recorded windmills, made in the Middle East over a thousand years ago, relied upon drag. A Savonius turbine also works mainly on this principle, so does a cup anemometer. In a different context, the spinnaker of a sailing boat, billowing out at the front of the boat with a following wind, pushes the boat forward by "catching" as much wind as possible. In all these cases, it is clear that the drag device—cup, scoop, or sail—can never move faster than the wind itself. The rotational speed of a turbine that operates on the drag principle is limited by the speed of the wind.

The second main type of wind turbine is the *lift machine*. Bladed windmills and wind turbines use lift to generate forces at right angles to the wind, producing a twisting force or *torque* on the main shaft and delivering useful power. The concept was well understood by the engineers who designed and built Europe's traditional windmills. Today's large HAWTs continue the tradition and are entirely driven by lift forces; drag forces, far from helping, are a hindrance. Returning to our sailing boat analogy, the mainsail and jib of a racing dinghy, unlike the spinnaker, are carefully designed to enhance lift and allow the boat to move at right angles (or even "closer") to the wind.

In such cases, there is no obvious relationship between the speed of the wind and that of the machine; indeed, the speed of a lift machine, although *dependent* on wind speed, is by no means *restricted* to it. An inspiring example is provided by British engineer Richard Jenkins who broke the world speed record for a wind-powered land vehicle in 2009 (see Figure 3.7). His racing car *Greenbird*, propelled by a rigid vertical "sail" with many of the characteristics of a turbine blade or aircraft wing, attained a speed of 126 mph (203 kph) on the dry bed of Ivanpah Lake in California.[3] And guess what—the average speed of the wind that propelled him was only about 35 mph! *Greenbird* admirably demonstrates that lift efficiently harnessed and drag effectively minimized are key to modern wind engineering.

Although large HAWTs rely on lift forces, the way in which they act on rotor blades is subtle and somewhat mysterious to many people. Fortunately, we can broach the topic gently by

Figure 3.7 Maximizing lift, minimizing drag: *Greenbird* on its record run (*Source:* With permission of Ecotricity).

considering a more familiar situation—lift-and-drag forces acting on the wing of an aircraft.

Figure 3.6(b) shows a flat plate inclined at an angle to the wind, generating both lift and drag. What it does not show is the turbulence and eddies produced on the plate's "downwind" side, which accentuate drag. For a machine designed to operate on the lift principle with minimum drag, special care must be taken to avoid turbulence by using a well-designed *airfoil section* instead of a simple flat plate.

All this has quite a history. By the time the Wright brothers made their powered flights at Kitty Hawk, North Carolina, in the first decade of the last century, they were fully aware of the crucial importance of wing shape for maximizing lift and minimizing drag (Figure 3.8). Experiments carried out by a number of investigators since the late 18th century, and continued by the Wrights themselves, had established the efficacy of double-surfaced airfoils resembling the wings of a bird. Furthermore, it was understood that a cambered wing, curved more on the upper surface than the lower, generates lift due to reduced pressure above and increased pressure below. Ever since the early days of powered flight, aeronautical engineers have striven to perfect airfoil sections, and their efforts have certainly influenced the modern renaissance of wind energy.

Figure 3.9 shows a typical airfoil section through the wing of an aircraft facing a horizontal airstream. There are several important features:

- The airstream divides between the upper and lower surfaces. *Since the upper surface is more curved (cambered), the air must travel further and faster over it, lowering the pressure.* Conversely, the lower airstream exerts positive pressure. Both effects contribute to lift.

Figure 3.8 Cambered wings with plenty of lift: the *Wright Flier 2* takes to the sky in 1904 (*Source:* The Library of Congress/Wikimedia Commons/Public domain).

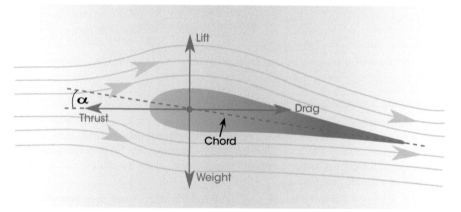

Figure 3.9 Forces acting on an aircraft wing in an horizontal airstream.

- The airflow over a well-designed smooth wing stays close to the wing surfaces, without forming eddies or turbulence. This reduces drag—although there is inevitably some residual drag due to surface friction.
- The line between the center of the airfoil's leading edge and the trailing edge is known as the *chord*. It is typically kept inclined to the airstream at a small angle (α) known as the *angle of attack*, to optimize lift.

The four red arrows in the figure indicate the directions (but not the relative strengths) of four key forces. Assuming the aircraft is in level flight at constant speed:

- Lift (also produced by the other wing, and probably a tailplane) is at right angles to the oncoming airstream and exactly balances the weight of the aircraft.
- Drag (also on the other wing and the aircraft's fuselage) is exactly counteracted by the thrust of the engine(s).

If the lift increases, the aircraft gains height; if the drag increases, it loses speed. In level flight at constant speed, *lift equals weight and drag equals thrust.*

An important measure of an airfoil's quality is its *lift-to-drag ratio*. Ideally, plenty of lift is provided to support the aircraft's weight, but little engine power is needed to overcome drag, leading to good fuel economy. The greater the lift-to-drag ratio, the better. We shall have more to say about the airfoils used in today's large HAWTs in Section 3.2.4.1.

There is one more important, and potentially disastrous, aspect of flight that should be mentioned here: the condition known as *stall*. If an aircraft gets into a "nose-up" attitude at low speed, the lift provided by its wing airfoils may decrease drastically, and the drag increases, causing it to crash—a pilot's nightmare. Stall is also very important to wind turbines and, interestingly, not only in a negative sense. Although it is generally undesirable for turbine blades to enter a stall condition during normal operation, in very strong winds *deliberate* stall may be used to limit the amount of power generated by a rotor, protecting the machine from damage. We will consider the positive and negative aspects of stall in Section 3.2.4.2.

The term "lift" is clearly appropriate for an aircraft, which is literally held up in the sky by it, but in other machines lift has a different function. For example, *Greenbird's* vertical "wing" (see Figure 3.7) generates a lift that acts horizontally, propelling the vehicle forward rather than skyward. A HAWT uses a lift to generate rotation around a horizontal axis. Regardless of how the lift is used, an efficient lift machine depends on a high lift-to-drag ratio.

We are now ready to explore the effects of lift and drag on a turbine blade in some detail. Imagine a section through the airfoil, fairly near the hub, as shown in Figure 3.10. The wind blows horizontally from left to right and, viewed from the front, the rotor is turning clockwise. Clearly, in this view, the airfoil section has a very different orientation from that of an aircraft wing in level flight, and this can be confusing. So, let us start by emphasizing that, viewed from the front, the airfoil presents its "underside" toward us—what, in an aircraft, would be its lower surface. The more curvy "topside" is out of sight and it follows that the lift force is acting somewhat away from us and toward the back of the turbine. With this in mind, what follows should be fairly straightforward!

There are other important distinctions between the aerodynamic environments of a wind turbine and an aircraft in level flight. As a turbine rotates, it *creates its own wind* at right angles to the "natural" wind, and it is the *resultant wind* that acts upon the blades. Rather than experience the wind coming from the left side of the figure, the blades operate in an airflow coming more from the direction of rotation. This is illustrated in the figure by blue vectors 1, 2, and 3 that indicate wind strength as well as direction. They are:

1. The "natural" wind at the turbine rotor, coming horizontally from the left.
2. The additional wind was caused by rotation.
3. The resultant wind acting on the airfoil is equal to the vector sum of 1 and 2.

We now consider the forces generated, shown by numbered red and orange vectors. The airfoil is inclined at a suitable angle of attack (α), generating plenty of lift (4) at right angles to the resultant wind. There is also some drag on the airfoil (5), and the net force it generates (6) is the vector sum of lift and drag (the amount of drag has been exaggerated here to show its effect clearly). And now comes a key point: we may resolve the net force (6) into two components: a desirable radial component (7) that produces blade rotation and an unwelcome axial component (8) that tries to bend the blade toward the back of the turbine.

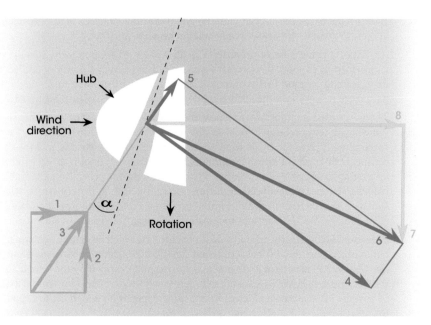

Figure 3.10 Wind speeds and forces on a HAWT airfoil.

Figure 3.10 encapsulates and explains ideas that many people find challenging. In particular, it shows that:

- To be effective, the blade airfoil must be inclined at a suitable angle to the *resultant* wind.
- The direction of the resultant wind depends on the ratio between the speed of the natural wind and the speed of the turbine.
- Drag on the blade causes its net force (6) to move slightly away from the direction of the lift force (4) toward the back of the turbine, reducing the useful radial force (7) and increasing the undesirable axial force (8). This emphasizes once again the value of a high lift-to-drag ratio.

There is one further major distinction to be made between the aerodynamic environments of a turbine blade and an aircraft wing. So far we have considered a point on the blade fairly near the hub. But the wind speed caused by rotation equals Ωr, where Ω is the angular velocity of the rotor in radians per second, and r is the radius considered (the distance from the center of the hub). So, at a given speed of rotation, and with a constant incoming wind speed, the speed and direction of the resultant wind vary according to the distance from the hub. This is quite different from an aircraft wing in level flight, which experiences the same wind speed and direction along its length. And the upshot is that, in order to optimize lift along the length of a turbine blade, it must be progressively *twisted* from hub to tip.

This important conclusion is illustrated by Figure 3.11, showing typical wind speeds and directions at three points along a blade. The right-hand diagram refers to a point near the blade tip at a distance R from the hub (R being the rotor radius). The "natural" wind, coming

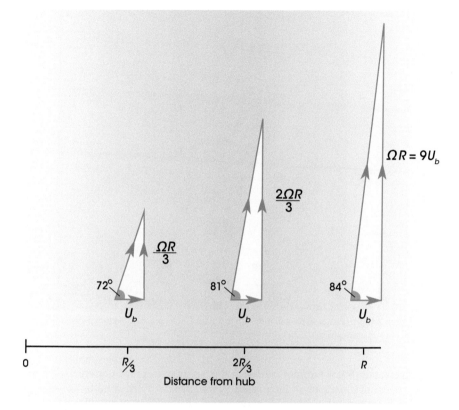

Figure 3.11 Wind speeds and directions at three distances from the hub.

in from the left, is denoted by U_b (see also vector no.1 in Figure 3.10). The wind speed caused by rotation, equal to ΩR, is here shown as equal to $9U_b$. The ratio between the rotational and natural wind speeds at the blade tip is therefore equal to 9 and is referred to as the *tip speed ratio*. It normally lies in the range 5 to 10 for large modern HAWTs and, as we shall see later, is one of the most significant parameters affecting turbine performance. Note that, near the blade tip, the resultant wind is almost entirely due to the rotational component, so the outer sections of the airfoil must perform well in an air speed many times greater than that of the incoming, "natural," wind.

The central diagram refers to a point two-thirds along the blade from the hub. The "natural" wind vector is the same, but the rotational wind speed is only two-thirds as great. The resultant wind is correspondingly reduced but is still dominated by the rotational component. The left-hand diagram takes these trends further, showing wind conditions at a point one-third along the blade from the hub. Even here, the rotational wind is considerably stronger than the natural wind. And notice how the direction of the resultant wind changes from hub to tip, making angles of 72°, 81°, and 84° with the rotor axis—the important effect that requires the blade to be progressively twisted (Figure 3.12).

Actually, it is more complicated than this because, as we have shown in Chapter 2, wind is highly variable. The speed of the "natural" wind impinging on a turbine rotor is always

Figure 3.12 Progressively twisted from hub to tip: the blades of a high-performance HAWT rotor (*Source:* With permission of Repower).

fluctuating. Some turbines are designed to run at more or less constant speed, but most large modern machines adapt their speed to the strength of the wind by swiveling the complete blade around its longitudinal axis using *pitch control*, improving efficiency over a wide range of wind speeds. We will explore these important topics in the following sections.

3.2.3 Rotor speed

How fast should a large HAWT rotate? The answer to this basic question, which depends mainly on aerodynamics, has important implications for the mechanical and electrical aspects of turbine design. Machines rated at several megawatts are now commonplace, and the industry is busy contemplating rotors rated at 10 MW for offshore wind farms, sweeping areas of the sky bigger than an international football pitch. The speed at which such giants rotate is obviously a major issue.

We have already seen that the Betz Limit places a theoretical ceiling of 59% on the efficiency of a wind turbine rotor, regardless of its detailed design or speed of rotation. We have also discussed lift-and-drag forces, explaining how they act on a turbine airfoil and have explained the need for a HAWT blade to be progressively twisted from hub to tip. It is now time to relate these ideas to rotor speed and the way in which it may be varied to suit the wind conditions.

It is helpful to start by recalling the simple relationship between the power, speed, and torque of a rotating machine:

$$P = \Omega T = 2\pi N T \tag{3.11}$$

where P is the power measured in Watts, Ω is the angular velocity in radians per second, N is the speed in revolutions per second, and T is the torque measured in Newton meters (Nm). For example, suppose a large turbine rotor is generating 3 MW of mechanical power while turning 12 times per minute, or once every 5 seconds. In this case, $P = 3 \times 10^6$ and $N = 0.2$, so the torque is given by:

$$T = P/2\pi N = 3 \times 10^6/0.4\pi = 2.39 \times 10^6 \ NM \qquad (3.12)$$

We should pause for a few moments to consider this enormous figure. A force of 1 Newton is close to one-tenth of a kilogram weight, so the twisting effort on the rotor shaft is about 240,000 kg m, thus 240 ton m. Imagine an Indian elephant weighing 2.4 tons standing on the end of a long (and light) horizontal lever, the lever would need to be 100 m long for the animal to produce this amount of torque!

This example emphasizes that large HAWT rotors are quintessentially high-torque, low-speed machines, unusual in today's technological world. By comparison, generators in conventional power plants usually run at 3000 revolutions per minute (rpm), or 3600 rpm in the USA. A machine running at 3600 rpm generates 300 times as much power as one running at 12 rpm, for the same amount of torque. Ask any electrical or mechanical designer, and he or she will confirm a general preference for high speed and low torque. Large low-speed HAWTs demand heavy shafts, expensive bearings, and high-ratio gearboxes (or specially designed low-speed electrical generators). This is bulky, heavy stuff requiring rotors and nacelles weighing hundreds of tons to be lifted and maintained on top of high towers.

Since power is proportional to the product of torque and speed, why not reduce the torque and increase the speed of megawatt rotors? The answer lies mainly in the aerodynamics of turbine blades, and we have already hinted at it in the previous section: the ratio between the rotational speed of the blade tips and the speed of the incoming wind—the *tip speed ratio*—has a major influence on rotor efficiency.

Figure 3.14 shows a typical plot of the power coefficient (C_p) against the tip speed ratio (λ) for a large HAWT with blades rigidly attached to its hub (as in many older, simpler, machines). We see that tip speed ratios between 5 and 10 are the most efficient at capturing the power in the wind. The famous Betz Limit of 59% remains beyond reach at the top of the figure, but values up to about 50% are possible provided the turbine speed is well-chosen in relation to the incoming wind speed. Ideally, the turbine should track the wind speed to maintain λ at, or close to, its optimum value. But, if a fixed-blade turbine is designed to run at a particular speed, it is clear that the optimum value of λ can only occur at one wind speed. In variable winds, fixed blades and a constant turbine speed mean that the power coefficient will often be well below its maximum value (Figure 3.13).

The precise form of the $Cp - \lambda$ curve depends on blade geometry and some complicated theory well covered elsewhere.[3, 4] Although intuition is a somewhat risky substitute for proper analysis, in this case, a simple "thought experiment" may help us understand what is going on. So let us imagine a large fixed-blade rotor, temporarily uncoupled from its gearbox and generator, and held stationary by a brake. The wind is blowing steadily at moderate speed. The brake is now released, and the rotor starts to turn, running freely in its bearings. Since we are not trying to generate any electricity, the torque is used entirely to accelerate the rotor; but both torque and acceleration are initially small because the airfoils' angle of attack is not well suited to the incoming wind direction—we are at the left-hand end of the $C_p - \lambda$ curve. As the rotor speed and tip speed ratio increase, the blades become more and more effective,

Figure 3.13 A large modern HAWT: high torque at low speed (*Source:* With permission of Repower).

generating plenty of torque and accelerating the rotor rapidly toward the peak of the curve. If we were generating electricity, this is where the rotor speed should ideally be held; but in this experiment, we are allowing it to run free, eager to see what happens—and hoping that it will not break up!

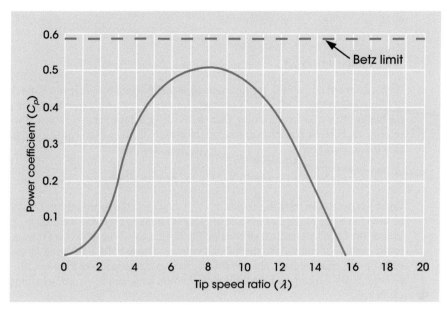

Figure 3.14 Power coefficient and tip speed ratio.

Unhindered by the need to generate electricity, the rotor's torque continues to provide acceleration and the tip speed ratio increases further, toward the right-hand end of the curve. The fixed blades now become less effective, their orientation less and less suited to the resultant wind which is now mainly due to the high speed of rotation. There is less lift and an increasing amount of drag on the blades, especially due to turbulence near the blade tips, which are now moving at very high speed. Eventually, we reach a speed at which diminishing lift is exactly counteracted by increasing drag, and there is no net torque available for further acceleration. The rotor has reached its maximum free-running speed at this particular wind strength.

We can now see, in broad terms, why the power coefficient C_p in Figure 3.14 behaves as it does. It is zero when λ is zero because power is proportional to the product of speed and torque, so "no speed, no power." It reduces to zero again at some high value of λ (typically between 15 and 20), because the rotor's diminishing torque is all being used to counteract drag. In between, it can generate useful power, especially if λ is close to optimum.

Thought experiments are cheap, risk-free, and may give valuable insights; but it is time to return to reality. So let us consider a 3-MW HAWT rotor, working steadily at half its rated power in a wind speed of 11 m/s. Typical figures might be:

- Rotor power: 1.5 MW
- Wind speed: 11 m/s
- Rotor diameter: 95 m

circumference: 298 m

rotational speed: 14 rpm

tip speed: 69 m/s

torque: 1.02×10^6 Nm

- Tip speed ratio 6.3

With a tip speed ratio of 6.3, the turbine is likely to be operating efficiently. Note the tip speed of 69 m/s = 248 kph = 155 mph: a veritable hurricane is blowing over the outer portions of the blades! It is also worth noting that the blade tips, rotating at a radius R with angular velocity Ω, have an acceleration toward the center (the hub axis) equal to $\Omega^2 R$. So in this case the acceleration is:

$$\Omega^2 R = (2\pi N)^2 R = (2\pi 14/60)^2 95/2 = 102 \text{ m/s}^2 = 10.4\,g \tag{3.13}$$

where g is the acceleration due to gravity (9.81 m/s^2). In other words, the radial force required to keep the outer portion of a blade in place is about 10 times its own weight, giving some idea of the huge longitudinal stresses on the blade as a whole, which are transmitted down to the blade *root* and then to the hub. Of course, there are also large wind forces producing rotation and bending the blade toward the back of the turbine (see Figure 3.10). The design of large HAWT rotors and blades is a mechanical engineering, as well as an aerodynamic, challenge.

So far we have considered a turbine with blades fixed rigidly to the hub. But the blades of today's powerful HAWTs *can generally be rotated* about their longitudinal axes, altering their *pitch* and allowing the angle of attack of the airfoils to be adjusted to the direction of the resultant wind. And, if the wind speed is more than needed to generate the rated output power, the blades can be *pitched* to spill the excess. We shall have more to say about pitch control later.

Many large machines also operate at variable speeds, allowing the tip speed ratio to be kept close to its optimum value as the wind speed varies. Speed control makes better use of the available wind resource, giving improved rotor efficiencies and higher annual electricity yields. This is illustrated by Figure 3.15, showing the electrical output achieved when turbine speed is matched to incoming wind speed, optimizing the tip speed ratio λ. In this example, the turbine achieves its rated power output (denoted as 100%) at an average wind speed of 12 m/s when running at a rated speed (100%) with the optimum value of λ. If the turbine speed is too low or too high the electrical output is reduced, as shown by the 12 m/s curve. This is similar to the $C_p - \lambda$ curve of Figure 3.14, except that we are now plotting turbine rotation speed instead of tip speed, and electrical power output instead of power coefficient (assuming a constant conversion efficiency between mechanical and electrical power).

At a lower average wind speed—say 9 m/s—the turbine speed must be reduced by about 25% to maximize power output; and, of course, the peak is considerably lower because there is less power in the wind at 9 m/s than at 12 m/s. If the wind drops further to 6 m/s, turbine speed must fall by about 50%, and the maximum power is only about one-eighth of its value at 12 m/s (as expected from the cubic relationship between wind speed and power, discussed in Section 2.1). The heavy red curve in the figure joins all the maximum power points, showing how the turbine speed should ideally be adjusted to the wind speed to achieve maximum power output.

Tracking wind speed in this way helps optimize power production over a wide range. It is most valuable at intermediate wind speeds—say between 5 and 12 m/s in this example. Below

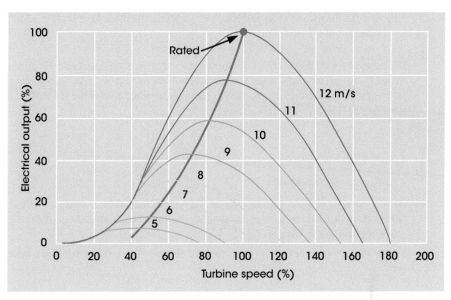

Figure 3.15 Tracking wind speed to obtain maximum turbine output.

5 m/s, there is very little power in the wind anyway, and above 12 m/s there is an excess. Generally, turbine speed control over a range of about 2.5:1 gives the required flexibility. A further advantage of speed variation is that it allows an increase in wind strength to be taken up by increased rotor speed rather than rotor torque and may be viewed as a form of torque control, preventing excessive torque on the drive train and electrical generator.

Rather than providing continuously variable-speed control, some simpler turbines have electrical generators that can operate at just two fixed speeds, with reduced flexibility. However, the current trend, especially with the largest machines, is toward continuous speed control.

The above analysis brings out many important points, but we must remember that wind is often highly variable. The picture painted in Figure 3.15, based on average wind speeds blowing steadily in strength and direction, is inevitably rather oversimplified. Complicating factors include the following:

- The presence of wind shear means that wind speeds are generally higher at the top of the rotor's swept area than at the bottom. Blades therefore experience cyclical fluctuations in tip speed ratio and aerodynamic loading as they rotate.

- Turbulence, often more serious at onshore sites than at offshore ones where the wind blows more steadily, can present rotors and blades with rapid changes in wind speed. The rotor's pitch and speed control systems may be unable to respond in time, forcing blade airfoils to operate well outside optimal design conditions, mechanical as well as aerodynamic.

- So far in this book, we have tended to assume that "wild wind" fluctuates in speed, but not in direction; or, if it does change in direction, the rotor may be turned to face it "head on" using special inbuilt *yaw motors*. But here again, the changes may be too fast for the rotor to respond effectively, leading the rotor to operate in wind that has a substantial off-axis component.

The net effect of all this is a more complicated wind regime than implied by Figure 3.15—and indeed by our previous discussions of turbine power output and annual electricity yield. And yet our main conclusions about rotor speed remain valid, even if the rotor often finds itself operating in less-than-ideal conditions.

3.2.4 Rotor blades

3.2.4.1 Choosing airfoils

We now turn to airfoil design, a topic that could fill hundreds of pages on its own! We will restrict ourselves here to a short review of its historical background and the main technical issues confronting the designer.[3, 4]

Several important issues relating to blade and airfoil design have been introduced in previous sections:

- However excellent is the design of its blades, and well-chosen its rotation speed, a rotor cannot capture more than 59% of the power in a steady airstream—a theoretical limit based on momentum theory that was established by Albert Betz back in the 1920s. In practice, large modern HAWT rotors achieve values up to around 50%.

- The diameter of a well-designed HAWT rotor, and hence the length (but not the detailed shape) of the blades, is essentially proportional to the square root of its rated power output—see Equation (3.4).

- The lift forces generated by blade airfoils have a radial component that translates into torque applied to a turbine's main shaft. There is also an unwanted axial component that tends to bend the blades backward, and this is aggravated by drag forces on the blades—see Figure 3.10. Therefore, blade airfoils need high lift-to-drag ratios.

- The blades of a HAWT must be progressively twisted from hub to tip to maintain a suitable angle of attack toward the resultant wind—see Figure 3.11.

- The tip speed ratio has a major effect on a rotor's aerodynamic efficiency, with optimum values typically in the range of 5 to 10. This in turn influences the range of suitable turbine speeds as wind strength varies.

In the early years of the wind energy renaissance, it was obviously sensible for blade designers to draw on the impressive body of information about aircraft airfoils that had been accumulated since the Wright brothers made their first powered flights in 1903. Aeronautical engineers spent almost a century developing airfoils for a huge variety of fixed-wing aircraft, military and civil, with a wide range of operational duties. They had also fine-tuned airfoil sections for aircraft propellers and helicopter rotors—spinning devices that share many of the aerodynamic features of wind turbines, even if they are designed to create wind rather than capture it!

However, it became clear in the 1980s that new thinking was required, and much research on wind turbine airfoils has since been done in the USA, the UK, Germany, Denmark, and the Netherlands. Several problems affecting wind turbine performance have received special attention:[3]

- Roughness on the leading edges of conventional airfoils caused by accumulations of insects, dirt, or ice can seriously degrade rotor power output, in extreme cases by as much as 40%. It is obviously difficult, and expensive, to clean the blades of a large turbine regularly.
- Traditional aircraft airfoils tend to produce far too much power at very high wind speeds, endangering turbine equipment, especially in machines with fixed blades.
- When turbine blades incorporate *pitch control* allowing excessive wind power to be "spilled," sudden gusts may occur too fast for the control system to respond, generating extreme forces.
- The very long blades of today's large HAWTs need special, thick airfoil sections near the hub for structural strength and support.

The last of these points leads to a related issue: the need to taper turbine blades from hub to tip. Figure 3.16 showed such tapering very clearly, and you may have previously noticed it in Figures 2.7 and 3.1. In fact, the blades of today's large HAWTs are invariably tapered (as well as twisted), reaching maximum width and thickness soon after leaving the hub and progressing to a much smaller, thinner section at the tip. From a mechanical point of view, this is certainly desirable because, as we noted in the previous section, longitudinal

Figure 3.16 Preparing for duty off the coast of Scotland: three blades for a large HAWT rotor (*Source:* With permission of Repower).

$(\Omega^2 R)$ stresses at the blade root caused by rotating mass—especially mass near the tip—can be very large. From this point of view, the lighter the outer sections of the blade, the better. And, of course, tapering reduces the amount of expensive material used in blade construction.

But, there are also compelling aerodynamic reasons for tapering a turbine blade. In a nutshell, theoretical approaches to blade design show that, for optimum performance, the chord of the airfoil section should be progressively reduced from hub to tip.[3, 4] Intuitively, we may imagine that a high lift-to-drag ratio at the tip requires a slim airfoil with a small chord because of the very high resultant wind speeds, which are mainly due to rotation (Figure 3.17). Certainly, drag forces at the tip are very serious for rotor efficiency and must be minimized. But as we go back toward the blade root, the lower resultant wind strength is adequately captured by a thicker, wider, airfoil.

To summarize, the tip of a blade is best fashioned as a relatively narrow, thin airfoil giving a high lift-to-drag ratio, whereas the root region, which accounts for a very small proportion of the rotor's swept area, must provide strength and structural support. In practice, the differing airfoil sections along the length of a blade are often chosen as members of an airfoil "family," specially tailored to address the above problems.[3]

The blade designer, in addition to choosing an appropriate family of airfoil sections, must consider the mechanical strength required to withstand static and dynamic forces, control strategies for survival in extreme wind conditions, potential noise problems, and material and manufacturing costs.

There is also the very important question of blade (and therefore rotor) efficiency in the wind regime for which the turbine is intended. We have already given this question some attention,

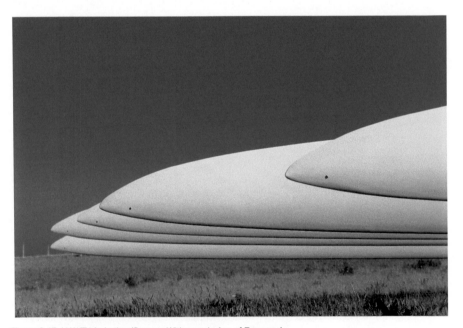

Figure 3.17 HAWT blade tips (*Source:* With permission of Repower).

explaining how speed control can be used to improve efficiency as wind speed varies (see Figure 3.15). Rather than focus on maximizing the power coefficient over a narrow range of wind speeds—an approach commonly adopted back in the 1980s—today's blade designers try to take the full range of wind speeds and conditions into account, maximizing annual energy capture.

A powerful weapon in the designer's armory is the so-called *blade element method* for predicting the torque and power generated by a particular blade geometry. Rather than consider the rotor as a circular disc that extracts kinetic energy from the wind uniformly over the whole swept area (as in Section 3.2.1), blade element theory divides the disc into a large number of thin annular sections at different radii from the hub. It then estimates the lift-and-drag forces on each portion of the blade as it moves around an annular section, as a function of wind speed, rotor speed, and the particular airfoil used at that radius. In this way, the designer can check and refine blade shapes for their aerodynamic performance and energy production. Details of the method are beyond the scope of this book but are well explained elsewhere.[3, 4]

3.2.4.2 Stall and pitch control

The aerodynamic condition known as *stall* can sometimes be catastrophic for an aircraft, but it has both positive and negative implications for wind turbines. In some machines, stall is deliberately used to limit the amount of power captured by the rotor in high winds, preventing damage. To understand how stall affects turbine operation, we need to explore the nature of the airflow pattern over an airfoil section.

Back in Section 3.2.2, we described the lift-and-drag forces acting on an aircraft wing in level flight and illustrated typical airflow over an airfoil in Figure 3.9. In this figure, the flow is "ell-behaved," with the airfoil inclined at an angle of attack (α) that generates plenty of lift and very little drag—as the designer intended. Typically, an airfoil section generates increasing amounts of lift up to angles of attack of about 10°; but beyond that the flow over the top surface tends to break up, becoming turbulent and causing a sudden loss of lift and increase in drag, as shown in Figure 3.18. This is the stall condition, and it is highly significant for turbine blades as well as aircraft wings.

In more detail, low values of α produce *attached flow* with the airstream remaining in close contact with the upper surface of the airfoil. There is a thin *boundary layer*, where the air speed increases from zero at the surface to that of the frictionless flow above. Flow in the boundary layer is smooth and steady at low angles of attack and is referred to as *laminar*; but if the angle of attack approaches a critical value, the boundary layer becomes increasingly turbulent toward the trailing edge, and stall begins to develop.

The magnitude of the lift-and-drag forces acting on an airfoil must clearly depend on its surface area—in other words on its chord (width) and span (length). The larger the surface presented to the oncoming wind, the larger the forces. So to compare the performance of different wings, blades, and sections, lift-and-drag forces are commonly expressed per unit of surface area and divided by the *dynamic force* of the wind to produce dimensionless *lift-and-drag coefficients* relevant to a particular wind speed.[3]

Figure 3.19 shows how the lift-and-drag coefficients of a typical airfoil vary with the angle of attack. As the angle is increased from zero up to around 10° the lift coefficient grows more or less in proportion, typically peaking between 1.0 and 1.5, but then declines rapidly as stall develops. The drag coefficient starts at a very low value—say around 0.01—but

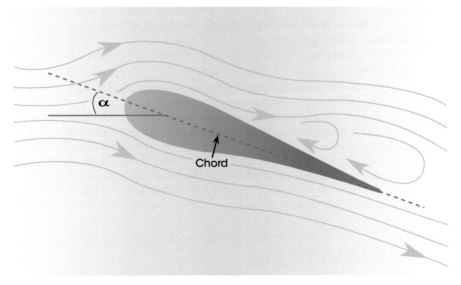

Figure 3.18 As the angle of attack increases, stall develops.

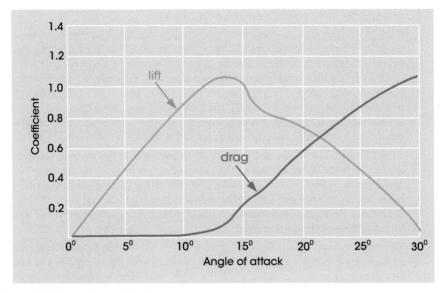

Figure 3.19 Lift and drag coefficients.

increases dramatically in stall. These figures suggest lift-to-drag ratios up to around 150 for a well-designed airfoil operating at optimum angles of attack. However, it must be emphasized that the details, especially in the deep stall region, depend heavily on the particular airfoil and wind speed considered.

How does all this relate to wind turbine performance? There are two main issues: the avoidance of stall conditions in normal operation and the deliberate use of stall to prevent a rotor from generating excessive power in very strong winds.

We will start by considering what happens when a rotor with fixed blades, running at constant speed, encounters a change in wind speed. The solid blue vectors in Figure 3.20 represent the "natural" wind, the wind due to rotation, and the resultant wind at three radii along a blade, when the tip speed ratio is 9.0. This is similar to Figure 3.11, which we used to explain why *a blade must be progressively twisted toward the tip* in order to maintain a suitable angle of attack. But in Figure 3.20, we have added dotted lines to show what happens when the natural wind strength U_b increases by 50%, assuming the rotor speed is unchanged. The direction of the resultant wind, and therefore the angle of attack of the fixed blades, increases by the angles colored red in the figure: 3.2° at the tip, 4.5° at two-thirds radius, and 8.2° at one-third radius from the hub. Of course, these are just representative figures, relevant to the tip speed ratio and the percentage change of wind speed we have chosen. But they demonstrate a crucial point: the blades' angle of attack increases as the wind gathers strength, and *the effect is much more pronounced near the hub*. Increasing the angle of attack by 8.2° is very likely to cause stall; but by 3.2°, much less so. We conclude that as the wind speed rises, the blades will tend to stall progressively, starting near the hub and moving out toward the tip.

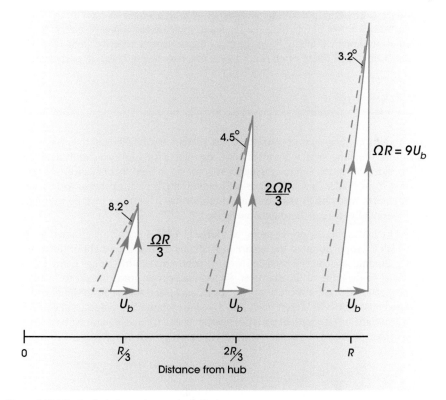

Figure 3.20 Effects of wind speed on angle of attack.

This seems a good moment for a historical diversion. The remarkable development of turbine technology in Denmark, including the Gedser machine of the 1950s (see Figure 1.12), convinced the Danes that some form of aerodynamic control mechanism should always be incorporated into their large turbines to prevent damage or even destruction in extreme winds. Danish machines imported into the USA during the Californian "wind rush" of the early 1980s benefited greatly from this design philosophy and were far more robust and reliable than their rivals. The commonest approach used *passive stall control*, designing the fixed blades to stall automatically and limit rotor power when the wind speed exceeded a certain value. Other techniques included deployment of control surfaces such as tip brakes, or movable flaps (ailerons) located near the trailing edges of the blades. Over many decades Danish wind turbines have been renowned for their effective use of aerodynamic control.

However, the use of a passive stall in a fixed-blade machine, although conceptually straightforward and cheap to implement, gives a very limited form of aerodynamic control. Large modern HAWTs are considerably more sophisticated. Start-up of a turbine may be achieved by pitching the blades to give substantial lift at a standstill. Blade pitch refers to turning the angle of attack of the blades of a rotor into or out of the wind to adjust the rotation speed and the generated power. There is a specific pitch angle for any given wind speed to optimize output power. Pitch angles greater or less than this value reduce power output, even to the point of zero rotation with high winds. Variable pitch allows very effective *active stall control* in which the blades are pitched to give a smooth transition from normal to stall-limited power generation, preventing overload in very high winds. In fact, excess wind power may be "shed" either by pitching the blades to increase the angle of attack (the normal stall condition) or by pitching them the other way toward a negative angle of attack (known as *feathering*). The latter, often faster to achieve, may be preferable in gusty conditions.

3.3   Mechanics

Modern multi-megawatt wind turbines present big challenges to mechanical engineers: rotors weighing over a hundred tons, nacelles with gearboxes and generators weighing considerably more, and towers sufficiently robust to hold them all aloft. As well as ensuring that rotor blades and hubs have sufficient strength and resistance to fatigue, mechanical engineers must design gearboxes and high-torque drive trains to cope with fluctuating, and occasionally extreme, aerodynamic loads.

As we have seen, the power rating of a large HAWT is mainly determined by its swept area, which is proportional to the square of its diameter. But the weight of the rotor depends on the amount of material used in its construction and, other things being equal, is proportional to the cube of the diameter. So we must expect rotor weight to increase more than proportionally to power rating—an unwelcome trend that is, however, already being countered in some of the largest machines by using carbon fiber materials in blade construction. The weight and strength of a nacelle also increase with machine rating because it must house a heavier, more powerful gearbox and generator, as well as support the rotor (Figure 3.21).

3.3.1   Gearboxes

To many people, a gearbox seems a commonplace object. Perhaps this is because we are used to vehicle gearboxes, which are nowadays so reliable that they tend to be taken for granted.

Figure 3.21 Up it goes: the nacelle of a 5-MW HAWT (*Source:* With permission of Repower).

But the gearboxes used in large HAWTs are unusual, and deserve our attention, for two main reasons:

- Power ratings are relatively large. For example, since 1 horsepower (hp) is equivalent to 746 W of electrical power, the gearbox of a 3-MW HAWT must be able to transfer up to $3 \times 10^6/746 = 4021$ hp from the rotor to the electrical generator. The 10-MW machines need gearboxes rated at over 13,000 hp.
- A HAWT gearbox must *increase* the speed from the rotor to the generator. For example, its input shaft may be rotating at 15 rpm and the generator at 1500 rpm—a 100:1 *step-up ratio*. Gearboxes used in other applications, including road vehicles, are generally *step-down* devices with smaller ratios.

We should just mention that an important class of large HAWT uses specially designed low-speed generators directly coupled to their rotors, avoiding the need for gearboxes. We will discuss them in Section 3.4.2.2.

The two gear configurations in Figure 3.22 illustrate some important general principles underlying gearboxes. Part (a) shows two *spur* gears mounted on separate parallel shafts. The larger gear is on the low-speed shaft, the smaller one is on the high-speed shaft, and the speed ratio is equal to the ratio between their diameters (or numbers of teeth). If we ignore the small power losses in well-designed gearboxes, we may equate the

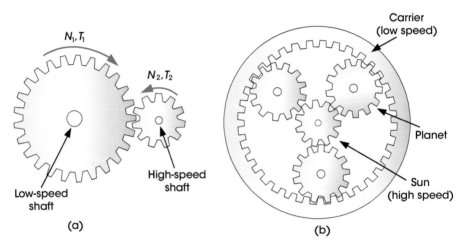

Figure 3.22 Gear arrangements for (a) a simple spur gearbox and (b) a planetary gearbox.

input and output powers, so that:

$$2\pi N_1 T_1 = 2\pi N_2 T_2 \quad \text{and} \quad N_1/N_2 = T_2/T_1 \tag{3.14}$$

where N_1 and N_2 are the input and output speeds, and T_1 and T_2 are the torques (see also Equation (3.11)). This shows that the torque ratio between the input and output shafts is the inverse of the speed ratio: if the speed is increased (as in a HAWT gearbox), the torque reduces in proportion. The electrical generator coupled to the output shaft is a relatively high-speed, low-torque unit compared with the rotor.

In practice, it is difficult to achieve speed ratios above six in a single-stage gearbox. Higher ratios require intermediate shafts and extra gears, referred to as a *gear train*. The overall ratio is then equal to the product of the individual ratios. For example, a ratio of 100:1 may in principle be obtained using three pairs of meshing gears with ratios of 5, 5, and 4.

However, it will probably not surprise you to learn that the simple arrangement of spur gears shown in the figure, or a multistage version of it, is not generally used in today's megawatt HAWTs. Designers tend to favor a *combined spur/planetary gearbox*, often in three stages, based upon the planetary principle illustrated in part (b) of the figure. The main components of a step-up planetary stage are:

- A large-diameter *carrier* with a ring gear on its inside circumference, driven directly by the low-speed (input) shaft.
- Intermediate *planet* gears, each free to rotate on its own shaft and driven by the ring gear.
- A central *sun* gear, driven by the planets, connected directly to the high-speed (output) shaft.

The planets have a smaller diameter than the carrier's ring gear and therefore rotate faster than the input shaft. The sun's diameter is smaller than that of the planets and therefore rotates faster still. The overall step-up speed ratio is given by:

$$S = 1 + \left(d_r/d_s\right) \tag{3.15}$$

where d_r and d_s are the diameters of the ring and sun gears, respectively.

There are two main advantages to this design:

- The input and output shafts are coaxial, giving a compact and comparatively light unit.
- Many pairs of gear teeth mesh together, reducing the loading on each gear.

The weight of a large HAWT gearbox, like that of the rotor, tends to scale roughly in line with the cube of the rotor radius, so the compactness and weight-saving of the planetary layout are very helpful. Even so, the huge torque on the rotor shaft of a large HAWT (discussed in Section 3.2.3) demands very robust, heavy gears at the low-speed end of a multistage gearbox. And, of course, ruggedness and complexity come at a cost.

With the advance of power electronics, low-speed multipole generators driven directly by turbine rotors without any need for gearboxes are of special interest; the development of gearless wind turbines is discussed in Section 3.4.2.2.

3.3.2 Towers

Towers for windmills and wind turbines have a long history. By the end of the 18th century, English and Dutch windmills had reached a high degree of sophistication, incorporating rotating caps on the top of wooden or stone towers (see Figure 1.9). The towers needed internal space, as well as the height and strength to support their rotors, because they housed milling or pumping equipment, stores, and people. In later years, huge numbers of steel lattice towers were manufactured for water-pumping windmills (see Figure 1.10), especially in North America, and lattice designs were still widely specified up to the 1980s. Today, we can occasionally spot an elderly tower or steel lattice windmill within sight of its modern counterpart: a poignant reminder of how far wind engineering has progressed over the centuries (Figure 3.23).

The new generation of megawatt HAWTs spearheading the current surge in wind energy capacity are almost all mounted on tubular steel towers. Fashioned from steel plates, rolled to produce the required curvature, and then welded, they give technical performance, ease of maintenance, longevity—and, many would say, visual elegance. They also provide a protected enclosure for access to the nacelle and for cabling and services (Figure 3.24). Their main technical requirements may be summarized as follows:

- *Strength and stiffness.* The tower must be strong enough to support the weight of nacelle and rotor (up to several hundred tons), and stiff enough to resist the large wind forces acting on the rotor.
- *Height.* Tower height is typically about one rotor diameter. However, the precise height depends on the wind regime at the site, including the amount of wind shear, and on potential visual intrusion. The top of the tower incorporates the interface with the nacelle and includes the stationary part of the yaw drive mechanism.

Figure 3.23 Towers old and new in Portugal and Canada (*Source:* (a) With permission of Repower. (b) With permission of Canadian Wind Energy Association).

- *Foundation.* Large onshore wind turbines are generally secured on reinforced concrete foundations that must be heavy enough to resist overturning in the most extreme wind conditions. Occasionally turbines are sited on rock and must be secured by steel rods inserted deep within it. Foundations for offshore turbines—generally a more challenging problem—will be discussed in the next chapter.
- *Resistance to corrosion.* Steel towers must be well-protected against corrosion, especially offshore. Special coatings/paints and/or steels with enhanced anti-corrosion properties are available.
- *Safety.* The inside of a tower must be designed for safe access, typically using guarded ladders or climbing bars and anti-fall harnesses.

As we shall see in the next section, the properties of a steel tower, especially its mass and stiffness, are very important because mechanical coupling with the rotor can produce significant vibrations.

3.3.3 Vibration and fatigue

A large wind turbine turning steadily in a strong breeze appears as a model of calm, dignified technology. Unlike a Formula 1 racing car powering through the gears or an airliner straining along the runway to get airborne, the turbine refuses to be hurried, generating clean electricity at its own pace.

Figure 3.24 About to go aloft a French wind turbine (*Source:* With permission of Repower).

However, external appearance is not always a good indicator of inner calm. A large HAWT has to endure many forces, static and dynamic, that are not directly concerned with the production of electricity. Static forces, for example, the load imposed by a steady wind on a stationary rotor, or the weight of a tower on its foundations, are fairly easy to visualize. More complicated are the cyclic forces that vary according to the position of the rotor, including the effects of weight and wind shear on individual blades as they rotate. Noncyclic dynamic forces, especially fluctuating wind forces acting on turbine blades and rotors, are the most awkward of all, both theoretically and in practice. Dynamic forces may set up unwelcome vibrations and, in the long term, cause fatigue and failure of mechanical components. And in the most extreme weather conditions, it is the unpredictable ones that threaten a turbine's survival.

Our exploration of "wild wind" in Chapter 2 emphasized that wind is generally variable in strength and direction. A turbine's annual energy output and capacity factor are enhanced by fluctuations in wind strength because of the cubic relationship between wind speed and power, making periods of above-average wind speed highly productive. In strong, gusty conditions, there are big variations in the power intercepted by a rotor and delivered by the blades, often over time scales measured in seconds rather than minutes. All this must be accommodated by the control systems that regulate rotor speed and blade pitching. It is hardly surprising if such a demanding operating environment sets up vibrations, and

Figure 3.25 Mechanical strength and resilience: a large modern HAWT (*Source:* With permission of Repower).

occasionally extreme loads, in the turbine's mechanical components—even though the stresses and strains are invisible to a casual onlooker (Figure 3.25).

The theory of structural vibrations, as applied to complicated mechanical components including turbine blades, rotors, and towers,[3] is beyond the scope of this book. However, we can illustrate some key ideas and terminologies with a few simple experiments and equations that illustrate, in broad outline, the issues facing designers of large wind turbines.

Imagine holding the long blade of a kitchen knife firmly down on a table as in part (a) of Figure 3.26, then deflecting the handle with the other hand and letting it go—an experiment

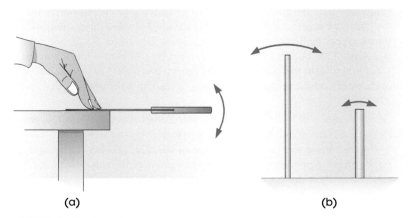

(a)   (b)

Figure 3.26 Simple vibrating systems.

that amuses almost every physics student. The knife vibrates up and down with a certain *frequency*, and with diminishing *amplitude*, eventually coming to rest. If the blade is now repositioned so that less of it overhangs the table, more strength is needed to deflect the handle and, on release, the vibrations are faster. The time course of vibration is a decaying *oscillation* whose frequency depends on *stiffness* and *mass*. Greater stiffness increases the frequency; greater mass decreases it.

And why do the vibrations decay with time? It is due to *damping* caused mainly by air resistance as the knife blade bends backward and forward. Without any damping, the vibrations would continue indefinitely with a frequency known as the *natural frequency*, but the vibrations die out with damping. Above a certain level of damping, a system displays no oscillations at all, returning to its starting position without any "overshoot."

Part (b) of the figure shows the second part of our experiment, in which two vertical wooden rods are held firmly in a wooden block. Imagine deflecting the top of the tall thin rod with a finger, then letting it go. Once again, assuming a small amount of damping, there will be vibrations. The shorter, thicker rod will be much harder to deflect because of its far greater stiffness; and, if any vibrations are visible, they will be at a higher frequency. Like the knife blade in part (a), each rod responds in a manner determined by its inherent stiffness, mass, and damping.

Although it is risky to stretch analogies too far, these experiments have considerable relevance to vibrations in wind turbines. After all, a turbine blade is essentially a long object held firmly, or *cantilevered*, out from the hub. It has certain mass, distributed along its length, and stiffness that depend on its construction and the direction of bending. Subjected to a sudden gust of wind, it will tend to vibrate, and the vibrations will fade with time depending on the amount of damping. A turbine tower is even more obviously a cantilevered structure, anchored firmly to the ground or seabed, and capable of swaying or vibrating. You may have noticed certain similarity with our wooden rods (even though we have not put model nacelles and rotors on the top!). With this motivation in mind, we will introduce a little mathematics.

Vibration theory is normally introduced by considering an idealized system comprising "lumped" mass, stiffness, and damping elements—the simplest type of *linear second-order system*. Such a system exhibits all the effects noted above and will serve well for our discussion. In particular, decaying oscillations take the form:

$$x = A \exp\left(-\xi \omega_n t\right) \sin\left(\omega_d t + \varphi\right) \tag{3.16}$$

where x is the displacement from the resting position, A is the initial amplitude, ξ is a damping constant, ω_n is the natural frequency measured in radians per second, ω_d is the vibration frequency, and φ is a phase angle. This shows that the amplitude of the oscillations decays according to a negative exponential at a rate determined by the damping constant ξ.

The vibration and natural frequencies are related by:

$$\omega_d = \omega_n \left(1 - \xi^2\right)^{0.5} \tag{3.17}$$

If $\xi = 0$, there is no damping, and the vibrations continue indefinitely at the natural frequency. If $0 < \xi < 1$, the frequency is reduced, and the vibrations decay. And if $\xi = 1$, the system is said to be *critically damped*, settling back to its starting point without oscillation. Higher values of ξ denote a system that is *overdamped* and increasingly "sluggish."

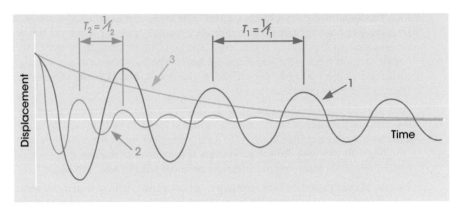

Figure 3.27 Vibrations in systems with mass, stiffness, and damping.

Figure 3.27 shows three typical responses of this type. Curve 1 reminds us of the kitchen knife: the vibration decays slowly, with many oscillations before it dies out. The time interval or *period* between successive peaks is T_1 and the frequency in cycles per second (Hz) is its inverse, thus $f_1 = 1/T_1$ (note that the frequency in radians per second equals $2\pi f_1$). Curve 2 denotes a stiffer system with more damping: the period T_2 is shorter, the frequency f_2 is higher, and it decays more rapidly. This might be like one of our wooden "towers." And, curve 3 shows the response of an overdamped system that returns to its starting position without any oscillations.

The concept of *natural frequency* is so important for wind turbines that it deserves more discussion. When a system is disturbed from rest and allowed to settle, any vibrations it displays are "natural" to it and show how it behaves when left "on its own." After the initial disturbance, and assuming no further external energy is supplied, the frequency of vibration is determined solely by the system's mechanical properties—its stiffness, mass, and damping. The response is *characteristic of the system*, not of the initial displacement or force that disturbs it.

The vibrations shown in Figure 3.27 relate to simple mechanical systems, whereas complex structures such as turbine rotors and towers generally possess many natural frequencies with different amounts of damping. It is also important to realize that deflections or vibrations set up in one part of a turbine may affect other parts due to *mechanical coupling*. The blades, hubs, gearboxes, generators, nacelles, and towers of a wind turbine are all interconnected mechanically, facilitating the transfer of vibration energy between them. It is quite possible for a mechanical disturbance originating in one place to produce a vibration elsewhere. It is rather like the opera singer who, as the story goes, produces a powerful note that shatters a wineglass on the other side of the room.

The above ideas are highly relevant to large HAWTs for several reasons:

- Vibrations may be set up by cyclic forces due to the rotation of the rotor and drive train, and, by random, dynamic forces including wind gusts.
- Mechanical structures including blades and towers have natural frequencies at which they "prefer" to vibrate. Sudden disturbances tend to set up vibrations at the preferred frequency (or frequencies), which decay at a rate determined by the amount of

damping. And, as we shall see below, repetitive forces close to a natural frequency can produce strong vibrations due to the phenomenon of *resonance*.

- Dynamic forces on one part of a turbine may set up vibrations in another part due to mechanical coupling.

So far we have considered vibrations caused by an initial disturbance, watching the system respond and eventually settle "on its own." But what if there are repeated disturbances that occur *at a frequency close to a natural frequency?* A good example is cyclic forces due to the rotation of a rotor. For example, suppose a three-bladed rotor is turning at 15 rpm and producing various cyclic forces on the blades. There will be force components that fluctuate at the *blade-passing frequency*, that is, $3 \times 15 = 45$ times per minute, corresponding to 0.75 Hz, and some of the pulsating energy will no doubt be transferred to the tower by mechanical coupling. It is not hard to imagine that unwelcome tower vibrations may be set up if 0.75 Hz is at, or close to, one of the tower's natural frequencies. And if the rotor keeps turning at 15 rpm, the vibrations will continue. This is known as *resonance*.

The classic approach to analyzing continuous vibrations in a linear system assumes that it is disturbed by a sinusoidal force of variable frequency. Sinusoidal inputs have the special property of producing sinusoidal responses at the same frequency, but with variable magnitude (and phase). The magnitude (amplitude) of the response of a second-order system with various degrees of damping is shown in Figure 3.28. Consider, for example, the red curve for which $\xi = 0.15$. As the frequency of the sinusoidal input is increased from zero, the response amplitude passes through a peak in the region of the system's natural frequency ω_n, then falls away as the frequency rises further. We see that excitation close to the natural frequency causes resonance. It is not hard to imagine that strong resonance in a lightly damped system can place big stresses on a component or structure, which may even cause its destruction in the long term. We are back again to the opera singer and the wine glass.

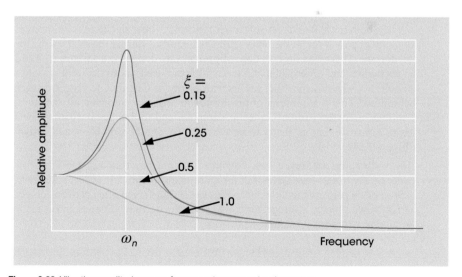

Figure 3.28 Vibration amplitude versus frequency in a second-order system.

As the amount of damping increases, the resonance becomes less marked, as shown by the orange and yellow curves in the figure. The resonant peaks are wider and occur at slightly lower frequencies—see Equation (3.17). If $\xi = 1$, there is no resonance at all (green curve), and the system is said to be *critically damped*.

As far as HAWT vibrations are concerned, we may expect the strongest ones to be caused by forces at or very close to a natural frequency, especially if the damping constant ξ is small. Of course, cyclic forces arising in practice are most unlikely to be precisely sinusoidal; but if they occur regularly at the right (or perhaps we should say wrong!) frequency, they have a similar effect. It is like a child on a swing: give a gentle push at just the right moments, and the swing goes higher and higher.

What does this imply for HAWT design? One of the most striking examples is provided by steel towers, which are essentially cantilevered structures with low damping that support large masses (nacelles and rotors) on the top. Of a tower's many natural frequencies, only the lowest or *fundamental* is normally of concern because it is most likely to coincide with cyclic forces set up by the rotor. These are mainly at the rotation and blade-passing frequencies. With this in mind, designers recognize three main classes of tower:[3]

- *Stiff tower*. The fundamental natural frequency is above the blade-passing frequency at the highest speed of the rotor. This normally avoids tower vibration problems.
- *Soft tower*. The fundamental natural frequency is between the rotor rotation frequency and the blade-passing frequency. In a three-bladed HAWT, these frequencies are in the ratio 1:3, and they vary with rotor speed. For example, if a rotor's working speed range is 12 to 24 rpm, corresponding to 0.2 and 0.4 Hz, the blade-passing frequency falls in the range of 0.6 to 1.2 Hz. A soft tower would therefore be designed to have its fundamental frequency at around 0.5 Hz.
- *Soft-soft tower*. The fundamental natural frequency is below both the lowest rotor rotation frequency and the corresponding blade-passing frequency—in the above example, less than 0.2 Hz.

Clearly, both the soft and soft-soft tower types may suffer significant vibrations as the rotor and blades pass through the main resonant frequency during the start-up or shut-down of the turbine. This raises the question: why not always design for a stiff tower? The answer lies in economics and tower installation—a stiff tower needs more steel and is more difficult to handle. So, why not construct towers with plenty of damping, to reduce or even suppress vibrations? This is a question of practicality as well as economics, for structural steel is a "springy" material with low inherent damping, and to dampen it down artificially would be an expensive option. Better to be aware of the vibration problem, and design around it (Figure 3.29).

Tower vibrations are fairly easy to visualize. They are important not only for the stresses they place on the towers themselves, but for unwanted aerodynamic effects that a swaying tower has on blade performance, and for the stresses it transfers to other turbine components. The situation with blades is altogether more complicated. Many different forces tend to deform or vibrate a blade as it rotates:

- *Gravity forces*, due to the blade's own weight, stretch or compress it when vertical, and bend it when horizontal.
- *Centrifugal forces* due to rotation stretch the blade regardless of its orientation.

Figure 3.29 Designed to withstand forces and vibrations: the tower and blades of a large HAWT (*Source:* With permission of Repower).

- *Bending forces* produced by the wind: a tangential component that generates useful torque but tends to bend the blade "edgewise"; and an unwanted, and normally the much larger, axial component bends the blade backward, causing "flapping."
- *Gyroscopic forces* caused by the yaw motion of the rotor. Whenever a spinning rotor is adjusted to face the wind, it acts like a gyroscope, producing bending and torsion forces according to the blade's position.

Many of the above forces are either cyclic or fluctuate in a random fashion. Wind forces have a steady component due to the average (mean) wind speed, with superimposed fluctuations. All this leads to a highly complex, ever-changing, force scenario for turbine blades and towers. Although distortions and vibrations may not degrade the short-term performance of a well-designed turbine, in the long term they threaten it through the process known as *fatigue*.

Fatigue occurs in engineering materials subjected to repeated cyclic loads and eventually—perhaps after millions of "cycles"—causes failure. Typically, small cracks start to develop in the material or component in the region of maximum stress and grow with repeated cycles. Susceptibility to fatigue depends very much on the material: try bending a thin strip of aluminum sheet backward and forward, and it soon fails; steel survives much better. Today's large HAWTs contain a variety of structural materials including steel, fiberglass reinforced plastic, and, increasingly in large blades, carbon fiber composites. All have their own distinctive fatigue characteristics.

The development of fatigue depends on the magnitude as well as the number of load cycles, making resonant conditions with low damping especially risky. Fatigue is also hastened when a component or structure, already stressed by a steady load, has a cyclic load superimposed on it—a common situation in wind turbines. And while few manufactured products experience more than a million load cycles in their working lives, turbine designers face a bigger challenge. For example, a three-bladed machine turning at an average of 15 rpm for 18 hours a day over 25 years produces $3 \times 15 \times 60 \times 18 \times 365 \times 25 = 4.43 \times 10^8$ load cycles at blade-passing frequency. We are in the hundreds of millions. And if the load cycles excite significant resonance in turbine components, we are also into dangerous territory. A severe fatigue failure in one blade could so unbalance the rotor that the whole thing shakes to pieces.

Progressive fatigue is to be expected in machines and structures subject to vibration and should be considered a normal part of wear and tear. But a large HAWT must also be able to withstand occasional extreme loads. Most of the time it operates within its design parameters, with average wind speeds described by a Rayleigh or Weibull distribution (see Section 2.2.2). It cuts in at one wind speed and starts generating, cuts out at a much higher speed to avoid damage, and is expected to withstand a reasonable amount of turbulence. But, it must also cope with extreme events—especially violent gusts superimposed upon storm-force winds. There may also be sudden changes in wind direction and shear which force the turbine's speed, pitch, and yaw control systems into rapid response. We introduced the theory behind the occurrence of extreme events, including the Gumbel distribution, in Section 2.2.2. So, when discussing the slow and steady onset of fatigue, essentially an aging process involving millions of load cycles, we should always bear in mind that a turbine is also threatened by single, "once in a lifetime," events.

3.4 Electrics

3.4.1 Alternating current (AC) electricity

The vast majority of today's large wind turbines feed their electricity into power grids based on *AC* electricity. The early years of the electrical power industry saw intense competition between AC systems championed by Nikola Tesla (1856–1943) and DC, or *direct current*, systems preferred by Thomas Edison (1847–1931) (Figure 3.30). However, the race was eventually won by AC, which came to dominate the generation, transmission, distribution, and utilization of electrical energy during the 20th century. Tesla, born in what is now Croatia, spent most of his life in New York and is particularly remembered for his invention of the AC induction motors that today power much of global industry—and are intimately related to a major class of wind turbine generators.

One of the principal advantages of AC electricity is that its voltage level may be easily changed using transformers. Typically, AC is generated at one voltage level, transformed up for long-distance transmission and distribution, and finally transformed down to a safe voltage level for use by consumers. Whenever the voltage is raised, currents are reduced, and vice-versa. Transmission at high voltage (and low current) reduces losses in power lines, allowing the use of much smaller and less expensive conductors. At the consumer end, AC is well suited to running cheap and efficient electric motors in a huge variety of applications, and even keeps our electric clocks running on time!

(a)   (b)

Figure 3.30 Pioneers of electrical power: Tesla and Edison (*Source:* (a) Napoleon Sarony/Wikimedia Commons/Public domain. (b) The Library of Congress/Wikimedia Commons/Public domain).

However, we must be careful not to dismiss DC power transmission. The power-carrying capacity of electric cables is generally greater when fed with DC rather than AC and DC is often favored for very-long-distance transmission of bulk power from one region or country to another. DC also has special advantages for long submarine links, including cables to connect offshore wind turbines to the land. We will return to this important topic in Section 4.5.

In an AC system, voltages and currents in linear circuits and devices vary *sinusoidally*. Currents flow backward and forward, changing direction at a rate determined by the grid frequency—for example, 50 cycles per second, or Hertz (Hz), in Europe and 60 Hz in North America. However, it is important to realize that the voltage and current waveforms, although sinusoidal in shape, are not generally *in phase* with each other, and this has profound implications for the generation and use of grid electricity. Nonlinear circuits, which include electronic switches used in wind turbine generation and control, introduce non-sinusoidal waveforms, and their analysis is more complicated. For the time being, we will focus on linear circuits because they illustrate and explain many key aspects of AC electricity.

In general, the time-varying waveform $v(t)$ of an AC voltage may be expressed as:

$$v(t) = V_p \sin(\omega t + \varphi) = V_p \sin(2\pi f t + \varphi) \tag{3.18}$$

where V_p is the peak value (amplitude), ω is the frequency in radians per second, f is the frequency in Hz, t is the time in seconds, and φ is the phase angle. If $\varphi = 0$ the function is a sine wave that passes through zero at $t = 0$. This may be thought of as a "reference" waveform. Other waveforms of voltage and current with different phase angles are shifted along the time axis with respect to the reference.

We now consider the voltage and current waveforms for three basic types of linear circuit elements: *resistors, inductors, and capacitors*. Figure 3.31(a) shows two cycles, or *periods*, of a continuous voltage waveform—a sine function with peak value V_p and period T (equal to

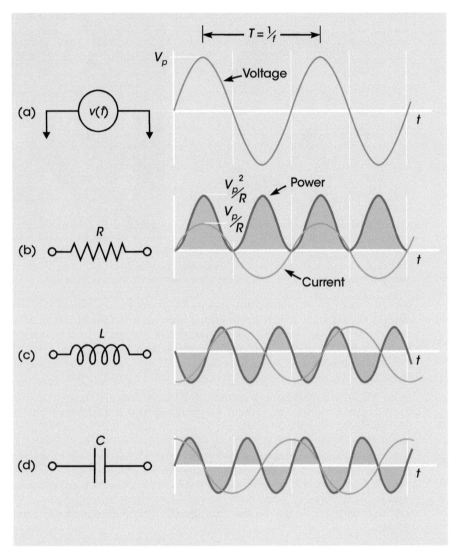

Figure 3.31 AC voltage, current, and power.

the reciprocal of the frequency in Hz). If this AC voltage is applied to a resistor of value R ohms, as in part (b), the current that flows is proportional to the voltage at every instant, and there is no phase shift. According to Ohm's law, its peak value is V_p/R. The instantaneous power dissipated in the resistor equals the product of voltage and current:

$$p_r(t) = V_p \sin(\omega t) V_p/R \sin(\omega t) = \left(V_p^2/2R\right)(1 - \cos(2\omega t)) \qquad (3.19)$$

We see that the power fluctuates cosinusoidally, but at twice the frequency of the voltage source and with an offset that keeps it positive. The area under the power curve, shaded red in the figure, represents the energy supplied by the AC source and dissipated in the resistor. In some ways, this is all rather surprising. After all, if we turn on an electric heater or an incandescent lamp—essentially both resistors—it is not obvious that the power is fluctuating at twice the supply frequency, but this is because any resulting "flicker" is too fast for our eyes and brains to follow.

It is clear from the figure that the *average* power dissipated in the resistor is half the peak power, that is, $V_p^2/2R$. The same amount of power would be dissipated if it were connected to a DC voltage of $V_p/\sqrt{2} = 0.7071\,V_p$. In other words, the heating effect of a sinusoidal voltage with peak value V_p is the same as that of a DC voltage of $V_p/\sqrt{2}$. Since heating is, in an important sense, what matters (and what we pay for when we turn on a heater!) the voltage of an AC supply is quoted as $V_p/\sqrt{2}$, referred to as the *rms (root mean square)* value. For example, the "220 V" of a European domestic supply refers to its rms value; the peak value is $220\sqrt{2} = 311$ V.

AC electricity becomes rather more interesting when we consider inductors and capacitors, because they produce phase shifts between voltage and current waveforms. The *inductor* shown in part (c) of the figure has the property of *inductance*, denoted by the letter L and measured in *Henrys (H)*. It may be thought of as an "ideal" coil of wire with no electrical resistance. It does not *dissipate* energy (get warm) like a resistor; it can only *store* energy in its magnetic field. To provide some motivation for the discussion that follows, we note that all rotating electrical machines, including wind turbine generators, depend on coils and magnetic fields. Inductance is central to their operation.

The figure shows current and power waveforms for an ideal inductor supplied with a sinusoidal voltage, and for the moment we will focus on phases, assuming for convenience that the waveforms have unit amplitude. The key point to note is that the current waveform, colored green, *lags* the voltage by a quarter of a period, in other words its phase angle is $-90°$ ($-\pi/2$ radians). Intuitively, this is because the magnetic field takes time to wax and wane, lagging behind the voltage waveform. Now $\sin(\omega t - \pi/2) = -\cos(\omega t)$, so the instantaneous power supplied to the inductor has the form:

$$p_L(t) = \sin(\omega t)\,(-\cos(\omega t)) = -\sin(\omega t)\cos(\omega t) = -0.5\sin(2\omega t) \tag{3.20}$$

This is shown as a red curve in part (c) of the figure. As with the resistor in part (b), the power fluctuates at twice the supply frequency; but it now goes negative as well as positive. The inductor alternately accepts power from the source and then returns it, so the average power is zero. It is not a net consumer of energy; rather it uses current from the source to establish an oscillating magnetic field.

To summarize, an "ideal" inductor or coil, when supplied with a sine wave of voltage, draws a sinusoidal current that lags the voltage by 90°. The source, although supplying voltage and current, does not provide net power. The product of an inductor's voltage and current is therefore referred to as *reactive power*—one of the most important concepts in AC electricity. Reactive power is measured in *volt-amperes reactive* to distinguish it from the *real power* dissipated by a resistor, which causes heating and is measured in *Watts (W)*.

So how is an AC generator affected by having to supply *reactive* power? Not much, you might think, because its main task is clearly to supply *real* power for heating or doing useful mechanical work. But there is a snag: although the current taken by a purely inductive load does not cause power consumption *in the load*, that same current must originate in the generator and flow along transmission lines and through transformer windings, all of which possess some electrical resistance and generate unwanted heat. So in practice a certain amount of energy is wasted getting reactive power to a reactive load, and power utilities tend to discourage reactive power demands which "use up" a substantial portion of their equipment's current-carrying capacity.

And now for the last in our trio of circuit elements—the capacitor. In its simplest form, a capacitor may be thought of as a pair of close-spaced metallic plates. It has the property of *capacitance*, measured in *Farads (F)*. We can deal with it relatively quickly because in many ways its behavior is the exact opposite of the inductor's. Whereas the AC current in an inductor *lags* the voltage by 90°, in a capacitor it *leads* by 90°. And, whereas the inductor stores energy in a *magnetic* field, the capacitor stores it in an *electric* field set up between the plates. Like the inductor, the capacitor alternately accepts power from the source and then returns it, and the net power flow is zero. The current and power waveforms are shown in part (d) of Figure 3.31, and the instantaneous power takes the form:

$$p_c(t) = \sin(\omega t)\cos(\omega t) = 0.5\,\sin(2\omega t) \tag{3.21}$$

A capacitor also requires reactive rather than real power, but with phase relationships opposite to those of an inductor. We will discuss several important ways in which capacitance affects AC power grids.

So far we have placed emphasis on phase relationships and the concept of reactive power. But how about the *amplitudes* of the currents that flow in inductors and capacitors? Specifically, since in their ideal form they contain no resistance, what limits the currents? The answer is their *reactance* which, like resistance, is measured in ohms (Ω); but, unlike resistance, reactance is frequency-dependent and involves the 90° phase shifts we have described above.

The reactance of an inductor is:

$$X_L = \omega L = 2\pi f L \quad \text{ohms} \tag{3.22}$$

where L is the inductance measured in Henrys. For example, in a 50-Hz AC system, an inductor of 1 Henry (1 H) has a reactance of $100\pi = 314$ ohms (Ω); in a 60-Hz system, the same inductor has a reactance of $120\pi = 377\,\Omega$. Inductive reactance is proportional to frequency.

The magnitude of AC current flowing in an inductor is simply equal to the AC voltage divided by the reactance. This is analogous to Ohm's law for a resistor—provided we remember the 90° phase shift. For example, a 1 H inductor, connected to a 50 Hz 220 V rms source, takes a current of 220/314 = 0.70 A rms. The peak current is $0.70\sqrt{2} = 1.0$ A.

Whereas the reactance of an inductor increases with frequency, that of a capacitor reduces. It is given by:

$$X_C = 1/\omega C = 1/2\pi f C \quad \text{ohms} \tag{3.23}$$

Once again, the magnitude of AC current that flows is equal to the voltage divided by the reactance. As an example, you may like to check that a 0.0001 F capacitor connected to a 60 Hz 120 V rms source takes a current of 4.52 A rms.

So far we have considered resistors, inductors, and capacitors as "ideal" elements. In practice, of course, they are not ideal; for example, the wire of an inductor coil invariably has

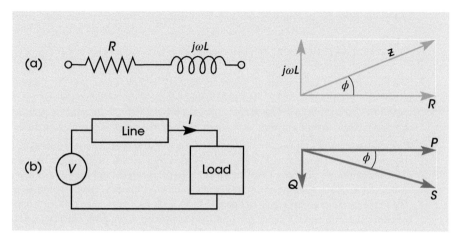

Figure 3.32 The j-notation for AC circuits.

some resistance. In other cases, we need to analyze circuits containing separate R, L, and C components. They all take sinusoidal currents when supplied with sinusoidal voltages, but amplitude and phase relationships are more complicated.

We can illustrate this with a simple example and at the same time introduce some further ideas and terminologies. Figure 3.32(a) shows a resistor and inductor connected in series. The resistor's value is R ohms, and we have labeled the inductor's reactance $j\omega L$, rather than just ωL. This simple change is in fact highly significant. The letter j (pure mathematicians use the letter i) denotes $\sqrt{-1}$ and implies that the reactance involves a 90° phase shift. Mathematically we are representing the reactance as an *imaginary* quantity, distinguishing it clearly from the resistance R which is represented as a *real* quantity. The *j-notation* allows linear AC circuits to be analyzed simply and elegantly using the rules of *complex arithmetic*.

A full account of the j-notation applied to AC circuits is a standard ingredient of electrical engineering textbooks, but for our purposes it may be reduced to a few key ideas. To start, we note that the resistor and inductor in Figure 3.32(a) are in series, and we may therefore add the resistance and reactance to give their combined *impedance*:

$$Z = R + j\omega L \quad \text{ohms} \tag{3.24}$$

The impedance, a complex quantity with real and imaginary parts, is represented in the figure. The resistance is drawn as a real number, the reactance as an imaginary number, and the impedance is found by vector addition. Alternatively, the impedance may be resolved into real and imaginary components. It is somewhat analogous to resolving the wind force on a turbine blade into axial and radial components—as we did in Figure 3.10.

The magnitude and phase of the impedance are given by:

$$|Z| = \sqrt{(R^2 + \omega^2 L^2)} \quad \text{and} \quad \varphi = \tan^{-1}(\omega L/R) \tag{3.25}$$

As an example, suppose that a coil with inductance 1 H and resistance 40 Ω is connected to a 50 Hz 220 V rms supply, and we wish to find the current that flows in the coil. The magnitude and phase of the impedance are

$$| Z |= \sqrt{\left(40^2 + (100\pi)^2\right)} = 317\Omega \quad and \quad \varphi = \tan^{-1}(100\pi/40) = 82.7° \qquad (3.26)$$

The magnitude of the current is therefore $220/317 = 0.69$ A rms, and its phase angle with respect to the voltage is $-82.7°$. Note that since we are *dividing* the voltage by the impedance to find the current, the *positive* phase angle associated with the impedance produces a *negative* phase angle for the current, in line with the normal rules of complex arithmetic.

A similar approach may be used to represent the real and reactive components of the power supplied by an AC generator. Figure 3.32(b) shows a generator supplying voltage and current via a transmission line to a load. In general, both line and load have reactance as well as resistance, and the generator's current is not in phase with its voltage. It supplies *real power (P)* to the resistive components of line and load, and *reactive power (Q)* to the inductive and capacitive components. The vector sum of these two powers gives the *apparent power (S)* supplied by the generator, equal to the product of its rms voltage and current and measured in *volt-amperes (VA)*. For example, a generator might supply 100 kVA of apparent power, consisting of 80 kW of real power and 60 kVAR of reactive power. Note that large quantities are expressed in kVA or MVA equivalent to the use of kW and MW for real power.

As previously mentioned, the supply of reactive power uses up part of the current-carrying capacity of a utility's generators and transmission lines and wastes real power in associated resistances. Commercial and industrial customers requiring substantial amounts of reactive power are often charged special tariffs, depending on the *power factor (F)* of their loads. The power factor is defined as the ratio between real and apparent power, and from Figure 3.32(b) we see that it is also equal to the cosine of the phase angle φ. Thus:

$$F = P/S = \cos \varphi \qquad (3.27)$$

If the power factor is unity, no reactive power is supplied, the phase angle is zero, and the load is purely resistive. When reactive power is demanded, the power factor reduces and the phase angle increases. A positive phase angle means the load is capacitive, giving a *leading* power factor; a negative angle denotes an inductive load, producing a *lagging* power factor—typical of industrial loads that include large electric motors. A power factor of 0.9 is typical of the level below which utilities start to charge large consumers for reactive power.

The negative effects of reactive power may be counteracted, or offset, using *power-factor correction*. Imagine, for example, that a load takes a current of 300 A at a lagging power factor of 0.8, equivalent to a phase angle of $-36.9°$. The current may be resolved into a real component of magnitude 300 cos 36.9° = 240 A, and a reactive component of magnitude 1000 sin 36.9° = 180 A. Now suppose capacitors, placed across the load, are designed to take a leading current of 180 A. This exactly counteracts the lagging current of the load and supplies the required amount of reactive power. The utility now provides just the real power via the transmission system, while the load and capacitor swap reactive power. This situation has, in fact, already been illustrated in Figure 3.31. Parts (c) and (d) show that the reactive current and power in the inductor and capacitor are in *antiphase* and will "cancel each other out" if they have equal magnitudes. For this to happen their reactances must be equal at the supply frequency, so that:

$$\omega L = 1/\omega C \quad and\ therefore \quad C = 1/\omega^2 L \qquad (3.28)$$

A smaller (and therefore cheaper) capacitor may be used to give partial power-factor correction—for example, raising it from 0.8 to 0.9. Large industrial and commercial users of electricity often choose to install parallel banks of capacitors, switching in more or less capacitance automatically as the power factor of their load changes.

Utilities often use correction equipment of their own because power factors close to unity tend to enhance system stability as well as reduce resistive losses in generation and transmission. So far we have only mentioned capacitors, but inductors may be used to correct leading power factors, for example, when feeding AC electricity into submarine cables possessing high capacitance.

Our final topic in this introduction to AC electricity concerns *three-phase systems*. So far we have dealt with a single sinusoidal voltage and the currents it produces in resistive and reactive loads. But in practice AC grids normally generate and transmit a set of three separate voltage *phases* (not to be confused with phase *angles*) displaced from one another by 120° ($2\pi/3$ radians). A set of such voltages is shown in Figure 3.33. Domestic customers with modest power requirements normally receive just one of the phases—a so-called *single-phase supply*. Commercial and industrial consumers are more likely to receive all three phases.

Three-phase AC has a number of advantages:

- Large three-phase generators, transformers, and motors are more efficient than single-phase versions.
- Three-phase is more economical to transmit than single-phase, using less conductor material for the same power and voltage levels.
- Power flow from a three-phase generator into a "balanced" linear load is constant (see below), reducing vibrations in the generator and three-phase motors supplied by it.
- Three-phase electricity can produce rotating magnetic fields of major importance in the design of generators and motors.

Figure 3.34(a) shows a three-phase system comprising generator, transmission lines, and load. The voltages for phases 1, 2, and 3 are produced by the generator's internal coils, or *windings*, which are arranged electrically as a "Y" or "star" and joined at the *neutral*

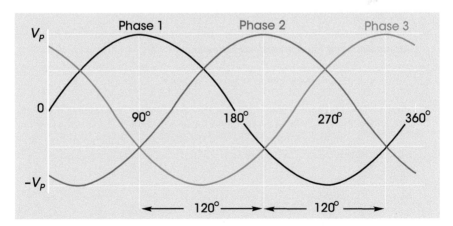

Figure 3.33 Three-phase voltages.

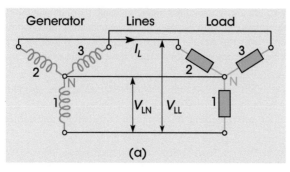 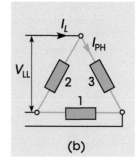

(a) (b)

Figure 3.34 Three-phase AC power.

point (N). Transmission is done by three-line conductors and one neutral conductor—a total of four cables. For simplicity, we have not shown any transformers but these, too, have three-phase windings. At the receiving end, the cables feed a three-phase load such as a large electric motor. In addition, low-power single-phase loads such as lighting may be supplied by one line and the neutral cable and distributed equally between the three phases. Households are normally supplied with one phase.

If the load is perfectly balanced with equal impedances in the three phases, the vector sum of currents at the neutral point is zero. In this situation, the neutral conductor carries no current—which explains why it is normally much thinner and lighter than the three-line conductors. In practice, loads are unlikely to be perfectly balanced, so the neutral conductor carries a small amount of current.

The voltages of the various phases, measured between line and neutral and denoted by V_{LN}, are out of step by 120°—see Figure 3.33. The voltage V_{LL} between any two lines equals the vector sum of two-phase voltages, giving $V_{LL} = \sqrt{3}\,V_{LN}$. The current in each line, I_L, equals the current in each phase, I_{PH}.

An alternative arrangement, known as "Δ" or "delta," is shown in part (b) of the figure. Transmission now requires only three cables, and if the three impedances are equal the load and line currents are again perfectly balanced. The line and phase voltages are now equal, but the line current is $\sqrt{3}$ times the phase current.

In a three-phase system, the total real and reactive power in the load may be expressed in terms of the rms line voltages and currents as:

$$P = \sqrt{3}\,V_{LL}I_L \cos\varphi \text{ and } Q = \sqrt{3}\,V_{LL}I_L \sin\varphi \tag{3.29}$$

Some of the ideas and terminologies in this section may seem strange if you are new to AC electricity; but they will be a great help for understanding the various types of generators used in large modern wind turbines, and the engineering challenge of integrating them into grid networks.

3.4.2 Generators

3.4.2.1 Introductory

When Michael Faraday (1791–1867) discovered *electromagnetic induction* in 1832, he may have had little inkling that, within a hundred years, the phenomenon would spawn a global electricity industry. A man of limited formal education, Faraday was appointed professor of chemistry at the Royal Institution in London in 1833 and is widely acknowledged as one of the greatest experimentalists of all time. He was fascinated by the relationship between electricity and magnetism and the possibility of making an electric motor. His theory of electromagnetic induction provided the essential insight; and today we realize that the fortuitous combination of Faraday's ideas, copper's conductivity, and iron's magnetism has formed the basis of electric motors and generators ever since.

From the point of view of electricity generation, the core principle of electromagnetic induction may be simply stated: a voltage is produced, or *induced*, in a conductor moving at right angles to a magnetic field. What matters is the *relative* movement—the conductor may be stationary and the field moving, or vice versa.

Figure 3.35 shows how this principle can be used to make a simple AC generator. A magnet with north and south poles rotates between the arms of a soft iron frame. Two slots in the frame carry an insulated copper wire in the form of a loop. The magnet forms the moving part—the *rotor*—and the frame and wire loop form the stationary part—the *stator*. As the magnetic field (green arrows) moves past the conductor equal but opposite voltages are induced in the two sides of the loop. Being in series, they add together and appear at the terminals.

Half a revolution later, the north and south poles of the magnet have swapped positions, the field has changed direction, and the induced voltages change sign. This is an AC device that generates a fluctuating voltage—positive then negative—each time the magnet goes through a complete revolution.

In electromagnetic induction, the magnetic field, the movement, and the induced voltage are always at right angles, or *orthogonal*, to each other. Looking carefully at the rotational system of Figure 3.35, we see that in this case the field is *radial*, the movement is *tangential*, and the conductor is arranged in *axial* slots.

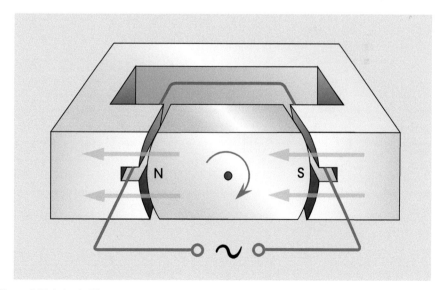

Figure 3.35 A simple AC generator.

How much voltage is generated as a magnetic field "sweeps past" a conductor? This depends on two factors: the strength of the field and the speed of movement—in other words, on the *rate of change of magnetic flux*. The field may be maximized by keeping the *reluctance* of the magnetic circuit as low as possible. In Figure 3.35, we have shown an iron frame that provides an "easy," low-reluctance path for the magnetic flux (apart from the small but necessary air gap between the frame and magnet). But even a strong magnetic field is unlikely to produce much voltage in a single conductor loop, so practical AC generators use coils with many turns of wire, referred to as *windings*. The instantaneous induced voltage in a winding is given by:

$$v(t) = -n\,(n\Phi/dt) \qquad (3.30)$$

where n is the number of turns, Φ is the magnetic flux measured in *Webers*, and $d\Phi/dt$ represents its rate of change.

Although this simple machine demonstrates several key principles relevant to wind turbine generators, it has its limitations. First, it is a single-phase device whereas large AC generators are normally three-phase. Second, its voltage waveform is unlikely to match the sinusoidal shapes shown in Figures 3.31 and 3.33. Although the magnet generates alternate positive and negative voltages as it rotates past the conductor slots, these are more likely to resemble "pulses" than a smoothly varying sine wave. In fact, the production of sinusoidal voltages requires careful attention to the profile of the magnetic field and the layout of the copper windings, part of the stock-in-trade of the design engineer. And finally, the magnetic field of a large machine is normally produced by windings on the rotor, supplied with DC, which acts as an electromagnet. This is referred to as *separate excitation*.

Figure 3.36 shows the main features of a high-power AC generator, this time producing three phases. The windings are spread in slots around the stator's circumference. For simplicity, we

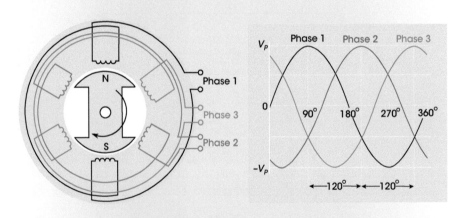

Figure 3.36 A three-phase AC generator.

have shown them well separated, but they actually overlap and the number of conductors in each slot is carefully devised to produce sinusoidal voltage waveforms. As the rotor turns at constant speed, the angular separation between the windings is translated into the required timing separation between the three voltage waveforms, shown on the right-hand side of the figure. One complete AC cycle is generated for each complete revolution. If the required frequency is 50 Hz, the rotor must turn 50 times per second, or at 3000 rpm; if 60 Hz is required, at 3600 rpm.

So far we have assumed that the rotor has a single pair of poles, and this is generally true of large AC generators driven by steam turbines in conventional power plants. However, generators for large wind turbines often have two-pole pairs, halving the rotational speed for the same output frequency—for example, 1500 rpm produces 50 Hz—with a corresponding reduction in the gearbox ratio. In principle there is nothing against even more pole pairs, giving further speed reductions; indeed, there is an important type of multipole, low-speed, wind turbine generator that dispenses with a gearbox altogether, allowing it to be directly driven by the wind turbine. In general, the speed of rotation in rpm is given by:

$$n = 60f/P \quad \text{giving} \quad P = 60f/n \tag{3.31}$$

where f is the frequency of the generated supply, and P is the number of pole pairs. For example, a direct-drive machine required to generate 60 Hz at a rotation speed of 20 rpm would need $60 \times 60/20 = 180$ pole pairs.

For reasons that will become clear shortly, the above machines are examples of *synchronous generators*, and the speed of rotation is known as the *synchronous speed*. When a machine's field is produced by separate excitation, it is known as a *wound-rotor synchronous generator* (*WRSG*); and when the field is produced by permanent magnets, it is referred to as a *permanent-magnet synchronous generator* (*PMSG*).

There is a second major type of AC machine, of great importance in modern wind engineering, known as the *asynchronous*, or *induction*, generator. Most people find it more difficult to understand than the synchronous generator, so it will be helpful to prepare the ground for a fuller description in later sections.

We have already seen how a rotating magnet generates a set of sinusoidal voltages in a three-phase stator winding. There is another side to the same coin: such a winding, if energized by an *external* three-phase supply, produces its own *rotating magnetic field*. For example, if we were to remove the rotor of the machine in Figure 3.36 and supply three-phase 50-Hz AC to its windings from an external source, the windings would act as electromagnets, producing a magnetic field rotating at 3000 rpm. Furthermore, it may be shown that if the individual fields produced by the three phases are sinusoidal in both time and spatial distribution, then their vector sum is constant in magnitude.[3] In other words, the field produced by a three-phase winding supplied with three-phase power is essentially like that produced by a rotating magnet. Back in the 1880s, it was Tesla's great insight to realize that this phenomenon could be used to make a new type of AC motor—the *induction motor*. During the 20th century, induction motors became the trusted workhorses of industry as well as powering many domestic appliances including washing machines and electric lawnmowers. And we shall see later that their first cousins, *induction generators*, are widely used in wind turbines to produce grid electricity.

What is the essential distinction between synchronous and asynchronous (induction) machines operating in a large electricity grid? A highly imaginative analogy based upon a

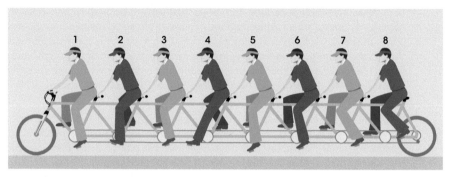

Figure 3.37 Cyclists on a "very long bike."

number of cyclists riding a "very long bike" has previously been described,[5] and we will use a modified and simplified version of it here.

Figure 3.37 shows cyclists on a very long bike that represents the electricity grid. Some cyclists act as generators, supplying power to the grid. Others act as motors (loads), demanding power from the grid. Each has pedals connected to a sprocket wheel that meshes with a long chain running the whole length of the bike. The overall aim is to keep the bicycle moving along a straight flat road at constant speed, equivalent to maintaining the grid frequency at its nominal value of 50 (or 60) Hz. Air resistance and rolling resistance are neglected.

The "generator" cyclists pedal actively, trying to push the bike forward, some with more force than others. The "motor" cyclists apply braking force to the pedals, trying to slow the bike down. If the bike speed is to remain constant the total forces produced by the two groups must be kept in balance at all times. This is equivalent to a grid in which electricity consumption must always be matched by generation.

As well as being a generator or motor, each cyclist is either *synchronous* or *asynchronous*, depending on how his pedals are linked to the driving chain:

- *Synchronous.* The pedals of a synchronous cyclist are connected to the sprocket wheel by a stiff, but slightly elastic, shaft. If the cyclist is a generator, his effort twists the shaft slightly forward by an amount proportional to the torque exerted. But if the cyclist is a motor, the shaft is twisted slightly backward. As a result, the angular positions of the various cyclists' pedals are not aligned—for example, they do not all pass the lowest point at the same instant. The pedals of a generator are always slightly in advance of those of a motor. But, it is important to realize that *synchronous cyclists must all pedal at the same speed*, dictated by the speed of the chain. In terms of our analogy, synchronous generators and motors connected to an electricity grid must rotate at a speed precisely determined by the grid frequency.

- *Asynchronous.* The pedals of an asynchronous cyclist are connected to the sprocket wheel by a *fluid coupling*. This consists of two small fan-like turbines facing each other in a fluid-filled housing. One turbine is rigidly connected to the pedals, the other to the sprocket wheel. The turbines are coupled together relatively "softly" by the viscosity of the fluid, and as one rotates it tends to drag the other with it. A generator cyclist transmits power by pushing the pedal turbine faster than the sprocket wheel turbine; a motorcyclist does the opposite. Power transfer in either direction depends on a certain

110

amount of *speed difference*, or *slip*—no power is transmitted if the turbines are rotating at the same speed. In terms of our analogy, an asynchronous generator must rotate somewhat faster, and an asynchronous motor somewhat slower, than the equivalent synchronous machine.

Let us now examine the cyclists in the figure more carefully. Suppose we are told that the four blue ones are all synchronous, and that one of them is acting as a generator and three as motors. Can we tell which is which? Indeed we can: cyclist 7 has his pedals slightly in advance of the others, so he must be the generator.

We are also told that the four red cyclists are asynchronous, with two generators and two motors. But this time we cannot tell which is which because a "snapshot" showing pedal positions gives us no information about rotation speeds and the amount of slip—the essential clue for distinguishing between generators and motors. We would need to know pedal *speeds*, not *angular displacements*.

The analogy offers further insights,[5] especially into the question of reactive power balance. As we saw in the previous section, a generator or grid must generally supply two types of power: active power that delivers heat or mechanical work and reactive power that supplies the out-of-phase AC currents demanded by inductive or capacitive loads. So far, our bicycle analogy has focused entirely on the real power generated or demanded by the cyclists. But the bicycle must also maintain balance and stability, not tipping over sideways and causing the cyclists to fall on the road. Forces trying to overturn the bicycle to the left may be considered analogous to producers of reactive power, for example, capacitors used for power-factor correction; forces acting to the right are analogous to consumers of reactive power, for example, motor inductances. Note that the cyclists' balancing act involves constant adjustments of their weight to left or right but does not require them to generate (or consume) real power. This is analogous to a grid network, where reactive power balance is important for stability and maintaining the system voltage at the required level.

In the following sections, we describe the main variants of synchronous and asynchronous generators used in large HAWTs. We will start with synchronous machines since they are in many ways easier to understand and relate to Figures 3.35 and 3.36. We will then move on to the asynchronous machines that dominated the early years of the wind energy renaissance and, supported by recent developments in power electronics, find increasing application in today's high-power, variable-speed turbines.

3.4.2.2 Synchronous generators and gearless wind turbines

Most large conventional power plants produce electricity using synchronous generators driven by steam or hydroelectric turbines. Two-pole machines coupled to high-speed steam turbines are often referred to as *turbo-alternators* and are accepted as the trusted workhorses of the electrical supply industry. In Europe, with its 50-Hz grids, they rotate at 3000 rpm; in North America, with 60-Hz grids, at 3600 rpm. We have already illustrated their main features in Figure 3.36. Hydroelectric generators are generally designed as multipole machines, allowing rotation at the much lower speeds of large water turbines. But in all cases, the generator is effectively tied to the grid it serves by a "stiff magnetic spring," with no option but to rotate at *synchronous speed*. As its power output changes, the rotor adopts a different *angle*, very much as the "synchronous cyclists' described in the previous section did with their pedals; but it cannot alter its rotation *speed*.

Another important feature should be mentioned. As we explained in the previous section, an electricity grid has to supply reactive as well as real power. One of the advantages of conventional synchronous generators is their ability to act as controlled sources of reactive power by varying the strength of their rotor fields,[3] allowing them to run at either lagging or leading power factors. Such flexibility is a great asset for controlling and stabilizing a large grid.

We may clarify the generation of real power by considering the interaction of magnetic fields. As already noted, the field of a WRSG is produced by feeding DC excitation current into the rotor's field windings; a PMSG uses permanent magnets instead. In either case, the rotor field rotates at synchronous speed and induces three-phase voltages in the stator windings (see Figure 3.36).

So far we have only mentioned induced voltages, but of course voltage on its own is not enough; the stator must also supply three-phase currents (and power) to the grid. This has an extremely important consequence: the currents produce their own rotating field that interacts with the field set up by the rotor. It is the "tension" between the two fields, rotating synchronously but with an angular offset from one another, that produces torque and delivers real power to the grid. This is analogous to the stiff but slightly flexible shaft linking the pedals of a "synchronous cyclist" to the chain of the bike shown in Figure 3.37.

Figure 3.38 illustrates the interaction of rotor and stator fields with a simple model. In this case, fields are produced by permanent bar magnets, rotating together at synchronous speed n, but offset from each other by an angle β. North and south poles attract, so the angular offset produces a constant torque on the magnets that try (but fail) to align them. In part (a), the torque on the rotor opposes its rotation, so the rotor must be supplying mechanical power and the machine is acting as a generator. But in part (b), with the offset angle reversed, the torque on the rotor aids its rotation, so the machine is acting as a motor. We see that positive values of β represent generation and negative values represent motoring—like most

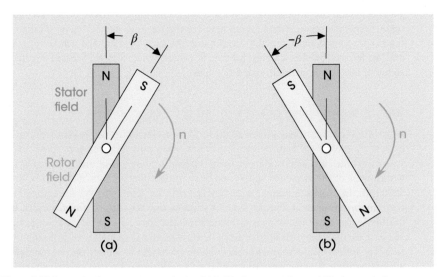

Figure 3.38 Interaction between stator and rotor fields (a) when generating and (b) when motoring.

electrical machines, a synchronous generator can also act as a motor. In either case, the rotor and stator are tied together by a "magnetic spring."

So what happens if a synchronous generator, coupled to a wind turbine via a suitable gearbox, is connected directly to a large electricity grid? The first point to note is that it is not self-starting; there has to be sufficient wind strength (or an auxiliary power source) to run the turbine up to synchronous speed before connecting it. The instant of connection must be finely judged, ensuring a suitable rotor angle as well as the correct speed. Thereafter, the wind turbine must turn at a constant speed dictated by the grid frequency. With a strong enough wind, it acts as a generator; but if the wind dies, the rotor tries to lag behind and starts to accept power from the grid. Our wind turbine wants to become a very large fan—hardly the desired effect, and certainly an expensive waste of grid electricity! Of course, it could be automatically disconnected from the grid; but in light and variable winds, there would be constant disconnection and reconnection, hardly the best way to treat a wind turbine.

There are two other major reasons why synchronous generators are unsuitable for connecting wind turbines directly to a grid:

- The connection is very "stiff." The grid demands a fixed rotation speed, allowing only small angular offsets of the generator's rotor. This is fine for synchronous machines driven by steam or water turbines, with power levels continuously under the control of supervisory engineers. But a wind turbine driven by variable and gusty winds is a very different proposition. Since the speed is fixed, any power fluctuations produced by the turbine rotor translate directly into torque variations and power surges into the grid. A stiff, inflexible connection places high stresses on mechanical components and may induce vibrations. Even wind turbines equipped with variable-pitch blades may be unable to react sufficiently fast to counteract such effects.
- Large modern HAWTs are increasingly designed as variable-speed machines. This allows the matching of rotation speed to wind speed, leading to enhanced efficiency and higher annual electricity yields (as discussed in Section 3.2.3), but it cannot be achieved with directly connected synchronous generators.

At this point, the prospects for synchronous generators in large HAWTs may seem distinctly unpromising. But in this field of electrical engineering, as in many others, the situation has been transformed in recent years by impressive developments in *power electronics*.

If a synchronous generator driven by a large wind turbine cannot be connected *directly* to a large grid, perhaps it can be connected *indirectly* using a power-electronic interface. This would make the generator independent of grid frequency, voltage, and phase. It may sound like an expensive option for a megawatt-scale HAWT, but if the high cost of the electronics can be offset by gains in turbine efficiency, and possibly also by omitting a heavy and expensive gearbox, it can prove very attractive.

Figure 3.39 shows the main elements of such a scheme. A wind turbine (W) drives a synchronous generator (SG) via a gearbox (GB). The gearbox is shown dotted to indicate that it can be omitted if a multipole low-speed generator is used. The generator delivers power to an electronic *power converter* (PC), also sometimes called a *frequency converter*, because its key function is to decouple the rotation speed of the wind turbine from the frequency of the grid. This is done by first converting the generator's AC output to DC with a *rectifier*, then converting it back to AC using an *inverter*. The DC link provides isolation between the two unsynchronized AC systems, allowing the wind turbine to operate over a wide speed

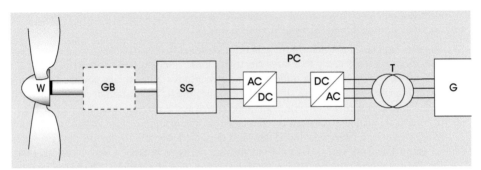

Figure 3.39 Main elements of a variable-speed wind turbine (W): GB, gearbox; SG, synchronous generator; PC, power converter; T, transformer; G, grid.

range. It also makes the grid connection much "softer," reducing sudden surges of torque and electrical power as the wind speed fluctuates. Finally, the AC output from the inverter is fed via a transformer (T) into the grid (G). The scheme is referred to as *full-scale* or *fully rated* power conversion because the full power output of the generator passes through the converter.

In addition to allowing generators to run at variable speeds, power converters can be designed to produce or absorb reactive power, and to provide a "soft-start" for running the turbine gently up to speed. In principle, the scheme of Figure 3.39 may also be used with asynchronous generators because the power converter decouples the generated frequency from that of the grid. However, as we shall see later, asynchronous generators are normally used with *partial-scale* converters, which handle only a portion of the output power and are considerably cheaper.

Rectifiers and inverters used in power converters are based on electronic switches such as diodes, transistors, and thyristors. Half a century ago, the pioneers of solid-state electronics could hardly have imagined that semiconductor devices would one day control megawatts of power. But huge advances in power-handling capacity have been made since the 1980s, opening electrical engineering up to a wide range of new ideas and possibilities. Wind engineering is certainly one of the beneficiaries.

We see that synchronous generators allied to full-scale power converters have found a valuable place in modern wind engineering. In principle, the decoupling of generator and grid frequencies allows a turbine to work over a wide range of speeds, referred to as *full variable-speed operation*; in practice, a range of about 3:1 is often chosen, sufficient to optimize wind energy capture between cut-in and rated wind speeds.

Low-speed multipole generators driven directly by turbine rotors without any need for gearboxes are of special interest. Figure 3.40 shows the nacelle of a gearless machine, its three-bladed rotor coupled directly to a low-speed generator (colored red). This is often referred to as an *annular* or *ring* generator because the stator and rotor windings are arranged in the form of annular rings. The main advantages of this system are the elimination of a costly gearbox with its associated losses, and the simplicity of the mechanical drive train, with a minimum number of bearings.

However, a directly coupled generator for a megawatt HAWT typically turns around 100 times slower than an equivalent four-pole machine driven via a gearbox. This means

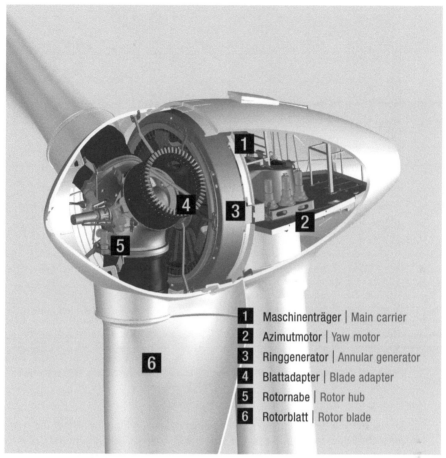

1 Maschinenträger | Main carrier
2 Azimutmotor | Yaw motor
3 Ringgenerator | Annular generator
4 Blattadapter | Blade adapter
5 Rotornabe | Rotor hub
6 Rotorblatt | Rotor blade

Figure 3.40 The rotor of this gearless wind turbine is directly coupled to a multipole, low-speed, synchronous generator (*Source:* ENERCON GmbH).

that its torque is about 100 times greater. As we noted in Section 3.2.3, the torque developed by a megawatt rotor can be equivalent to an Indian elephant standing at the end of a 100-m lever! Very high-torque, low-speed, electrical machines are comparatively rare these days and demand special design and construction skills—a point well illustrated by the stators being fitted with three-phase windings in Figure 3.41. In general, the physical size and losses of a low-speed generator depend more on torque than rated power, and efficiency can be improved by designing with a large diameter. Although this increases the size of the nacelle and makes transport more difficult, the overall simplicity and reliability of modern direct-drive machines have secured a substantial proportion of the current market, with clear signs that they will be increasingly favored for offshore wind farms.

In principle, the rotor fields of gearless synchronous generators can be produced by permanent magnets rather than by separate excitation. Permanent magnets avoid the power losses that occur when current is passed through copper windings, known as *copper losses* or i^2R

Figure 3.41 Assembling stators for high-power gearless wind turbines (*Source:* With permission of ENERCON GmbH).

losses, and can be made lighter and more compact. On the other hand, the strength of the magnetic field is fixed, unlike that of a wound rotor which can be adjusted to control reactive power and output voltage—flexibility that is admittedly less important when the generator is connected to the grid via a power converter. These and other issues are well discussed elsewhere.[5, 6] Megawatt HAWTs using permanent magnets are currently in production and seem likely to make a considerable impact in the coming decade.

3.4.2.3   Asynchronous generators

3.4.2.3.1   *Squirrel-cage rotor and wound-rotor induction machines*

Asynchronous generators connected to an electricity grid do not rotate at exactly synchronous speed because the generation of real power requires a certain amount of *slip*. Asynchronous operation is analogous to the action of a cyclist on a "very long bike" whose pedals transmit power to the drive chain via a fluid coupling—a concept already discussed and illustrated in Figure 3.37. Wind turbines coupled to asynchronous generators, also referred to as *induction generators*, can cushion sudden power fluctuations by allowing their rotors to speed up or slow down slightly, storing or releasing kinetic energy in the turbine and drive train and "softening" the grid connection. Furthermore, as we shall see in the next section, recent developments in power converters allow induction generators to operate over a wide speed range. And finally, turbines with induction generators may be self-started and run up to speed using their generators as motors.

In its basic form, an asynchronous (or induction) generator is cheaper and much simpler in construction than a synchronous machine with the same power rating. Yet the details of its operation are, by common consent, harder to visualize. Since most people are more familiar with motors than generators, and since the two are intimately related, we will start by explaining the action of an *induction motor*.

A three-phase induction motor has windings arranged around its stator in much the same way as a synchronous machine and, when supplied with a three-phase AC supply, produces a magnetic field that rotates at synchronous speed. It was Tesla's genius in the late 1800s to realize that the rotating field could be used to *induce* voltages in a simple form of rotor and that the resulting currents would produce torque, tending to "drag" the rotor around with the field. And so, the induction motor was born.

The simplest type of rotor is known as a *squirrel-cage rotor*, shown in its basic form in Figure 3.42. Iron laminations are built up to form a solid core, with a set of axial slots filled with cast aluminum bars that take the place of the more familiar (but more expensive) copper conductors. In practice, there are generally many more bars than shown in the figure, and they are skewed slightly to reduce torque fluctuations. The bars are short-circuited together by two *end rings* that complete the "cage." Squirrel-cage induction motors are produced in vast quantities for applications ranging from industrial drives to domestic appliances and have gained a reputation for robust reliability.

What happens when a stationary three-phase squirrel-cage motor is first connected to an AC supply? We may visualize the following sequence of events:

- A magnetic field, rotating at synchronous speed, is set up by the stator windings.
- Three-phase voltages are induced in the rotor bars, and since they are short-circuited by the end rings, currents flow. Initially, the rotor is stationary, the currents have the same frequency as the supply, and the rotor produces its own magnetic field that also rotates at synchronous speed.

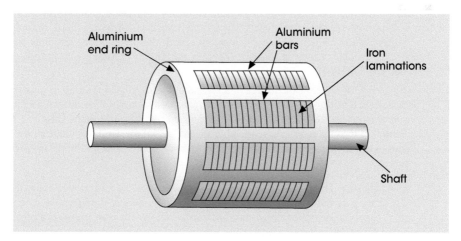

Figure 3.42 A squirrel-cage rotor.

- Interaction between the stator and rotor fields produces torque, and the rotor starts to turn. As it gathers speed, attempting to "catch up" with the stator field, the frequency of the induced currents in the rotor bars decreases.

- The rotor can never reach synchronous speed. If it did, there would be no induced currents in the rotor bars, no torque, and the machine could not act as a motor delivering mechanical power. For torque and power to be produced, there must be a certain amount of *slip*. This contrasts with a synchronous machine, which rotates at precisely synchronous speed.

We see that the key to motor operation is the *induction* of rotor currents by the stator's magnetic field, which tries to "drag" the rotor with it. The slip is defined as:

$$s = \left(n_s - n \right) / n_s \tag{3.32}$$

where n is the rotor speed and n_s is the synchronous speed. Slip is generally quoted as a percentage of the synchronous speed, with values between about 0.5% and 3% typical for motors when delivering their rated power. Large high-power motors generally run at lower slip values than small ones.

As an example, suppose a large three-phase two-pole induction motor connected to a 50-Hz supply is running at 1% slip. Its rotation speed is 99% of synchronous speed, in other words, $0.99 \times 3000 = 2970$ rpm. Note that the rotor is "slipping back" through the stator field at 30 rpm, or 0.5 revolutions per second. It follows that the frequency of the induced rotor currents has fallen right down to 0.5 Hz and that the field they produce is rotating at just 30 rpm *relative to the rotor*. But since the rotor is itself turning at 2970 rpm, the field is rotating at 3000 rpm *relative to the stator*. The same argument holds for any other value of slip; the stator field and induced rotor field both rotate synchronously, interacting to produce a steady torque, even though the rotor body is moving at less than synchronous speed.

So how does an induction machine work as a generator? The answer is "rather easily." As a general principle, rotating electrical machines can act either as motors or generators. If they accept electrical power and deliver torque, they are motoring; but if they accept torque and produce electricity, they are generating. We should also note that, to a reasonable approximation, induction machines may be considered *linear*. An induction motor running just below synchronous speed can be "pushed" up to and beyond synchronous speed by applying external torque to its shaft. Above synchronous speed, the slip and torque simply change sign, and the machine starts to generate. There is nothing to prevent a smooth transition from motor to generator.

Figure 3.43 shows how the torque of a typical induction machine varies as it works up from zero speed (*start-up*) toward synchronous speed as a motor and becomes a generator above synchronous speed. Speed values are normalized to the synchronous speed n_s. The motor portion of the characteristic, shaded blue, is important for wind turbines that use their generators to provide motorized run-up, and we may identify several important points on the curve:

- *Start-up*. The torque and currents at zero speed tend to be large, well above the nominal rating of the machine. Therefore, rather than connect the machine suddenly to the supply, a controller known as a *soft-starter* is normally used to ensure a gentle start-up.

- *Pull-up*. The pull-up, or minimum, torque must comfortably exceed the resisting torque for the run-up to continue.

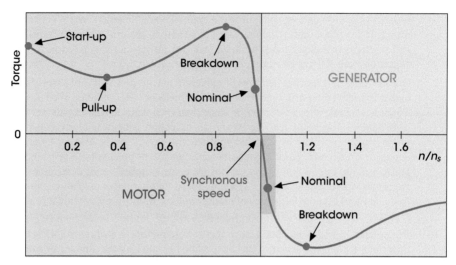

Figure 3.43 Typical torque-speed characteristic of an induction machine.

- *Breakdown.* The breakdown torque is the peak value available when the machine is running as a motor with low values of slip. If exceeded, the motor stalls.
- *Nominal.* The nominal torque is the value at which the machine is designed to operate for long periods.

Following a motorized run-up and assuming the wind is sufficiently strong, a turbine can start generating power. Torque and slip change sign, and the generator runs slightly above synchronous speed. For example, if the slip is −2%, the normalized speed $n/n_s = 1.02$. As wind strength varies, the torque fluctuates around its nominal value, producing small changes in slip. Note that only a very small portion of the complete generator characteristic is used during normal operation—indicated by dark green shading in the figure. The machine generates within a narrow speed range—so much so that a turbine of this type is often described, rather confusingly, as *fixed speed*.

Figure 3.44 illustrates the main elements of such a turbine. The rotor (W) and gearbox (GB) drive an induction generator (IG), shown by two circles representing its stator and

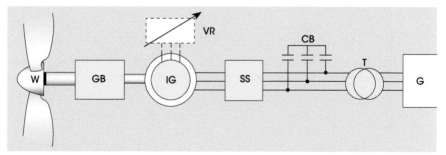

Figure 3.44 Main elements of a "fixed speed" wind turbine: W, turbine rotor; GB, gearbox; IG, induction generator; VR, variable resistance; SS, soft-starter; CB, capacitor bank; T, transformer; G, grid.

rotor (for the moment, we will ignore the optional variable resistance (VR) connected to the rotor). Three-phase stator windings feed electricity to the grid (G) via a transformer (T). A soft-starter is included to aid motorized run-up. Power-factor correction is provided by a capacitor bank (CB), a point that deserves some explanation.

One of the most important features of induction machines is that (not surprisingly) they are highly inductive, requiring reactive power to build up and maintain their magnetic fields. Whereas synchronous generators produce their own fields and have inherent flexibility for controlling voltage and reactive power, an induction generator always runs at a lagging power factor and must be supplied with reactive power by the grid, or by local capacitor banks. From a circuit point of view, its stator and rotor possess both resistance and inductance, and the interaction of the various elements as speed varies determines the shape of the torque-speed characteristic.[3] When we recall that inductive reactance is proportional to frequency and that the frequency of induced rotor currents is proportional to the amount of slip, it is not surprising that the characteristic shown in Figure 3.43 has a complicated form.

In any case, it is important to realize that the characteristic in Figure 3.43 is typical rather than definitive; in particular, it can be modified by altering the amount of rotor resistance. A squirrel-cage rotor, with its cast bars and end rings, has fixed resistance; but an alternative form of rotor uses copper windings instead of aluminum bars, producing a *wound-rotor induction generator* (*WRIG*). The ends of the windings are brought out to *slip rings* on the rotor shaft, and electrical contact is made by *carbon brushes* that bear upon the slip rings (note that the word *slip* is here being used in a different sense). Although slip rings and brushes make WRIGs less robust than simple squirrel-cage machines, a wound rotor introduces some very important options—including the addition of VR to the rotor, shown by the dotted lines in Figure 3.44.

Typical effects of adding external resistance are shown in Figure 3.45, which focuses on the region close to the synchronous speed that is relevant to a machine working as a generator. The main effect is that the slip for a given output torque (and power) increases as the

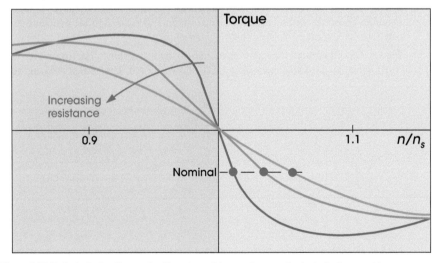

Figure 3.45 Typical effects of increasing the rotor resistance of a wound-rotor induction machine.

rotor resistance increases. Starting with the variable resistance set to zero (red curve), the rotor windings are effectively short-circuited, and the rotor behaves like a squirrel-cage rotor. Adding resistance (orange and green curves) reduces the slope of the torque-speed characteristic, increasing the operating speed range from, say, 1% to 5% of synchronous speed. A given torque fluctuation produces a larger speed change. This allows sudden power surges to be partially absorbed by changes in turbine speed, producing a "softer" grid connection and reducing mechanical stresses on turbine components. It is sometimes referred to as *high-slip* operation.

The addition of rotor resistance causes power losses, which would normally be considered a disadvantage. However, in high winds when the wind turbine rotor is producing excess shaft power, the introduction of rotor resistance allows the excess to be dissipated in the rotor circuit, keeping the stator power within its rated value. This is a valuable alternative to shedding power by pitching the blades, which may be too slow in gusty conditions.

A further option provided by a wound rotor is known as *slip power recovery*.[3] The power losses incurred by adding rotor resistance reduce the generator's overall conversion efficiency—and the higher the slip, the greater the losses. It would be attractive if, instead of dissipating power as unwanted heat, slip power could be "recovered" and put to good use. Any AC power taken from the rotor is at a variable frequency and cannot be fed directly into the grid. However, it can be rectified and then inverted by a power converter to make it grid-compatible, adding to the flexibility of high-slip operation.

Twenty years ago, it was common practice to design large wind turbines with asynchronous generators and fixed blades, giving robust and relatively inexpensive "fixed-speed" machines. Unable to vary their speed by more than a few percent, or to pitch their blades to accommodate changes in wind speed, such machines were relatively inefficient at converting wind power to electricity and relied on *passive stalls* to protect them from damage in storm conditions. Today, the megawatt HAWT market has swung decisively toward variable-speed machines and pitched blades—using either synchronous generators and full-scale power converters, as described in the previous section, or, as we shall see in the following pages, induction generators and partial-scale converters.

3.4.2.3.2 Doubly fed induction generators

In the previous section, we shifted attention gradually from the simplest form of an induction machine with its squirrel-cage rotor to WRIGs. In particular, we saw how the rotor of a WRIG is not restricted to accepting mechanical power from the wind and delivering electrical power to the stator via rotating magnetic fields. Electrical power may also be *extracted* from the rotor via slip rings and brushes, modifying the torque-speed characteristics. We now take this important idea a stage further by explaining how electrical power may also be *fed into* the rotor, producing one of the most important recent advances in wind turbine technology—the *doubly fed induction generator* (*DFIG*).

The DFIG concept has several key advantages:

- It produces wind turbines that operate over typical speed ranges of about 2:1, or slightly greater, sufficient to give high overall efficiency. This is sometimes referred to as *limited variable-speed* operation, distinguishing it from the fuller variable-speed operation of synchronous generators described in Section 3.4.2.2.

- It incorporates a *partial-scale* power converter which is considerably cheaper and has lower losses than a full-scale converter.

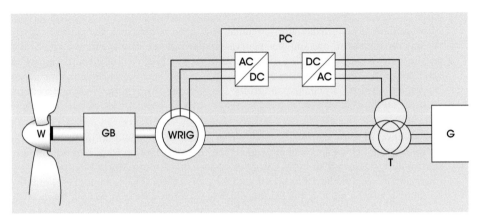

Figure 3.46 Main elements of a variable-speed turbine using a doubly-fed induction generator: W, wind turbine rotor; GB, gearbox; WRIG, wound-rotor induction generator; PC, power converter; T, transformer; G, grid.

- The power converter may be designed to include facilities such as soft-start, voltage and torque control, and power-factor correction.

For all these reasons, DFIGs have been increasingly favored as the generators for megawatt HAWTs in recent years, a trend that seems very likely to continue.

Figure 3.46 shows the main components of a DFIG scheme. The wind turbine and gearbox drive the shaft of a WRIG. The generator's stator is connected to the grid (G) via a transformer (T). In addition, its rotor can supply electrical power to, or receive power from, the grid via a partial-scale PC. The term *doubly fed* refers to the fact that the stator and rotor can both be supplied with AC voltages. In this arrangement, the power converter must clearly be able to transfer power in either direction, in other words, it is *bi-directional*.

If you are familiar with conventional electrical generators, this probably seems a curious system. A generator normally takes in mechanical power on the rotor and delivers electrical power from the stator. The idea—even the possibility—of the rotor accepting and delivering *electrical* power seems strange to most people. So, we will start with a few general observations.

The WRIG at the center of the scheme has two three-phase windings, one on the rotor and one on the stator. In principle, both can generate rotating magnetic fields and transfer power across the air gap. The machine's characteristics depend on the *relative* speed difference between the rotor and stator—the amount of slip. It is even possible to imagine clamping the rotor and allowing the stator to spin! All this points to an essential equivalence between stator and rotor. A stator normally delivers real power, and accepts reactive power, from the grid. But there is no reason in principle why the rotor should not also send or receive electrical power, using brushes and slip rings on its shaft, even if it is turning and receiving mechanical power from the outside world at the same time. This insight, and its practical application, give extraordinary flexibility to the turbine designer, perhaps above all for developing cost-efficient machines that match their speed to the wind.

So where does the wind power captured by the turbine actually go? The *power balance* in Figure 3.47 illustrates the flow of real (but not reactive) power through the system,[6] assuming electrical power is being delivered to the grid by the rotor as well as the stator. On the

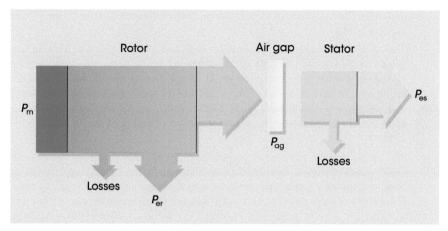

Figure 3.47 The flow of real power in a doubly-fed induction generator.

left-hand side, the input mechanical power to the rotor is denoted by P_m. The rotor delivers electrical power P_{er} to the grid (via the power converter). There are small power losses in the rotor, due mainly to resistance in its windings, and the residual input power is transferred across the air gap to the stator. There are also small stator losses, and the remainder is delivered to the grid as stator electrical power P_{es}.

The alternative is for electrical power to be *fed into* the rotor by the power converter. The arrow denoting rotor power now points the other way, but the power balance of the system must still be maintained. For example, the extra power might be used to increase the torque and/or speed of the turbine rotor, or it might be fed across the air gap to the stator.

But how can electrical power be fed into the rotor? In the previous section, we saw that power can be taken out of a wound rotor via slip rings and brushes and dissipated as heat in a variable resistance. That may seem fairly straightforward, but feeding power is harder to visualize, especially since the frequency of voltages and currents in the rotor varies with the amount of slip. But the situation may be summarized by saying that power can be "pushed back" provided the power converter injects AC voltage into the rotor at the right frequency, with appropriate amplitude and phase. This is certainly possible using today's sophisticated power electronics.

Reactive power can also be manipulated in various ways. As we pointed out previously, an induction machine always runs with a lagging power factor and requires reactive power to set up and maintain its magnetic field. This reactive power must come from somewhere—in the case of a "fixed-speed" turbine, it is supplied from the grid or a local capacitor bank (see Figure 3.44). But an added advantage of the DFIG concept is that the reactive power required by the generator can be supplied from the power converter.

It is clear that DFIG systems depend crucially on their power converters—not only for the ability to convert from AC to DC and back at high efficiency but also for the technical sophistication of the algorithms and control circuits that allow real and reactive power to be taken from or sent to the rotor. As Figure 3.46 indicates, this requires two bi-directional AC/DC units (rectifier-inverters) together with a DC link (shown by red and blue lines) that decouples the frequency of the rotor from that of the grid and allows the turbine to run at variable-speed.

The AC/DC unit connected to the rotor is known as the *rotor-side converter* and that connected to the transformer and grid is known as the *network-side converter*. They are controlled independently of one another. Although there is considerable flexibility when allocating roles to the two units, in many cases, the rotor-side converter provides torque control and voltage or power-factor control for the DFIG; the network-side converter controls the voltage of the DC link and transfers rotor power to and from the AC system at unity power factor.

The DFIG concept offers a comprehensive range of control options, but understanding the details requires expertise in electrical machines and circuit theory beyond the scope of this book. Fortunately, good accounts may be found elsewhere.[5, 6] In the rest of this section, we focus on some operational aspects of DFIGs, as they affect the wind turbines that drive them.

As already noted, probably the most valuable attribute of the DFIG is its variable speed. Back in Section 3.2.3 and Figure 3.15, we explained that a rotor speed range of about 2:1 is sufficient to give good turbine efficiency over a wide range of wind speeds. This is less than the 3:1 range typical of synchronous generators used with full-scale power converters, but it is considerably cheaper. It turns out that a DFIG can produce a 2:1 speed variation with a converter rated at about 30% of the maximum power produced by the wind turbine. For example, a HAWT rated at 3 MW needs a *partial-scale* power converter rated at about 1 MW. Not only is this a lot cheaper than a full-scale converter to handle 3 MW, but it also saves power losses in the converter itself.

The speed variation is obtained partly by operating the DFIG *super-synchronously*, in other words above synchronous speed, as with the "fixed-speed" squirrel-cage and wound-rotor machines described in the previous section and partly by operating it *sub-synchronously*, or below synchronous speed. Sub-synchronous power generation may seem surprising because we have previously assumed that an induction machine acts as a motor below synchronous speed (with positive slip) and as a generator above synchronous speed (with negative slip). But in fact, a DFIG can operate as a generator both sub- and super-synchronously. Basically, this is because electrical power may be fed into, as well as out of, the rotor. In either case, the stator feeds energy into the grid.

The situation is summarized in part (a) of Figure 3.48. At the top, real electrical power P_{er} is taken out of the rotor, which runs above synchronous speed ($n > n_s$). But electrical power may also be supplied to the rotor, which then runs at less than synchronous speed ($n < n_s$). The desired speed ratio of around 2:1 is achieved by a seamless transition from sub- to super-synchronous operation. Typically, the speed can be made to vary between 60% and 130% of synchronous speed using a partial-scale power converter that can handle about a third of the full-rated power of the wind generator. This corresponds to a range of slip values between 40 and −30%, far greater than with conventional induction generators.

Part (b) of the figure shows how the torque-speed characteristic of a typical induction machine begins to change as real power is fed into or out of the rotor. In a DFIG system, this is achieved by injecting a voltage with suitable frequency and phase from the rotor-side converter. As in Figure 3.43, positive torque corresponds to motoring, negative torque to generation. The red curve shows the characteristic of the basic machine, which changes from motoring to generation as the speed passes through synchronous speed ($n/n_s = 1.0$). The yellow curve shows what happens with negative voltage injection; the machine only

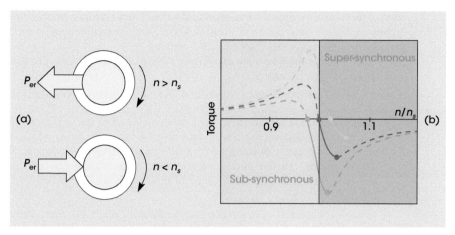

Figure 3.48 Sub- and super-synchronous operation of a DFIG.

starts to generate at super-synchronous speeds. Conversely, with positive voltage injection, indicated by the orange curve, generation starts at sub-synchronous speeds. The solid portion of each curve between the two colored dots indicates the region for real power generation. If reactive power is required, the rotor-side converter must again inject voltage at the correct frequency, but with a 90° phase shift.[6]

The above discussion may imply that the DFIG and power converter are operating fairly steadily, adjusting from time to time to accommodate changes in average wind speed. But in practice, they operate in a highly dynamic environment, adapting to sudden gusts and working with the systems that control pitching of the blades. The fast response of electronic circuits compared with mechanical blade pitching is yet another important advantage of the DFIG concept, smoothing power flow into the grid and protecting the turbine and its drive train from excessive stresses in gusty or extreme conditions.

3.5 Turbine control

Effective control of large wind turbines poses big challenges, perhaps above all because their rotors face an unruly power source—what we have previously called "wild wind." Typically, a large HAWT generates electricity over a 6:1 range in average wind speed, equivalent to a power range of more than 200:1. It must cut in and cut out appropriately and survive sudden gusts. Control functions include

- Optimization of wind capture by the turbine rotor and efficient conversion into electrical energy.
- Protection from mechanical and electrical overloads.
- Supervision and sequencing of turbine operations.
- Monitoring performance and health of system components.
- Interfacing with other turbines in a wind farm and supplying electricity in a form acceptable to the grid operator.

Some of these objectives require engineering solutions based on subtle combinations of aerodynamic, electrical, and electronic control. Others are essentially supervisory. The whole package must be integrated into a reliable working system; and since large turbines are expected to operate for long periods without detailed human supervision, this demands sophisticated computer control.

It is worth recalling the three main types of configurations used in today's megawatt HAWTs:

- *Fixed speed* using an induction generator connected to the grid (see Figure 3.44). In many older designs, the blades were bolted to the hub, giving simple and robust machines protected from damage by *passive stalls*. More recent designs incorporate pitched blades, valuable during start-up, and for controlling power by *active stall*. However, the minimal speed variation of these machines limits overall aerodynamic efficiency.

- *Variable speed based upon a DFIG and partial-scale power converter*. This configuration (see Figure 3.46) has become very popular for megawatt HAWTs in recent years, offering turbine speed ranges of about 2:1, or slightly greater. Active control of pitched blades is used to limit power production above rated wind speed and offers a full complement of safety features.

- *Variable speed using a synchronous generator and full-scale power converter*. This configuration (see Figure 3.39) is the other principal contender for today's high-performance machines. Speed ranges up to about 3:1 are typical. Generators incorporate either wound or permanent-magnet rotors and many are low-speed machines that dispense with the need for gearboxes. These turbines also make extensive use of blade pitching for control and protection.

Modern variable-speed turbines, whether based on induction or synchronous generators, achieve impressive technical performance by careful control of two parameters: *rotor speed* and *blade pitch*. We may use the power curve for a 2-MW turbine shown in Figure 3.50 (see also Figure 2.2) to describe a typical control strategy when operating between cut-in and cut-out wind speeds (Figure 3.49):

- *Region A-B*. In this region, from just above cut-in to just below rated power, the turbine's available speed range is used to keep the tip speed ratio close to optimum as wind speed varies, with the aim of maximizing the power coefficient (see Figure 3.15). In the example shown, the wind speed range is about 7 to 13 m/s, implying a turbine speed range of around 2:1. Efficient speed control at low to moderate wind speeds is crucial to the turbine's overall performance (note that below point *A* there is very little power in the wind; above point *B* there is a power surplus).

- *Region B-C*. The turbine is approaching its maximum speed, and control changes toward a constant-speed mode while allowing the power (and therefore torque) to fluctuate. However, a careful watch is kept on rotor torque, which must be held below its rated value to prevent excessive stress on the drive train, gearbox (if fitted), and generator.

- *Region C-D*. There is excess power in the wind, but the generator's rated power (2 MW in this case) must not be exceeded, so the control system changes to a constant-power mode. This is the region where active blade pitch control is essential. As wind speed increases, the blades are pitched to shed more and more power, either by increasing

Figure 3.49 These HAWTs are in the 2-MW class but have different generator configurations. On the left, an asynchronous generator with gearbox (Vestas A/S); on the right, a synchronous generator without gearbox (*Source:* (a) With permission of VESTAS. (b) With permission of ENERCON GmbH).

the angle of attack (active stall) or by pitching them in the other direction (feathering). The working turbine is at full stretch in this region and increasingly reliant on blade pitching for control and protection as it approaches cut-out.

As wind strength varies, there are often transitions between the three regions, and the aim of a well-designed control system is to enable smooth changes while ensuring that the turbine's rated power, speed, and torque are not exceeded.

Figure 3.50 focuses on speed and power, but torque is also very important. To understand how torque control is achieved, we need to consider the relationship between generator

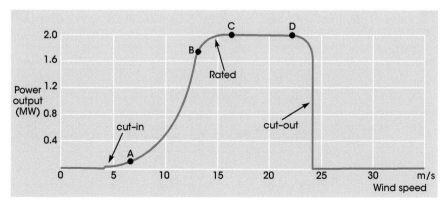

Figure 3.50 Typical power curve for a 2-MW turbine.

speed, torque, voltage, and current. As a rule, the *voltage produced by an electrical generator is proportional to speed, and current is proportional to torque*. Now the mechanical power P_m driving a generator is proportional to both speed and torque (see Equation (3.11)), and the real electrical power P_e it produces is given by the product of voltage and in-phase current. So if we assume negligible losses in the machine itself, the mechanical input power equals the electrical output power, and to a reasonable approximation:

$$P_m = 2\pi NT = P_e = VI \quad \text{watts} \tag{3.33}$$

where N, T, V, and I are the speed, torque, voltage, and in-phase current, respectively.

Equation (3.33) suggests that *rotor speed and torque may in principle be managed either mechanically or electrically*. For example, rotor torque can be modified either by direct adjustments of blade pitch or by controlling the generated current. More generally, the mechanical operation of a turbine can be monitored and controlled from the generator end of the drive train—an important option widely used in control schemes for variable-speed megawatt HAWTs.

Although we have already used the word "control" many times, we have not so far commented on its fundamental principles. In the case of large wind turbines, it is clear that a wide range of control actions is needed—not only during normal operation but also for start-up and shut-down, in emergencies, for ensuring the safety of equipment and personnel, and so on. Most of these actions are allocated to individual *subsystems* under overall supervisory control. The basic concept involved is one that permeates the whole of classical control theory—that of *feedback*.

Figure 3.51 illustrates the use of feedback in a simple control loop that is designed to ensure an "actual output" matches a "desired output" with acceptable accuracy. By "output" it means the value of a variable that we wish to control—it might be a voltage, current, torque, speed, angle, or temperature. A signal representing the actual output is *fed back* and subtracted from the desired value to produce an *error signal*. Clearly, if the actual and desired outputs are identical, the error is zero and no control action is needed. However, a nonzero error activates the forward path comprising the controller, power amplifier, and actuator to bring the actual output into line. It is often said that, in such a control loop, *the error drives the system*. And since the actual value is *subtracted* from the desired value to calculate the error, the feedback is described as *negative*. Negative feedback may be viewed as the guiding principle of automatic control.

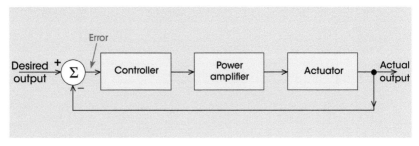

Figure 3.51 A simple feedback control loop.

Large wind turbines incorporate many subsystems of this type, controlling parameters such as:

- Rotor speed and torque
- Blade angle of attack
- Yaw rate and position
- Electrical output

The box labeled *controller* in the figure is normally electrical or electronic (typically a microprocessor or computer). Depending on the particular application, the *power amplifier* that increases the power level to that needed by the *actuator* may be electrical/electronic or hydraulic and the actuator may be electrical (motor, switch, etc.) or hydraulic. For example, the actuators that provide pitch control by rotating turbine blades about their longitudinal axes are electric motors or hydraulic pistons, with pitch angle as the control variable.

An extremely important advantage of negative feedback is that it makes the subsystem more or less impervious to external disturbances. Again, taking blade pitch control as an example, gusty wind conditions subject the blades to unwanted external forces that tend to alter their pitch setting—the "actual output" in Figure 3.51. This information is fed back and produces an error signal that restores the blade pitch to the desired value.

However, a word of caution is in order. The simplicity of Figure 3.51 may imply that matching "actual" and "desired" outputs is a straightforward matter: give an order, and it is immediately obeyed! But things are not that easy, mainly because the units in the forward path (controller, power amplifier, and actuator) are all limited by their own *dynamic responses*. In a nutshell, they cannot respond immediately to requests for control action, and if the power amplifier is turned up in an attempt to accelerate the response, there may be an overshoot or even oscillation. Such effects are beyond the scope of this book, but it is important to realize that they are central to control system design.

When the control subsystems of a large turbine have been carefully designed and tested, they must all be brought together to create an efficient, safe, and reliable working system. Control is at several levels and is essentially *hierarchical*.

In Figure 3.52, we assume that a large HAWT forms part of a wind farm that integrates and manages the electrical outputs from all its turbines and delivers power to the grid. The *wind farm controller* is at the top of the command chain, responsible for management issues affecting the farm as a whole (we shall have more to say about this in the next section).

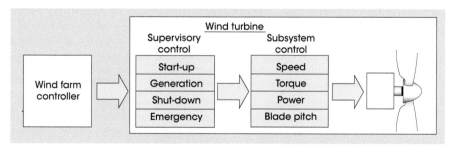

Figure 3.52 Hierarchical control of a large HAWT.

Next in the chain comes *supervisory control* of the individual turbine, typically sequencing start-up, shut-down, and emergency procedures, and overseeing operation during electricity generation. And finally, we have the turbine *subsystem controllers*, often based on the type of negative feedback loop shown in Figure 3.51, that control settings of individual parameters such as turbine speed and blade pitch. The picture is necessarily simplified, showing only a proportion of the turbine's control requirements, but it does underline the hierarchical nature of a highly sophisticated control structure.

The orange arrows in the figure suggest a one-way flow of command and control from wind farm to individual wind turbine. But as we have already seen, feedback is an essential aspect of effective control, allowing actual outputs to be compared with desired ones and taking remedial action when necessary. In fact, there is extensive feedback within the various "boxes" in the figure, and from turbine back to the wind farm, which we have not attempted to show in the figure. Supervisors need to know that orders issued have actually been carried out and that the system is working according to plan.

3.6   Onshore wind farms

3.6.1   Introductory

Large wind turbines are generally placed in groups to form *wind farms*, also known as *wind parks*. These may contain anything from two or three individual machines to more than a hundred machines producing hundreds of megawatts. Gigawatt onshore wind farms are firmly in prospect for the coming decade, especially in countries with large, remote, windswept areas of limited landscape value.

The development of wind farms got underway in California in the late 1970s, notably in the Altamont Pass, San Gorgonio Pass, and Tehachapi Mountains. By the 1990s, many wind farms were being installed in Europe, especially in Denmark, Germany, and Spain, and since 2000 there has been a worldwide explosion in capacity (see Section 1.5) that shows no signs of abating. The ratings of individual turbines have grown systematically over the years, initially from tens to hundreds of kilowatts and then on toward the megawatt machines that dominate today's utility-scale market.

The benefits of scale ensure that large wind farms based on megawatt HAWTs are very attractive to developers and electric utilities. Many of the largest operational wind farms are in the USA, with installed capacities in the range of 300–700 MW. The difficulty, of course, is to find sites extensive enough to accommodate large numbers of turbines, taking account of public acceptability as well as physical and geographical limitations. Densely populated European countries such as Denmark, Germany, and the Netherlands that have already invested heavily in onshore wind have increasing difficulty locating new sites—one of the main reasons for the current surge of interest in offshore wind farms, which we cover in the next chapter. But other countries with extensive, sparsely populated, flat plains or uplands have a long way to go before exhausting their onshore potential.

A prime example of a region that is already heavily committed to wind energy is provided by northwest Germany. Figure 3.53 shows wind farm locations in the neighborhoods of Hamburg and Bremen up to the North Sea coast. Most of this countryside is flat and windswept—part of the North German Plain—and has proved a magnet for wind farm developers over the past 20 years. Older installations are generally small by today's

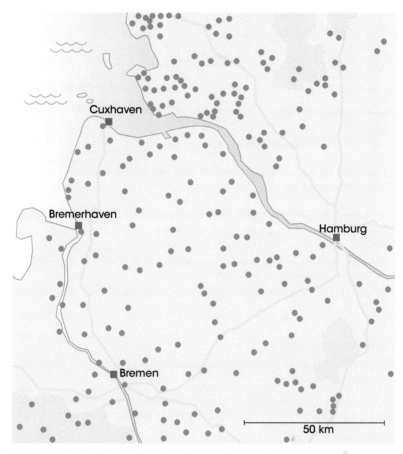

Figure 3.53 Wind farm locations in north-western Germany (*Source:* Data source: www.thewindpower.net).

standards, typically comprising a few machines rated at well under 1 MW; but there are also plenty of new wind farms based on the current generation of megawatt HAWTs. Germany has been, and still is, admirably active in wind power, with one of the highest total installed capacities in the world, plus one of the highest per head of population (see Figure 1.19). However, it is easy to appreciate that turbine installation on this scale may be approaching saturation in terms of suitable sites, public acceptability, and the challenge of integrating large amounts of renewable electricity into local grid networks.

Canada is a good example of a different scenario: a vast country with low overall population density where the penetration of wind energy is currently far lower than in many European countries, offering great scope for future development. Many other nations enjoy high wind resources that are so far largely untapped. However, it is important to bear in mind that the proximity of wind farms to grid networks and centers of population has an important influence on wind energy economics (Figure 3.54).

From a technical point of view, the biggest challenge for wind farm developers is to find accessible sites with high average wind speeds, preferably located reasonably close to existing

Figure 3.54 Plenty of scope for onshore wind: scenes from Canada (*Source:* With permission of Canadian Wind Energy Association).

grid networks. Potential sites must also be assessed for environmental and social impact, and legal issues such as land ownership, permissions, and access to power lines cleared. Once a site has been selected the exact number and locations of the wind turbines must be carefully planned, with the aim of maximizing annual energy yield and reducing, as far as possible, the costs of access and installation.

When all permissions are in place, it is time to prepare the site for turbine installation. Typically this involves

- Building or upgrading access roads (which may be time-consuming and costly, especially in rugged terrain).
- Constructing turbine foundations (which tend to be very site-specific, depending on subsoil, bedrock, etc.).
- Installing power and data transmission cables to interconnect the various turbines and control center, plus any transformers and switchgear necessary for grid connection.

The erection of megawatt HAWTs demands special skills and large cranes that have their own transportation problems. Blades must be maneuvered with great care, preferably in calm weather to avoid damage and the risk of danger to personnel. It is always a great relief for installation engineers to see their charges safely "up and running" (Figure 3.55).

3.6.2 Siting and spacing

We have already discussed wind resources in Chapter 2, including the use of wind atlases and specialized software to predict the wind regime in a local area, preferably supplemented by careful measurement of wind speeds at candidate wind farm sites—especially those with complex geography or obstructions. It is now time to consider the siting of individual turbines, referred to as *micro siting*, which forms a key part of the planning process for a successful wind farm. Effective placing of turbines in relation to neighboring machines, prevailing winds, and variations in the local terrain is essential to optimize energy production.

Things are fairly straightforward when a wind farm is to be built on an extensive flat site, well clear of significant obstructions in all directions. Needless to say, such sites are comparatively rare, but good examples may be found on the Great Plains of North America, the North German Plain, some coastal sites, and arid desert-like regions of the world. In such cases, wind farm designers are untroubled by variations in local terrain and can site and space an

Figure 3.55 Heavy work in high places (*Source:* With permission of Repower).

array of machines in an optimum configuration, provided there are no serious subsoil or rock problems to complicate turbine foundations.

As we have seen in Section 3.2.1, the basic task of a wind turbine rotor is to extract as much kinetic energy from the moving airstream as possible. In doing so, it creates a wake and produces turbulence that can adversely affect the performance of neighboring machines. Repeated over a wind farm as a whole, such effects give rise to *array losses* that reduce annual energy production. So, the question arises: how far should turbines be spaced apart? Clearly, generous spacing helps minimize array losses, but it also restricts the number of turbines that can be fitted on a given area of land and increases the costs of access roads and cabling. A compromise is needed.

Figure 3.56 shows a regular array of turbines on an extensive flat site in relation to the prevailing wind direction. There are two spacings to be considered: *downwind* and *crosswind*. Generous downwind spacing minimizes interference by a turbine's wake on its downstream neighbors and, although there are no hard and fast rules, a value between 8 and 10 rotor diameters is often recommended to limit total array losses to below 10%. If we take a 2-MW HAWT with a rotor diameter of around 90 m as typical of large, modern onshore wind farms, this implies a downwind spacing of around 800 m.

The need for crosswind spacing is less clear. On the face of it, a spacing of just over one rotor diameter might seem enough to avoid expensive arguments between adjacent machines! However, the situation is complicated because most sites do not have a single, dominant, prevailing wind direction; turbines use their yaw motors to face different wind directions at different times of day and in different seasons of the year. Clearly, a 90° change of wind direction converts crosswind into downwind, and downwind into crosswind. In practice, a spacing of about five rotor diameters is often chosen for the *nominal* crosswind direction

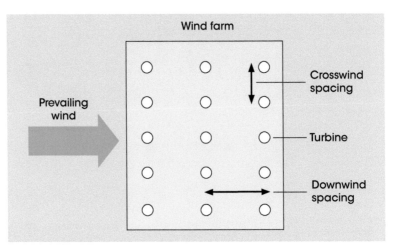

Figure 3.56 An array of wind turbines on an extensive flat site.

which combined with a downwind spacing of 800 m implies a land area of about 0.4 km² per 2-MW turbine. However, the actual value depends on turbine size and the wind regime at a particular site.

A useful aid to visualizing variations in wind direction is the *wind rose*. The example in Figure 3.57 summarizes 10 years of measurements at a site near Plymouth in the southwest of England that is exposed to prevailing winds from the Atlantic Ocean. The length of each colored bar indicates the relative amount of time that the wind blows within a certain speed range in a particular direction. It is clear that, in this case, the wind is most likely to blow strongly from the southwest, regarded as the prevailing direction. But winds from the northwest and east, although normally gentler, are also quite common during winter and spring months. In such cases, we cannot define "downwind" and "crosswind" precisely, and sensible judgments must be made about turbine spacing. Some locations, for example, the San Gorgonio Pass in California, have predominantly unidirectional wind roses; others, including

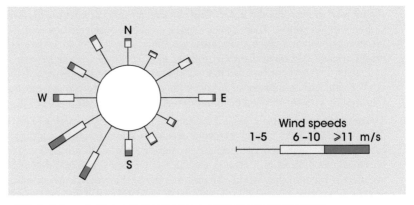

Figure 3.57 A wind rose for Plymouth, England (*Source:* Data source: UK Met Office).

the northwest coast of Spain, tend to be bi-directional (e.g. northeast and southwest). In these cases, high annual energy yields may be obtained by increasing downwind spacing and reducing crosswind spacing.

Array losses depend to a considerable extent on the wind rose pattern. If predominantly unidirectional or bi-directional, the wind spends most of its time blowing in line with the nominal downwind direction and a generous downwind spacing is desirable. But if the wind rose is well spread, crosswind spacing is also important. Another important factor is the amount of turbulence in the air stream. The energy loss behind a turbine, represented by its wake, is replenished over a certain distance by the exchange of kinetic energy with the surrounding wind field. The exchange is more rapid and effective when the wind field is turbulent, leading to smaller array losses in downstream turbines. Such effects have been extensively studied in practice and by computer modeling.[3]

The power curve of a wind farm—its power production as a function of wind speed—has a different shape from that of an individual turbine (see, for example, Figure 3.50). This is because array losses and turbulence affect turbines according to their position in the array. As the wind speed approaching a wind farm increases from zero, the first row of turbines starts operating, followed by those further downwind. At any given wind speed, downwind machines tend to produce less power than upwind ones, and the wind speed must exceed the rated value by some margin before all turbines reach their rated power. Furthermore, turbulence in the wind causes the power output from turbines in different parts of the wind farm to fluctuate differently, and some turbines may be temporarily out of service. The effect of all this is to modify the farm's power production, which is generally lower than that expected from multiple isolated turbines.

It is important to note that turbine spacing is not solely concerned with energy production. Tight spacing in the downwind direction means that turbines are more affected by turbulence caused by upstream neighbors. This may produce unacceptably large mechanical stresses on blades and rotors, eventually producing fatigue and affecting the warranties given by turbine manufacturers. Similarly, tight spacing in the nominal crosswind direction can produce unacceptable stresses if the wind blows for substantial periods in "nonprevailing" directions, and it may be necessary to shut down certain turbines to avoid damage.

So far, we have assumed an extensive flat site without significant obstructions to wind flow, with turbines arranged in a regular array. But, of course, things are rarely so simple. You may like to refer back to photographs showing a variety of sites with features and irregularities that require careful, individual, micro siting:

- Figure 2.4(b): Essentially a flat site, but with high trees in certain directions.
- Figure 2.7: A site in a hill country with complex terrain.
- Figure 3.49(b): A coastal site, clear toward the sea but with significant obstructions inland.
- Figure 3.54(a): A fairly flat site containing clumps of trees.
- Figure 3.54(b): A site with a single line of turbines strung out along a ridge, spaced by about two rotor diameters.
- Figure 3.58: A restricted but fairly flat site containing some obstructions, with a long line of close-spaced turbines.

Figure 3.58 Presenting a united front to the prevailing wind (*Source:* With permission of Georges Moreau).

In the last two examples, array losses and turbulence effects are small when the wind direction is perpendicular to the line of turbines but may become serious if it changes substantially. Such locations would normally have substantially unidirectional or bi-directional wind roses. Further discussion on wind turbine spacing is included in Chapter 5.

The technical objectives of turbine siting include maximizing annual energy production, avoiding excessive turbulence, and minimizing installation costs. But there are also important environmental and social considerations. From the point of view of the general public, especially those living close to wind farms, visual intrusion and turbine noise are generally of most concern. We shall discuss these and related topics in Chapter 8.

3.6.3   Monitoring and control

Wind farm operators need to monitor the performance of individual turbines, and of the wind farm as a whole, from a central location and take corrective action if necessary. They must also check that the electricity supplied to the grid meets contractual agreements. The computer system that acts as a "nerve center" for this complex task is generally referred to as a *supervisory control and data acquisition (SCADA) system*. It may be supplied by the turbine manufacturer, or alternatively by an independent company whose software is designed to work with a variety of turbine types and sizes within the same wind farm.

SCADA systems display appropriate monitoring and control data on computer screens. A typical view showing the layout and operating status of a wind farm is shown in Figure 3.59. In this example, the farm comprises a total of 18 turbines arranged in three groups along low hill ridges. The power output from each group is metered before entering a substation where the grid connection is made. Meteorological masts support anemometers and wind

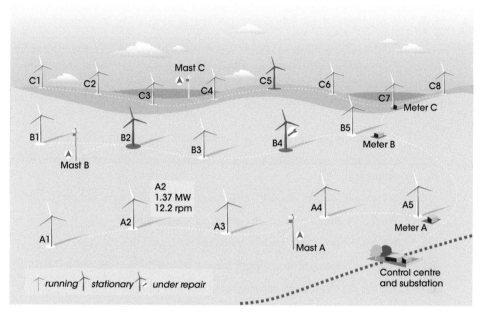

Figure 3.59 A typical screen showing wind farm layout and operational status.

direction indicators. Data from the masts and turbines is fed by cable or optical fiber to a control center from where orders are issued to control the overall operation of the farm. The display indicates the operational status of each turbine, and selecting an individual machine (e.g. A2) reveals its power output and speed of rotation.

The SCADA system of a large modern wind farm allows operators to select various screens, giving a comprehensive range of incoming data, such as:

- Each turbine's operational status, plus parameters such as wind speed, blade pitch angle, and gearbox oil temperature.
- Power output from each turbine, or turbine group.
- Atmospheric temperature, pressure, and air density.
- Real and reactive power delivered to the grid, wind farm capacity factor, and estimates of revenue based on energy production and utility tariffs.
- Grid voltage and frequency at the substation.

The incoming information is used to initiate supervisory and control actions that may include

- Monitoring turbines for safe operation and choosing optimal operating states.
- Start-up and shut-down of individual turbines, or groups of turbines, including emergency shut-down and grid disconnection.
- Fault diagnosis, repair, and maintenance.
- Keeping records of turbine and wind farm performance.

In our discussion of turbine control in Section 3.5, we noted that a large modern HAWT incorporates many subsystems devoted to controlling individual aspects of performance—rotor speed, generator torque, blade pitch angle, and so on—and is in many respects a self-regulating machine. The SCADA system moves the overall control function forward by integrating all turbines on a wind farm into a safe and efficient production unit.

Self-assessment questions

Q3.1 What are the two basic forces that wind exerts on the wind turbine blades?

Q3.2 How Richard Jekins' Greenbird was able to move much faster than the wind propelling it?

Q3.3 What positive effect can the "stall" condition have on a wind turbine?

Q3.4 What is the function of "pitch control" in wind turbine?

Q3.5 What is the reason for modern wind turbines having three blades?

Q3.6 What is the max fraction of wind energy that can be captured by wind turbine?

Q3.7 What is the relationship between rotor diameter and rated power?

Q3.8 How is the "rated power" of wind turbine been determined?

Q3.9 Why is HAWT preferred over VAWT?

Q3.10 Why do large HAWTs are low-speed machines?

Q3.11 Why does wind turbine have a cut-off wind speed?

Q3.12 Why wind turbine blades and aircraft wings are more curved on one of their two sides?

Q3.13 Why dirt or ice accumulation can degrade rotor power output?

Q3.14 What is the role of gearbox and why direct drive may be preferred in offshore wind turbine?

Problems

3.1 The blades of a wind turbine are ready to be added on a turbine mast on the ground with 12-degree slope when a storm front arrives. The turbine crew huddles in their truck as the rain starts and the wind picks up, and the engineer drives his truck upwind of the blades to disrupt the airflow around the blades, in an attempt to prevent them from being lifted by the wind and damaged.

The blades are 4.6 m (15 feet) long, 0.61 m (2 feet) wide, and have a mass of 45.36 kg. As the front arrives, the temperature drops to 15.6°C (60°F). Assume that the blades are approximately symmetric airfoils, and that the center of both the lift and the drag force is on the center of mass of the blade, and that the leading edge is facing the direction of the wind. Assume the air density is 1.25 kg/m³ and that the lift coefficient of a symmetric airfoil is approximately: $C_l = 2\pi \sin\alpha$).

(a) Was there a reason to be concerned? At what wind speed will the blades be lifted by the wind, assuming that there is no drag?

(b) If they are lifted by a 22.35 m/s (50 mph) wind, how fast will they be accelerated horizontally, if the blade's lift-to-drag ratio, C_d/C_l, is 0.035?

3.2 The following table lists the operating conditions at two different points of a wind turbine blade:

Location r/R	Wind velocity at blade (m/s)	Wind velocity at blade (ft/s)	Chord (m)	Chord (ft)	Angle of attack (degrees)
0.2	20.98	68.83	1.35	4.43	5.20
0.9	70.32	230.71	0.42	1.38	7.41

These conditions were determined at 15°C (59°F), for which the kinematic viscosity is 1.46×10^{-5} m²/s.

What is the Reynolds number at each blade section?

3.3 Assume that a two-bladed wind turbine experiences 25 mph and 40 mph winds. At 25 mph (tip speed ratio = 10), a blade section of 1.5 m experienced an angle of impact of 7.19 degrees. At 40 mph (tip speed ratio = 6) with an angle of impact of 20.46 degrees on the same 1.52 m section, the blade is starting to stall. Given the following operating conditions and geometric data, determine the relative wind velocities, the lift and drag forces, and the tangential and normal forces developed by the blade section at the two different tip speed ratios. Also, determine the relative contribution (the fraction of the total) of the lift-and-drag forces to the tangential and normal forces developed by the blade section at two different tip speed ratios.

How do the relative velocities and the lift-and-drag forces compare? How do the tangential and normal forces compare? How do the effects of lift-and-drag change between the two operating conditions?

Operating conditions are listed in the following table:

λ	α	φ	a
6	20.46	21.71	0.072
10	7.38	8.63	0.375

This particular blade section is 1.52 m long, has a chord length (c) of 0.762 m, and has a center radius of 4.88 m. The following lift-and-drag coefficients are valid for $\alpha < 21°$, where α is in degrees:

$$C_l = 0.42625 + 0.11628\alpha - 0.00063973\alpha^2 - 8.712 \times 10^{-5}\alpha^3 - 4.2576 \times 10^{-6}\alpha^4$$
$$C_d = 0.011954 + 0.00019972\alpha + 0.00010332\alpha^2$$

3.4 A wind turbine with a 30 m diameter rotor produces 278 kW output in a 12 m/s wind. What is its efficiency?

3.5 A wind turbine with an efficiency of 40% produces 1 MW output at a wind velocity of 14 m/s. What is the turbine rotor diameter?

3.6 Wind velocities are not constant throughout the day. The daily average power produced by a wind turbine is the power averaged over the wind velocity for the day.

Calculate the average power for a turbine with a diameter of 25 m and an efficiency of 40% if, during a 24-hour period, the wind velocity is:

3 m/s for 3 hours;
8 m/s for 16 hours;
12 m/s for 4 hours;
15 m/s for 1 hour.

3.7 A wind turbine rotor turning at 60 rpm is brought to a stop by a mechanical brake. The rotor inertia is 15,343 kg·m².

    (a) What is the kinetic energy in the rotor before it is stopped? How much energy does the brake absorb during the stop?

    (b) Suppose that all the energy is absorbed in a steel brake disc with a mass of 24 kg. Ignoring losses, how much does the temperature of the steel brake disc rise during the stop? Assume a specific heat for steel of 0.46 kJ/kg-C.

3.8 A cantilevered 2.1 m long main shaft of a wind turbine holds a 1600 kg hub and rotor at its end. At rated power, the turbine develops 325 kW and rotates at 60 rpm. The shaft is a 0.18 m in diameter cylindrical steel shaft.

    (a) How much does the shaft bend down at its end as a result of the load of the rotor and hub?

    (b) How much does the shaft twist when the turbine is operating at rated power? What is the maximum shear stress in the shaft?

3.9 A wind turbine on a 30.48 m tower is subject to a thrust load of 22.24 kN during operation at 300 kW, the rated power of the turbine. In a 53.6 m/s hurricane, the thrust load on the stopped turbine is expected to be 80.07 kN.

    (a) If the tower is a steel tube 1.52 m in outer diameter (O.D.) with a 0.0318 m thick wall, how much will the top of the tower move during rated operation and in the hurricane force winds?

    (b) Suppose the tower were a three-legged lattice tower with the specifications given in the following figure, how much will the top of the tower move during rated operation and in the hurricane force winds? Ignore any effect of cross bracing on the tower.

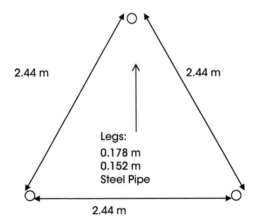

2.44 m   2.44 m

Legs:
0.178 m
0.152 m
Steel Pipe

2.44 m

3.10 The wind turbine rotor shown in the figure below has a rotor rotation velocity Ω of 1 Hz (60 rpm) and is yawing at an angular velocity ω of 12 degrees per second. The polar mass moment of inertia of the rotor is 16,297 kg m². The rotor weighs 1751 kg and is 3.66 m from the center of the bed plate bearing support. The shaft bearings, centered over the same support, are 1.22 m apart. The directions of positive moments and rotation are indicated in the figure.

(a) What are the bearing loads when the turbine is not yawing?

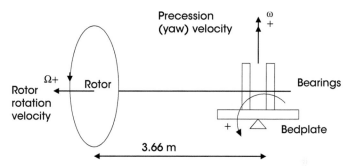

(b) What are the bearing loads when the turbine is yawing?

3.11 A 21.64 m long turbine blade has a nonrotating natural frequency of 1.6 Hz. In operation, rotating at 50 rpm, it has a natural frequency of 1.78 Hz. If the blade has a mass of 1006 kg, what hinge-spring stiffness and offset would be used to model the blade dynamics with the simple dynamics model?

Answers to questions

Q3.1 Lift and drag.

Q3.2 It creates a strong horizontal lift force.

Q3.3 It can protect the wind turbine from damage in extreme wind gust conditions.

Q3.4 The pitch system regulates the power output of the wind turbine by adjusting the angle of rotor blades according to wind velocity; it also functions as the main brake.

Q3.5 You would expect that more blades would offer a higher harvesting of the wind's kinetic energy as long as the movement of one does not create low-pressure areas around the other. However, blades are expensive, and it was found that 3-blade turbines offer a good combination of cost, technical performance and long-term reliability. Old windmills have four sails but increasing the number of blades from 3 to 4 in modern WT is not cost-effective. Esthetics play a role too; a 3-blade WT looks nice.

Q3.6 About 50% the theoretical limit is 59% (Betz Limit).

Q3.7 The higher the rotor diameter, the higher the swept area and the rated power.

Q3.8 At a wind speed of 12 m/s.

Q3.9 HAWT can employ bigger rotor diameters and are easier to adjust to changes in wind speed and direction.

Q3.10 They are low-speed but high-torque machines, and they can produce high speeds at the generator with gearboxes or permanent magnets.

Q3.11 To protect from very strong winds when drag can extend lift force.

Q3.12 To induce lift as the air velocity on the curved side is higher and the pressure is lower.

Q3.13 Surface roughness can reduce air velocity and lift.

Q3.14 To adjust speed from rotor to generator. It has many parts that may be susceptible to corrosion in marine environments and/or need frequent maintenance/repair.

References

1. German Wind Energy Association (BWE). www.wind-energie.de (Accessed on May 4, 2024).
2. P. Gipe. *Wind Energy Basics: A Guide to Home and Community-scale Wind Energy Systems*, 2nd edition. Chelsea Green Publishing: Vermont (2009).
3. J.F. Manwell *et al*. *Wind Energy Explained: Theory, Design and Application*, 2nd edition. John Wiley & Sons Ltd: Chichester (2009).
4. S. Heier. *Grid Integration of Wind Energy Conversion Systems*, 2nd edition. John Wiley & Sons Ltd: Chichester (2006).
5. T. Ackermann (ed.). *Wind Power in Power Systems*. John Wiley & Sons Ltd: Chichester (2005).
6. O. Anaya-Lara *et al*. *Wind Energy Generation: Modelling and Control*. John Wiley & Sons Ltd: Chichester (2009).

4 Fundamentals of offshore system

4.1   Introduction

The world's first offshore wind farm was a modest affair by today's standards. Located in shallow water just off the coast of Denmark, the Vindeby farm started in 1992 with eleven turbines and an installed capacity of 5 MW. Since then, Denmark has been joined by other countries with North Sea coastlines—the UK, Germany, the Netherlands, and Belgium—in pioneering the development of utility-scale offshore wind farms. China made significant progress in the last decade. Many other nations around the globe including the USA, Taiwan and Japan, are now planning to "move offshore," (Figures 4.1 and 4.2) presenting their wind energy industries with challenging new horizons.

Why this sudden surge of interest in generating renewable electricity offshore? And why have the countries of northwest Europe been leading the way, especially in the North Sea? We may identify a number of contributing factors:

- In recent years, European countries with high population densities have found it increasingly hard to locate new sites for onshore wind farms, especially very large ones that benefit from the economies of scale.

- Developments in wind turbine technology are constantly increasing the power ratings, rotor sizes, and efficiency of the largest turbines. The Vindeby machines mentioned earlier were each rated at 450 kW. Today, 2–6 MW turbines are commonplace, and new 10 MW and greater capacity turbines are installed. Very large HAWTs with rotor diameters over 100 m are well suited to offshore duty where there is plenty of space and visual intrusion is less problematic.

- Although offshore turbines are comparatively costly to install and maintain, wind resources are superior to those onshore, with higher average wind speeds and lower turbulence.

Onshore and Offshore Wind Energy: Evolution, Grid Integration, and Impact, Second Edition.
Vasilis Fthenakis, Subhamoy Bhattacharya, and Paul A. Lynn.
© 2025 John Wiley & Sons Ltd. Published 2025 by John Wiley & Sons Ltd.
Companion website: www.wiley.com/go/fthenakis/windenergy2e

Figure 4.1 Moving offshore (Figure 4.3 1st edition) (Source: With permission of Repower).

Figure 4.2 Moving offshore: new horizons for wind energy (Figure 4.1 1st edition) (Source: With permission of Orsted).

- As far as the North Sea is concerned, several countries that have amassed vast experience in wind energy—Denmark and Germany in particular—have a dwindling supply of land suitable for large onshore turbines. The UK, especially in Scotland, has huge offshore engineering experience thanks to North Sea oil (which is now running out), leaving a reservoir of skilled personnel keen to transfer to the wind engineering sector. And the extra costs and difficulties of offshore installation are manageable in the North Sea, which is shallow in many places.

Add to all this, the enthusiasm of governments to encourage wind energy and the undoubted problems of working offshore are increasingly seen as challenges rather than limitations.

Figure 4.3 shows the locations of operating, approved, and planned offshore wind farms in northwest Europe, including the all-important North Sea. Most sites are quite close to shore. Given the rapidly developing situation, it is inevitable that such a snapshot will date fairly quickly. Some planned farms may not materialize (and others are too closely spaced to be shown individually on a map of this scale). But the figure gives a useful overview, confirming that Europe's operating wind farms are soon to be joined by many more, especially in England (E), Denmark (D), and Germany (G); followed, a few years later, by a further set that are firmly planned, including many in Scotland (S) and along the northern and western coasts of France (F).

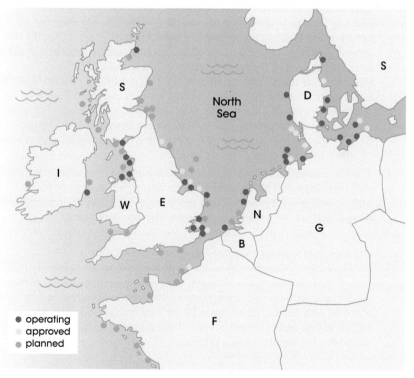

Figure 4.3 Offshore wind farm locations in northwest Europe (Figure 4.2 1st edition).

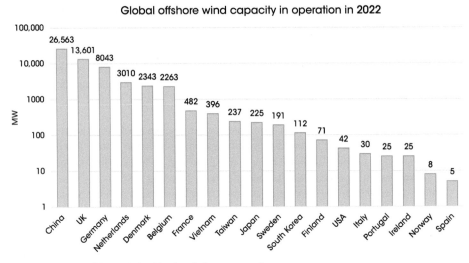

Figure 4.4 Cumulative offshore wind turbine installations per country.

How about moving further offshore? If we refer back to Figure 2.19 illustrating the North Sea's wind resources, it is clear that average wind speeds—the basic technical criterion for selecting wind farm sites—increase markedly with distance from the coastline. And visual intrusion becomes less problematic. However, there are two offsetting disadvantages. The water is generally deeper far from shore, making turbine installation more costly and difficult, and it is more expensive to bring the generated electricity ashore by undersea cable. We will revisit these topics in later sections. Not only are offshore wind farms increasing rapidly in number, but so are their average installed capacities. Figure 4.4 shows the cumulative capacity from 19 countries. Recently commissioned offshore wind farms are often based on turbines rated between 6 and 16 MW. We can design a wind farm generating 1 GW with 65–170 turbines. And it is reasonably certain that Europe will be generating tens of gigawatts of offshore electricity by 2025.

There is a remarkable contrast between the average installed capacity of onshore and offshore installations. This is surely where the true significance of offshore wind lies: very large utility-scale installations based on multi-megawatt HAWTs. It promises to revolutionize many aspects of the wind energy industry in the coming decade.

4.2   Offshore wind power fundamentals

As discussed in the Introduction section, offshore wind power is among the established scalable renewable energy technologies. They are relatively easy to construct due to the sea routes and vessels available to transport parts from manufacturing sites to turbine locations. There are other advantages for going offshore:

(1) The first and foremost advantage of offshore wind power is that the average wind speed in the sea is generally higher and more consistent than onshore, which means that the offshore wind turbines will be able to work more efficiently than the onshore

ones. Also, since there are no obstructions over the sea, the wind is less turbulent than onshore, which means that the fatigue effects on the turbine generator are less and therefore the lifespan of the turbines can be greater if corrosion is prevented.

(2) The second advantage is that there are less layout constraints. As there is more free space available in the sea, offshore wind farms can be more versatile and scalable. That means that larger installations can be built. Assuming that they do not have the transport limitations that transport over land may have, wind turbine generators can be larger in capacity than the onshore ones, achieving this way more energy production per turbine. Also, as the wind turbines are placed far from the coast, public opposition due to potential visual and noise issues is lessened.

(3) Offshore wind can directly support the hydrogen development (e.g., hydrogen hub for filling hydrogen-powered ships; hydrogen powered industrial hubs) and significantly contribute to a transition to low-carbon energy.

ASIDE: Oldest and largest machine

Wind power has an established track record for centuries in the sense that it has been used for centuries for sailing ships, sawing wood, grinding grains, and many more. It is one of the oldest sources of "machine" power which fits the definition of a machine (namely. a piece of equipment with several moving parts that uses wind power to carry out a particular type of work such as sawing wood). The invention of wind-powered sawmills by Dutchman Cornelis Corneliszoon van Uitgeest in the late 16th century helped Holland increase ship production through automated wood cutting, outcompeting their European rivals who were relying on slow manual processes. It can also be said that offshore wind turbine machines are also the largest moving machine. A typical 10-MW turbine will have three 80 m long blades and each weighing circa 35 tons turning at average of 8–10 rpm.

A typical offshore wind farm can generate 1 GW of power, equivalent to one or two standard nuclear power plants. Figures 4.5 and 4.6 illustrate schematically typical layouts of

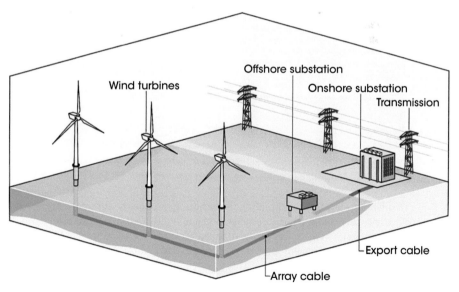

Figure 4.5 Layout of an offshore wind farm.

147

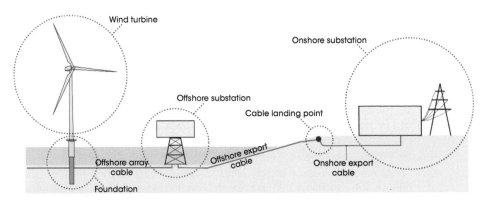

Figure 4.6 Schematic of wind farm grid integration.

Figure 4.7 Aerial view of Dudgeon Wind Farm showing turbines and substation (Source: Photo by Jan Arne Wold/Equinor).

an offshore wind farm, annotating the various components: turbine generator through the electricity cables to the offshore substation and finally to the onshore power grid. Figure 4.7 shows an aerial view of the Dudgeon wind farm, with a capacity 400 MW, located off the coast of Norfolk showing some of the turbines and a substation.

The first offshore wind farm was installed in Denmark in 1991 and had a capacity of 4.95 MW. In the following years, the growth was slow with only a small number of projects being developed in Denmark and the Netherlands. It was not until early 2000s that the offshore wind capacity has started increasing significantly. In 2018, the worldwide capacity of offshore installations was 23 GW and since then it increased at an annual rate of 15% reaching 35 GW in 2021. The world leaders in offshore wind power production (in 2022) are China, UK, and Germany with 26.5 GW, 13.6 GW, and 8 GW, respectively (Figure 4.4). Over the next 5 years, more than 300 GW of new capacity is expected to be installed. By then, China will have the largest share in offshore wind power, and at the same time, a new market will be created in North America, where the first offshore wind farms will be installed. The target of USA is to generate 30 GW by 2030.

Figure 4.8 shows wind farms planned in the USA.

Offshore wind farm locations in China

Also in China, about 15 GW of offshore wind farms are either operating or under construction or proposed; it is noted that one of the proposed wind farms will have a capacity of 6 GW (Figure 4.9).

4.3   Tackling intermittency of wind power

4.3.1   The sun, wind, and wind-resource map

In a nutshell, wind will blow as long as the sun is shining and therefore never-ending resource. The ultimate source of energy responsible for the creation of wind is the sun. Sun bombards the surface of the earth with radiation energy leading to a continuous warming. Of this energy, a large amount is sent back to space and what remains is transformed into heat energy. Due to the shape of the earth, its surface is not heated evenly resulting in the equator getting more energy than the poles. As a result, there is a continuous heat transfer from the equator to the poles. The atmospheric air consists of nitrogen and oxygen (totaling 99%), and like all gasses, it expands when heated and contracts when cooled. When solar radiation hits the surface of the earth, it causes the air to get warmer, which means that the air gets lighter and less dense. This also results in the rise of warm air in higher altitudes and the creation of areas of lower atmospheric pressure, where the air is warmer. Due to the variance in pressure, the air will move from a high pressure to a lower pressure area, to reach equilibrium. Wind can therefore be defined as the movement of atmospheric air masses from a region of high atmospheric pressure to a region of low pressure.

However, the wind circulation system is even more complex. At 30° and 60° latitude, there is a major change in atmospheric pressure creating zones of high and low pressure, respectively. The air circulation within these zones is known as cells, and the major winds created as a result are known as trade winds. Due to the Coriolis effect induced by the rotation of the earth, the winds deflect to the right in the Northern hemisphere and to the left in the Southern hemisphere one, leading to a spiral movement of the air mass. Figure 4.10 shows this upper altitude global wind system. In addition to the global wind system, there are localized influences as well that are mostly related to the terrain of an area and its proximity to water bodies.

Wind characteristics, such as the speed and direction, are not regular and can have a spatial and temporal variance. More specifically, wind speed is lower around the equator, and it increases closer to the poles. The highest wind speeds are created within the region of

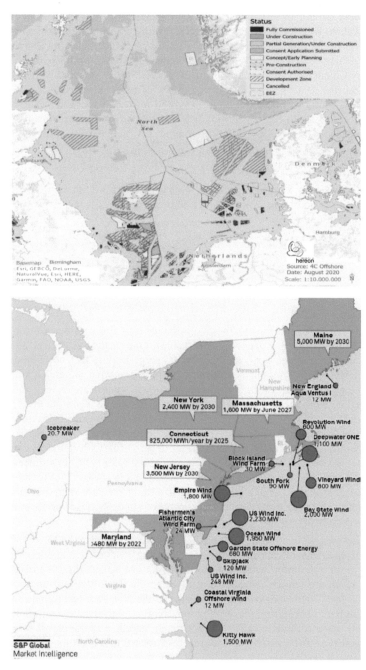

US offshore wind targets and projects. Source: S&P Global Market Intelligence.

Figure 4.8 Offshore wind farm locations in the USA.

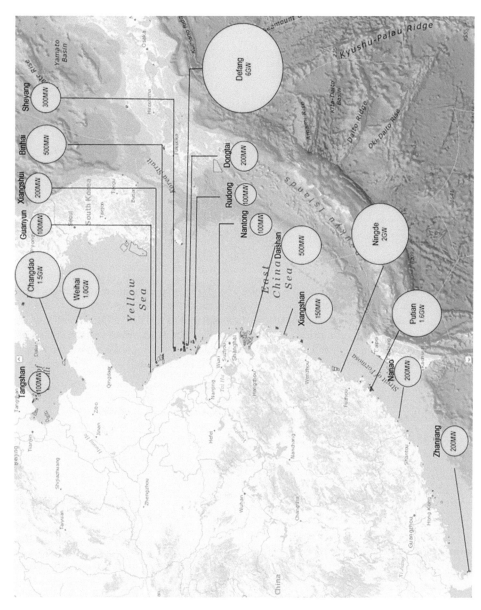

Figure 4.9 Offshore wind farm locations in China.

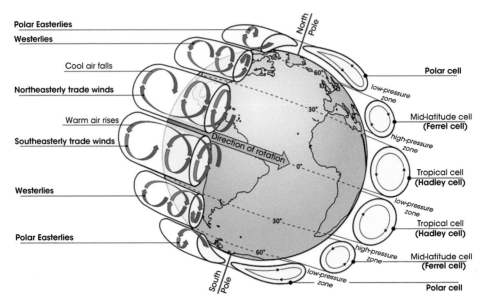

Figure 4.10 Global circulation of wind.

30° and 60° latitude, mostly because of a global wind circulation mechanism. However, the Southern hemisphere has less land obstructions than the Northern one, resulting in higher wind speeds. Therefore, the distribution of land has a major influence on the wind speed as it can obstruct the free movement of air. In other words, wind speed can be higher in the open sea than on land. Also, the wind speed is dependent on the time of the year, i.e., in the Northern hemisphere, the mean wind speed is higher during winter. Further details on wind energy can be found in Letcher (2023).[8]

ASIDE: Limits to wind power utilization

Gustavson[1] carried out a study to estimate the limits to wind power utilization. It has been suggested that approximately 2% of the solar energy (solar flux) striking the Earth's surface is converted into kinetic energy in wind. The average solar flux considering the whole of the earth's surface is $350\,W/m^2$, and the total solar energy is 1.8×10^8 GW. Using 2% value provides the total wind energy over an earth element of $7\,W/m^2$. Out of the total wind energy, circa 35% is dissipated within 1 km of the earth's surface and which provides an average of $2.5\,W/m^2$. This kinetic energy of the wind is converted by turbines, and the suggested 10% extraction rate limit provides a value of $0.25\,W/m^2$. It is important to note that the spatial distribution of wind energy both across the surface of the Earth and vertically through the atmosphere. As a rule of thumb, average annual wind speeds of 6 m/s and above at hub height are considered a safe investment.

4.3.2   Wind and waves

Offshore wind turbines and support structures generally face a harsher environment than their onshore counterparts. Winds are stronger, waves impose unwelcome dynamic loads, corrosive sea spray, and saltwater test materials and components. It is a wild world out there, as any oil rig worker or ocean sailor will confirm.

We have already discussed many features of the wind in Chapter 2, and two of them deserve further attention in relation to offshore turbines: *wind shear* and *turbulence*.

Wind shear describes the increase in average wind speed with height above a surface and depends mainly on surface roughness. A simple power law is often used to quantify the effect (see also Equation 2.4):

$$U/U_0 = \left(h/h_0\right)^a \tag{4.1}$$

where U is the speed at height h, U_0 is the speed at a reference height h_0, and a is a *power law exponent* that depends on the roughness of the surface. Onshore, typical values of a range between 0.1 for smooth surfaces and 0.3 for rough ones. A value of 0.14, or 1/7, long regarded as typical of grassland areas on the Great Plains of the USA, is often used as a default value for other sites in the absence of specific data and gives rise to the so-called *1/7th power law*. Offshore, the same value is often used for sea surfaces—although in practice the roughness must obviously depend on the height and pattern of the waves.

Variations of wind speed with height are important, but turbine output is more directly related to the power in the wind, which is proportional to the *cube* of the speed. We may modify Equation (4.1) to give the equivalent variation of *power* with height (see also Equation (2.5)):

$$P/P_0 = \left(h/h_0\right)^{3a} \tag{4.2}$$

This equation allows us to predict the benefit, in power terms, of raising the hub height h of a turbine above a reference height h_0. Curves for several values of a have already been drawn in Figure 2.5.

Our primary interest here is in large offshore HAWTs rated between (say) 2 and 10 MW, with rotor diameters between about 80 and 140 m and similar hub heights. Figure 4.11 addresses this situation by taking a hub height of 100 m as a reference, then using Equations (4.1) and (4.2) to estimate the effects of raising or lowering it by various amounts. Yellow is used for wind speed, red for wind power, and both are normalized to unity at 100 m. Curves are drawn for $a = 0.14$ (the 1/7th power law), and the spread around each covers the range $0.1 < a < 0.2$, which includes most offshore sites, in most sea states. For example, if $a = 0.14$, raising the hub height of a turbine from 100 to 120 m is expected to increase wind speed by about 2.5% and wind power by about 7.5%. If $a = 0.2$, denoting a rough sea surface, speed increases by about 4% and power by about 12%.

The benefit of increasing hub height is undoubted, but against it must be set the extra cost of a higher tower and stronger foundations. The balance of advantage may be hard to strike because it is not easy to predict the wind shear properties of a particular site. Complicating factors include:

- The roughness of a sea surface (and therefore the value of a) is greatly affected by wave conditions, which in turn depend on wind conditions, both locally and at a distance.
- There is often a substantial time lag between changes in wind speed and changes in wave height and pattern.

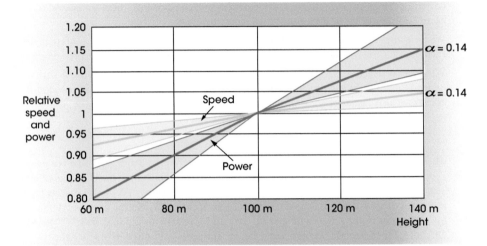

Figure 4.11 Variations of wind speed and power due to wind shear.

- Wind shear is affected by seasonal temperature differences between air and water.
- Many offshore wind farms are fairly close to the shore. Winds blowing from the shore, which may develop over relatively rough land surfaces including hills and cliffs, tend to produce higher values of α than winds blowing over long stretches of sea. The effects of wind shear on turbine performance, measured over a complete year, therefore depend on variations of surface roughness with direction and on the wind rose pattern for the site.

Another feature mentioned in Chapter 2 is *turbulence*. This refers to random short-term fluctuations in wind speed and direction, including the strong gusts known by sailors as *squalls*. It can exert heavy and even dangerous dynamic loads on turbine blades and structures. The intensity of turbulence at sea is generally less than on land because of lower surface roughness and less variation of air temperature with height. However, it is worth noting that turbulence is not entirely counterproductive because, on a wind farm, it tends to accelerate energy exchange between turbine wakes and the prevailing wind, reducing wake lengths. Shorter wakes in turn reduce array losses caused by interaction between adjacent turbines. We should also note that in conditions of low *atmospheric* turbulence, individual turbines may experience plenty of *localized* turbulence caused by upstream neighbors. Indeed, on a large offshore wind farm, the major contributors to turbulence are often the turbines themselves (Figure 4.12).

How about that other great influence on offshore turbines and structures—ocean waves? The first point to make is that surface waves on oceans, seas, and lakes are caused by the wind. Indeed, wave energy—another renewable resource with great potential—may be thought of as a child of wind energy. Unfortunately, waves are unwelcome in the context of offshore wind turbines for two main reasons: they have a major influence on surface roughness and wind shear, and they push and buffet turbine support structures, producing unwanted stresses and vibrations.

Figure 4.12 Catching the wind, resisting the waves (Source: With permission of Orsted).

Surface waves are not formed instantly but build up with time and with distance from the nearest windward shore, known as the *fetch*. Locally generated waves in shallow and medium-depth water are often short and choppy, whereas waves arriving from afar over deep water tend to produce a regular *swell*. A well-developed wave pattern may take hours or days to form and travel hundreds or even thousands of kilometers across an ocean. The wave heights and patterns experienced by an offshore wind farm often bear little relationship to how the wind is blowing locally, and wave direction may not coincide with wind direction. So, although surface waves are wind generated, their relationship with the wind that blows at a particular place and time is complex.

Wave patterns are often divided into three broad classes, illustrated by Figure 4.12:

- *Regular waves*. Also known as *Airy waves* after George Airy (1801–1892), English mathematician and astronomer, these waves are sinusoidal, or close to sinusoidal, in the form and are typical of long-distance ocean swell.
- *Irregular waves*. These have more complicated shapes and lack a single, well-defined height or wavelength (distance between adjacent crests). They may be formed by inter-action of regular waves arriving from different directions.
- *Random waves*. Typical of choppy seas produced by strong and variable local winds, random waves display a wide and unpredictable range of heights and wavelengths (Figure 4.13).

The designers of offshore turbines need to consider the dynamic forces exerted by waves and whether they occur at, or close to, the natural frequencies of towers and support structures. In the case of regular (Airy) waves impinging on single tubular supports known as *monopiles*,

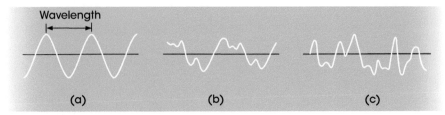

Figure 4.13 Wave patterns: (a) regular, (b) irregular, and (c) random.

theoretical predictions are relatively straightforward;[2] but with irregular or random waves, far less so. Additional problems can arise with *breaking* waves, produced when the bases of waves can no longer support their tops, causing collapse. This is most likely to happen as waves enter shallow water close to shore. And last but not least, the designer needs to consider rogue waves that occasionally place extreme loads on support structures and may even threaten rotating blades at the low point of their travel.

The statistics of waves[2] are analogous to those of the wind that we covered in Section 2.2.2. In particular, the frequency of occurrence of random waves of different heights may be modeled using a *Rayleigh distribution*, and the most dangerous wave that should be allowed for over the lifetime of a wind farm may be predicted using a *Gumbel distribution*. It seems rather satisfying that wind and waves, so intimately connected in nature, can be characterized by similar theoretical models.

4.3.3 Wind-resource maps

In order to use wind as a reliable source of energy, it is essential to have the wind speed and direction variance data. Such a tool is a wind map, showing how the mean wind speed is distributed in an area. Figure 4.14 shows the global wind map for a period between 1979 and 2012, at an altitude of 100 m developed by Nikitas (2020).[9] Figure 4.15 shows an offshore wind map developed by The World Bank and the Technical University of Denmark (DTU) (https://globalwindatlas.info/en).

4.3.4 Minimum wind speed to produce power and power curve of a turbine

Figure 4.16 shows a series of power curves for larger turbines (8–12 MW) and is essentially power production of a turbine plotted with respect to wind speeds. The power curve figure shows that a turbine can generate power from a wind speed starting from as low as 4 m/s and reach the rated power at about 12–15 m/s. It is therefore clear that high wind speed is not needed to produce power. For the power generation equation, see Section 3.2.1 "Rotor efficiency and the Betz limit."

4.3.5 Storage of energy from offshore wind

At some level of grid penetration, electricity storage would be needed to accommodate the variability of the wind-resource variability. The combination of offshore wind and with storage systems such as batteries is schematically shown in Figure 4.17. Another option is to produce hydrogen with the excess wind power and storing it as a dispatchable fuel (Figure 4.18).

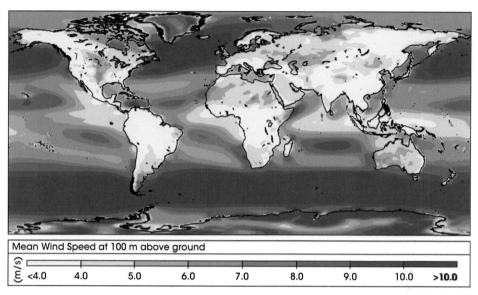

Figure 4.14 Global mean wind speeds for 1979–2012.

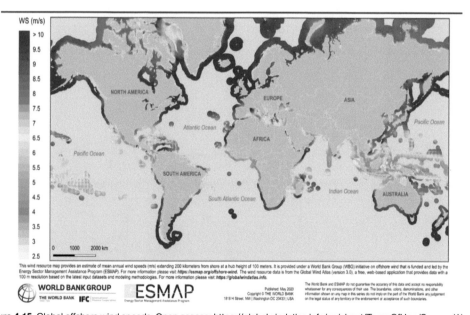

Figure 4.15 Global offshore wind speeds. Open access: https://globalwindatlas.info/en/about/TermsOfUse (Source: World Bank Initiative, 2020/CC-BY 4.0).

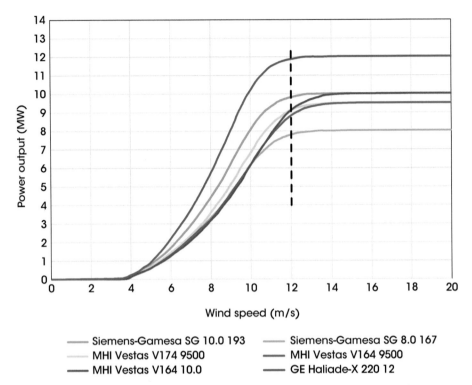

Figure 4.16 Power curves for turbines (Source: Compiled by the author from manufacturer's data).

Legend:

- Siemens-Gamesa SG 10.0 193
- MHI Vestas V174 9500
- MHI Vestas V164 10.0
- Siemens-Gamesa SG 8.0 167
- MHI Vestas V164 9500
- GE Haliade-X 220 12

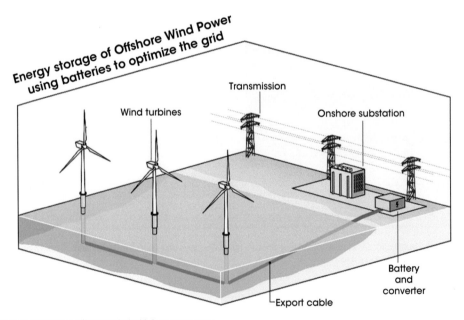

Figure 4.17 Combining offshore wind with battery storage.

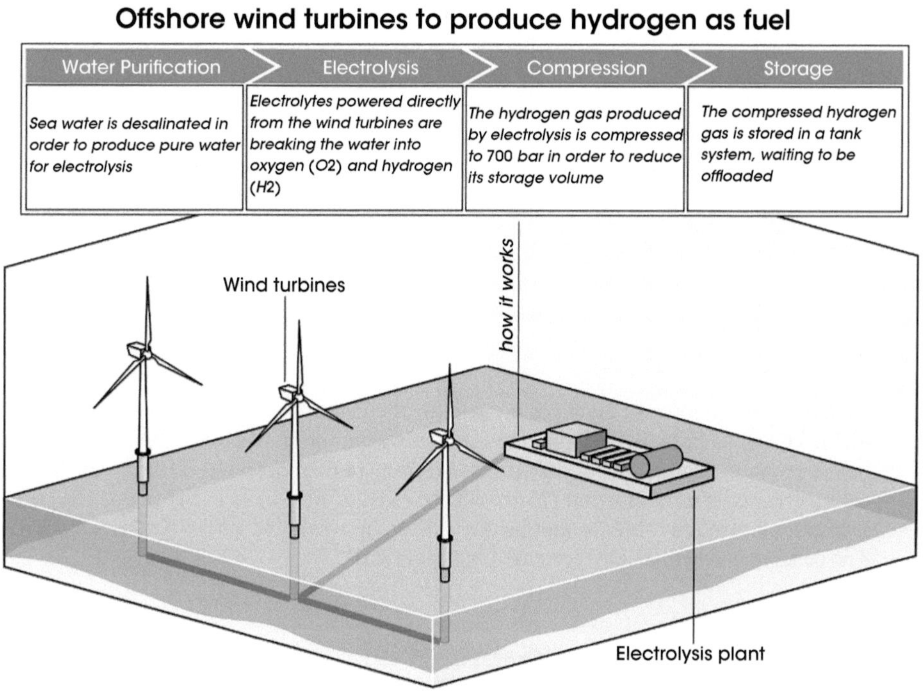

Offshore wind turbines to produce hydrogen as fuel

Water Purification	Electrolysis	Compression	Storage
Sea water is desalinated in order to produce pure water for electrolysis	Electrolytes powered directly from the wind turbines are breaking the water into oxygen (O2) and hydrogen (H2)	The hydrogen gas produced by electrolysis is compressed to 700 bar in order to reduce its storage volume	The compressed hydrogen gas is stored in a tank system, waiting to be offloaded

how it works

Wind turbines

Electrolysis plant

Figure 4.18 Use of offshore wind to produce hydrogen.

4.3.5.1 Hydrogen production using offshore wind—the Japanese "Jidai" concept

The 2011 Fukushima Daiichi Nuclear Power Plant (NPP) disaster was a devastating moment in the history of mankind where cascading event due to earthquake led to failure of the Nuclear Power Plant. Following the disaster, many countries such as Germany and Japan reduced their reliance on nuclear power and compensated with fossil fuels and renewables. Within the framework of global warming, and among others such as energy security, Japan aims to become a carbon-free country through a "hydrogen society."

The main idea is to generate hydrogen from water through renewable energy sources, such as wind, solar, and hydroelectricity, and the Japanese named it the "Jidai concept," see Figure 4.18. The Jidai concept is a four-step process: (i) seawater is desalinated; (ii) electrolysis is used to produce hydrogen and oxygen from water; (iii) hydrogen gas is compressed to 700 bar to reduce storage volume; and (iv) high-pressure hydrogen gas is stored in a module-based tank system. Similar attempts are also ongoing in major European economies. These attempts have been boosted by other technology developments, i.e., the invention of hydrogen-powered cars, trains, ships, and even aircrafts.

4.3.5.2 ASIDE: potential of green hydrogen

(1) The potential of hydrogen is huge in achieving net zero: through the existing off-shore infrastructure of pipe networks, hydrogen can be transported for distribution.

With the advent of hydrogen cars and trains, the economy can be transformed without the need for expensive metals that are needed for battery production.

(2) Studies are being conducted to demonstrate that a 100% hydrogen gas network is equally as safe as the existing natural gas. It is worth noting that burning natural gas to heat homes and businesses accounts for approximately a third of the UK's carbon emissions. Hydrogen powered commuter trains are available, and it has been reported in New Civil Engineer that 30% of the UK rail fleet could be suitable for running hydrogen powered trains. In summary, wind power has the potential to carry the transition to low-carbon energy, transforming the fossil-fuel energy landscape to a more sustainable energy future.

4.3.5.3 BATWIND project

Equinor, the developer of Hywind, is one of the first companies to develop storage solution and is known as Batwind and the promotors are Equinor and Masdar. When Batwind will be in operation, it will store energy produced from an offshore wind farm by which electrical power produced at Hywind (25 km off the coast of Peterhead in Scotland, UK) will be transported through cables to an onshore substation where the 1 MW batteries are placed. The concept is schematically presented in Figure 4.17.

4.4   Offshore systems and choice of foundations

4.4.1 Turbines and foundations

The current surge of interest in offshore wind energy is stimulating the design and development of ever-larger turbines, with 18 MW machines in prospect. The benefits of scale will always tend to favor large machines over small ones; but what is the best design configuration for these turbines, and will new types soon appear with even higher power ratings?

Onshore, the wind energy industry is dominated by the three-bladed horizontal-axis machines that have undergone continuous development and improvement since the 1980s. A huge amount of operating experience has been gained worldwide with these HAWTs. From time to time, new designs of wind turbine are proposed, and the offshore market is no exception. Large vertical axis machines are currently being considered, principally because their rotors and generators can be installed relatively close to the sea surface and the rotors do not need to yaw into the wind. But whether they can ever rival the efficiency, reliability, and cost-effectiveness of today's large HAWTs seem doubtful and will remain so until prototypes have been installed and tested over a lengthy period. In the meantime, today's impressive HAWT designs will no doubt see further incremental improvements, consolidating their hold on the burgeoning offshore market.

However, we should not assume that HAWT machines originally designed for use onshore can be shipped out to sea without modification. The basic design philosophy may be sound, but offshore turbines face a tough environment and need special features:

- Extremely high reliability, because of the difficulties and costs of access, maintenance, and repair at sea.

- Resistance to corrosion and wear in the marine environment.
- The ability to withstand very high wind speeds, squalls, and heavy seas.
- Offshore turbines are often designed with higher cut-out wind speeds than their onshore counterparts, taking advantage of the stronger wind regimes. This adds to electricity generation but requires extra robustness.

As far as installation is concerned, the major difference between onshore and offshore turbines lies in their foundations. Onshore, securing a turbine is normally fairly straightforward using a heavy reinforced concrete pad set in the ground, group of piles or by rods grouted into holes drilled in bedrock. Offshore, it is generally more challenging, and foundations must take account of turbine weight and height, the nature of the sea bed, the depth of water, and expected storm conditions.

In shallow and medium-depth waters, three main types of structure are used, as illustrated in 19.

- *Monopile*. This is a steel tube driven into the sea bed using a heavy-duty pile driver, to a typical depth between 20 and 40 m. A yellow *transition piece* including a platform (see also Figures 4.21 and 4.22) connects the monopile to the base of the turbine's tower. A monopile is normally the simplest and cheapest type of foundation in shallow water, including much of the North Sea.

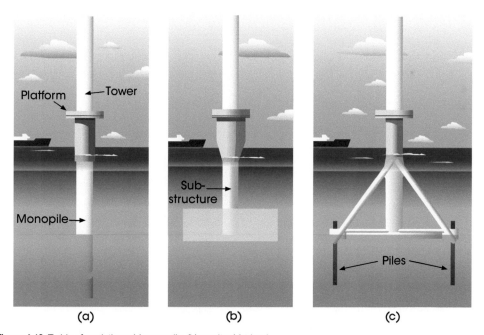

Figure 4.19 Turbine foundations: (a) monopile; (b) gravity; (c) tripod.

Figure 4.20 On the move: transition pieces with platforms (Source: With permission of Orsted).

Figure 4.21 A jackup barge, its long legs extended down onto the sea floor, begins turbine installation (Source: With permission of Orsted).

Figure 4.22 A 6 MW turbine supported by a "jacket" off the coast of Scotland (Source: With permission of Repower).

- *Gravity foundation.* An alternative to the monopile, a gravity foundation relies on its own weight to secure and stabilize the turbine. It needs a large base area to counteract the overturning moment produced by the turbine, and the sea floor must be dredged uniform and level to receive it. A gravity foundation for a large turbine may weigh in excess of 1000 tons including ballast and is normally fabricated in reinforced concrete.

- *Tripod or "jacket".* A welded triangular or four-legged structure made from tubular steel is often the preferred option in medium-depth waters, secured to the sea bed by piles. Less sea floor preparation is needed than for gravity foundations. Structures known as "jackets," similar to lattice towers, are often suitable since they have been used in the offshore oil and gas industries for many years and their design and performance is well understood (see also Figure 4.22).

In water depths above 60 m, the size and cost of rigid underwater structures tend to be prohibitive, and it is hardly surprising that current offshore wind farms are almost all located in fairly shallow water. But as the offshore industry develops, there is increasing interest in floating turbines for deep-water locations. In principle, a floating turbine can be assembled close to shore, towed out to sea, and anchored to the sea bed with relatively light, simple, equipment. It can subsequently be moved if required and eventually decommissioned leaving little

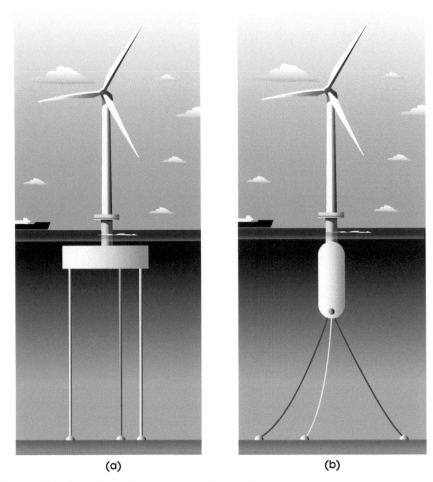

(a)   (b)

Figure 4.23 Floating turbines: (a) tension-legged platform and (b) spar buoy.

or no trace behind. Of various floating turbine concepts proposed over the years, Figure 4.23 illustrates two concepts:

- *Submerged tension-legged platform.* This system consists of a floating platform held just below the sea surface by slim tension legs or *tendons* anchored to the seabed. Clearly, the platform must have sufficient buoyancy to support the turbine while maintaining tension in the legs under all sea and wind conditions. Similar platforms are used by the offshore oil and gas industries on some of their rigs. A prototype supporting a relatively small wind turbine was placed in a water depth of about 110 m off the coast of southern Italy in 2007.[3]

- *Spar buoy*. Kept in place by mooring lines and anchors, a spar buoy is essentially a long closed cylinder that floats upright below the sea surface and is stabilized by ballast in its base. Buoyancy requirements are comparable with those of a submerged tension-legged platform. A prototype system of this type, supporting a 2.3 MW HAWT and with its floating element extending 100 m below the sea surface, was installed in about 200 m of water off the coast of Norway in 2009.[4]

The detailed design and construction of such systems are only hinted at by the illustrations in Figure 4.23 and will no doubt evolve with experience. Some very interesting and challenging issues must be addressed if floating turbines are to match the reliability and operating efficiency of machines fixed to the sea floor by rigid structures. Perhaps most importantly:

- *Turbine stability*. Floating structures cannot easily provide highly stable platforms for mounting wind turbines. How is the turbine controlled in yaw so that it faces the oncoming wind? How does any pitching or rocking of the float in heavy seas impinge on turbine aerodynamics and efficiency?
- *Stresses and fatigue*. Interactions between wind and waves mean that hydrodynamic stresses on the undersea structure interact with aerodynamic and mechanical stresses on the turbine, producing a complex fatigue environment.

As the offshore wind industry develops, moving into deeper waters, and experiencing more severe wind conditions, it is clear that there will be plenty of employment opportunities for innovative design engineers and skilled turbine installers.

As mentioned earlier, typically, for up to 60 m of water depth, the foundations are placed directly on the seabed and are of a fixed type—known as bottom-fixed wind turbine. However, for deeper waters, the wind turbines are placed on top of floating structures. Figure 4.24 presents the two types of systems (Grounded and Floating) together with main types of foundations commonly used today at different depths of water.

Figure 4.25 shows four types of floating system: Semi-submersible; Tension Leg Platform (TLP) system; (c) barge type; (d) spar type.

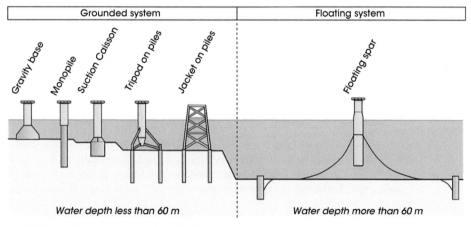

Figure 4.24 Types of systems (grounded and floating).

165

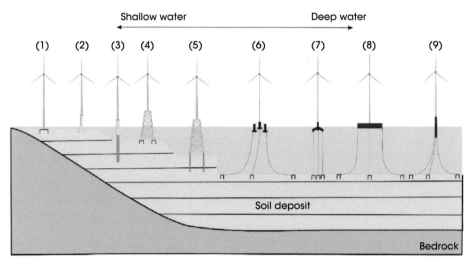

Figure 4.25 Typical bottom fixed and floating foundations. (1) Suction bucket caisson, (2) gravity-based foundation, (3) monopile, (4) jacket on suction caissons, (5) jacket on long piles (also known as pin piles), (6) semi-submersible, (7) tension leg platform (TLP), (8) barge, (9) spar.

The choice of foundation will depend on the following: site conditions (wind, wave, current, seabed condition, ground profile, water depth, etc.), available fabrication and installation expertise, operation and maintenance, decommissioning laws of the land, and finally economics. Below is a description of an ideal foundation to support offshore wind turbines following Bhattacharya (2019)[3]:

(1) A foundation is capacity or "*rated power*" specific (i.e., 6 MW or 10 MW rated power) but not turbine manufacturer specific. In other words, a foundation designed to support a 5 MW turbine can support turbines of any make. There are other advantages in the sense that turbines can easily be replaced even if a particular manufacturer stops manufacturing them.

(2) Method of foundation installation which is not necessarily weather sensitive, i.e., not dependent on having a calm sea or a particular wind condition. The installation of the first offshore wind farm in the USA took more time than that was expected as a result of unsuitable weather conditions.

(3) Low maintenance and operational costs, i.e., need least amount of inspection. For example, a jacket-type foundation needs inspection at the weld joints.

Figure 4.26 shows choice of foundations based on turbine size and associated sea depths. Figure 4.27 shows examples of turbine foundations. It is advantageous to have a large number of turbines in a wind farm to achieve economies of scale. If the continental shelf is very steep, grounded (fixed) turbines are not economically viable, and a floating system is desirable.

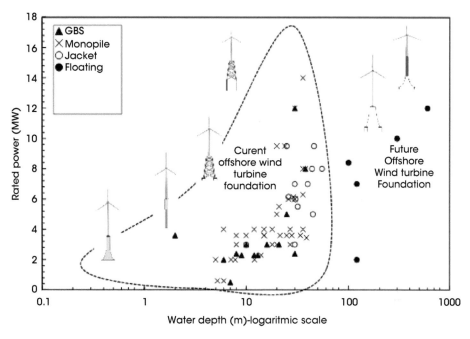

Figure 4.26 Global trends in offshore wind farm systems.

Figure 4.27 Examples of bottom-fixed jacket on suction caisson foundation, Aberdeen Offshore Wind Farm, Alpha Ventus (Source: (a) With permission of Navingo. (b) With permission of Orsted. (c) Courtesy of AREVA Wind).

4.5 Offshore wind farm case studies

4.5.1 Case study: *Horns Rev 2*

We previously noted that the world's first offshore wind farm entered service at Vindeby off the coast of Denmark in 1992. The "offshore club" was soon joined by Sweden,

The Netherlands, and the UK. But true to its reputation as a small but powerful engine room of modern wind technology, Denmark retained the record for wind farm size (in terms of installed capacity) for the next 17 years. The record book looks for the first 20 years looks like this:

Wind farm	Country	Year	No. of turbines	Installed capacity (MW)
Vindeby	Denmark	1992	11	5
Middelgrunden	Denmark	2001	20	40
Horns Rev 1	Denmark	2002	80	160
Nysted	Denmark	2004	72	166
Lynn and inner dowsing	UK	2009	54	194
Horns Rev 2	Denmark	2009	91	209
Thanet	UK	2010	100	300

We focus here on *Horns Rev 2* which began spinning its turbines in 2009. It attracts a great deal of international attention because of its size, technical features, and Danish pedigree and follows on from the well-known *Horns Rev 1* wind farm, sited in the same area but closer to the shore, which held the record in 2002–2004.

Rev is Danish for reef and reminds us that *Horns Rev 2* is sited on a reef in shallow water about 30 km off the westernmost point of Denmark, close to the city of Esbjerg. There is an excellent wind resource with prevailing westerly winds blowing unhindered across the North Sea, giving an average wind speed close to 10 m/s at the hub height of the turbines. The water depth and sea bed are suitable for monopiles with a diameter of 3.9 m and various lengths between 28 m and 40 m, set in the sea bed with the help of a large jackup barge (Figure 4.28).

Details of the turbines[5] and wind farm[6] include:

- Installed capacity   209 MW
- Number of turbines   91
- Turbine rated power   2.3 MW
- Rated wind speed   13–14 m/s
- Cut-in and cut-out wind speeds   4 m/s and 25 m/s
- Rotor diameter   93 m
- Hub height   68 m
- Rotor speed   6–16 rpm
- Weight (rotor and nacelle)   142 tons
- Turbine layout   13 rows in 7 radial circles
- Water depth   9–17 m
- Wind farm area   33 km^2
- In-farm cable length   70 km
- Cable length to shore   42 km

The electricity generated by the turbines is collected by undersea cables and delivered to a substation located within the wind farm (see Figure 4.29). The main power cable to shore includes optical fibers for data transmission.

Figure 4.28 One of the 25 support vessels used during installation of *Horns Rev 2* (Source: With permission of Orsted).

Figure 4.29 The *Poseidon* accommodation platform being installed next to the substation by a large jackup barge (Source: With permission of Orsted).

Figure 4.30 Installation complete at *Horns Rev 2* (Source: With permission of Orsted).

The annual energy production of *Horns Rev 2* is estimated as equivalent to the consumption of about 200,000 Danish households. Back in Section 1.4, we discussed the concept of "equivalent households," widely used by developers and electric utilities to inform the public about the amount of electricity generated by wind farms, and explained how important it is to understand the underlying assumptions. The completed *Horns Rev 2* is shown in Figure 4.30.

A pioneering feature of *Horns Rev 2* is its accommodation platform, known as *Poseidon*, which allows personnel to stay "on site" rather than traveling 2 hours by boat from Esbjerg harbor. It has 24 rooms including kitchen, dining room, study room, and gym, with a total area of 750 m² on three decks. *Poseidon* weighs 420 tons, supported by its own monopile, and is joined by a walkway to a substation platform incorporating a helipad.

The installation of *Horns Rev 2* took 20 months, employed 600 construction staff, and required 25 support vessels. It created quite a stir when it started operation in 2009 and will no doubt contribute handsomely to Denmark's offshore electricity generation for many years to come.

4.5.2 Case study: *London Array*

Construction work for *London Array*[7] began in 2009. The UK had already gained considerable experience with large offshore wind farms, including the *Thanet* farm sited off the Kent coast that raised the bar to 300 MW. *London Array* moves offshore installations a stage further, aiming to be the first wind farm in the world to reach the milestone figure of 1000 MW, or 1 GW.

It is always an advantage to generate electricity close to population centers, curtailing the inevitable power losses associated with long-distance transmission. This makes a large wind farm sited in the Thames Estuary an attractive proposition. There is a very good wind resource with an average wind speed at turbine hub height of about 9 m/s (see Figure 2.19). When completed *London Array* is predicted to supply enough electricity, over a full year, to satisfy the needs of about 750,000 households in a heavily populated region of southeast England.

Figure 4.31(a) shows the wind farm site, which is 20–30 km offshore from Margate (Kent) and Clacton (Essex) and about 120 km from central London. The total site area is about 250 km^2. Electricity is taken ashore by submarine cables to a new substation at Cleve Hill near the city of Canterbury. The substation boosts the voltage from 150 to 400 kV and feeds the electricity into the national high-voltage transmission grid.

London Array is being achieved in two phases and we focus here on *Phase 1*, the initial installation of 175 turbines with a combined capacity of 630 MW. Part (b) of Figure 4.31 shows the turbine layout, emphasizing the importance of seabed topology in the design of offshore systems. There are two major issues in the case of *London Array*: sand banks in the estuary and shipping lanes.

Phase 1 occupies an area of about 100 km^2, covering part of the *Long Sand* and *Deep Knock* sandbanks plus the shallow channel separating them. Water depths and the composition of sandbanks and seabed are well suited to monopile foundations topped by transition pieces. It is essential to avoid *Black Deep*, a much deeper channel along the northwest margin of the site, which is used as a main shipping lane by vessels going to and from the Port of London.

The turbine array forms a rectangular grid pattern. Prevailing winds blow mainly from the southwest so the grid is slanted in this direction. It also suits the other main winds—east to northeast—that tend to blow across the North Sea from continental Europe in the winter and early spring. Downwind spacing between adjacent turbines is about 8.5 rotor diameters, and crosswind spacing is about 5.5 rotor diameters (as discussed in Section 3.6.2).

Technical details of the turbines[5] and array[7] for Phase 1 may be summarized as follows:

- Installed capacity   630 MW
- Number of turbines   175
- Turbine-rated power   3.6 MW
- Rated wind speed   13 m/s
- Cut-in and cut-out wind speeds   3 m/s and 25 m/s
- Rotor speed   5–13 rpm
- Rotor diameter   120 m
- Hub height above mean sea level   87 m
- Weight of rotor and nacelle   225 tons
- Array area   100 km^2
- In-farm cable length   209 km

The 209 km of in-farm cabling collects electricity from the individual turbines and delivers it at 33 kV to two in-farm substations supported by their own monopiles (not shown in the

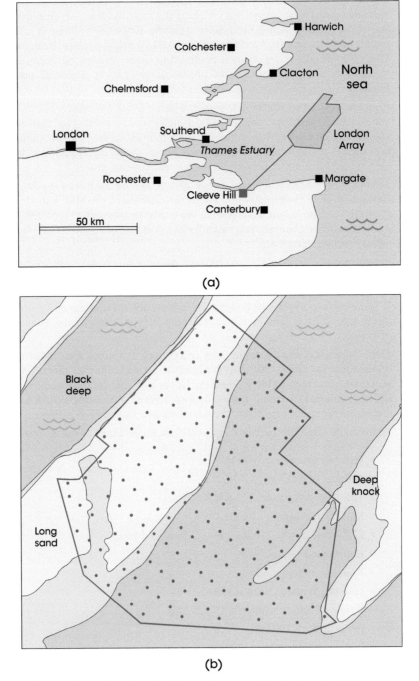

(a)

(b)

Figure 4.31 *London Array*: (a) location in the Thames Estuary and (b) turbine layout for Phase 1.

Figure 4.32 Onshore work in progress: groundworks for the Cleve Hill substation (Source: With permission of London Array Ltd).

figure). These substations transform the AC voltage to 150 kV for submarine transmission to the onshore substation at Cleve Hill using specialized cables that include optical fibers for data transmission. Although it is always tempting to focus on the most dramatic and visible parts of an offshore wind farm—turbines standing majestically above the ocean waves—we should always remember the design effort and engineering expertise needed to bring the precious electricity ashore (Figure 4.32). Finally, Figure 4.33 shows *London Array's* first stage of offshore installation that began in 2011.

4.5.3 Case study: World's first floating wind farm

Hywind Scotland Pilot Park is the first floating wind farm with rated capacity of 30 MW with five turbines, located 30 km off the coast of Scotland, see Figure 4.34. In October 2017, Hywind started delivering power to Scottish Electricity grid. The water depth is 95–120 m, and the mean waves (Hs) are 1.8 m. The average wind speed at 100 m height is 10.1 m/s, and a spar type of concept is used. The turbine structure is anchored using conventional three-line mooring as is used in FPSO (floating production storage and offloading) anchoring. The anchor piles are having a dimension of 7 m diameter and are 16 m long. The control system that is used to dampen out motions is the blade pitch control. The whole system is full scale tested through Hywind Demo 2.3 MW turbine installed in 2009 having 5 years of operational data. Figure 4.35 shows some of the installation photos of the floating system.

Figure 4.33 Offshore installation begins (Source: With permission of London Array Ltd).

Hywind Scotland Pilot Park

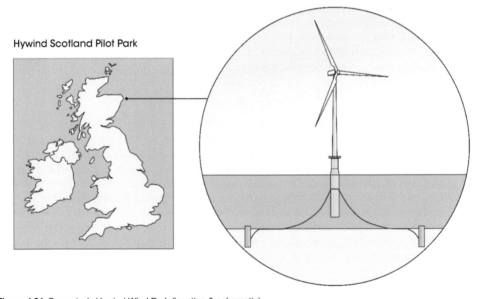

Figure 4.34 Case study Hywind Wind Park (location & schematic).

Figure 4.35 Installation of the turbine: (a) The substructure being pulled to the location; (b) lowering of the substructure; (c) anchored to the seabed with chains; (d) the RNA Assembly together with the tower is being transported for connecting to the substructure (Source: Equinor).

4.6 Bringing the power ashore

The design of a cable system for a large offshore wind farm raises important technical and economic issues. Within the farm itself, the AC electricity generated by individual turbines must be collected by submarine cables and delivered to one or more substations for onward transmission to shore. Up to this point, the system voltage is typically limited to 33 kV by the cost and size of transformers and associated switchgear.[2] But now comes the big question—how far away is the *onshore* substation, and which technology is most suitable for bringing the power ashore?

For cable lengths up to about 50 km conventional three-phase high-voltage AC transmission is the most economical solution. The transmission voltage is chosen to strike a balance between the capital cost of equipment and the expected power losses in transformers and cables over the projected life of the system. For example, *London Array* uses 150 kV for the final run to shore, transformed up to 400 kV in an onshore substation (Figure 4.36) before feeding into the UK's main transmission grid.[7] The system is having AC throughout.

For distances to shore above 50 km, an alternative technical solution becomes increasingly attractive—*high-voltage direct current* (*HVDC*) transmission. For distances greater than 100 km, it is virtually imperative. In view of our comments in Section 3.4.1 about the historical ascendancy of AC over DC, this may seem rather surprising. But since the 1930s, HVDC has quite often been chosen for bulk transmission of electrical power over long distances, and today there are onshore HVDC links over 2000 km long in China and Brazil.

Figure 4.36 An offshore substation for a large wind farm (Source: With permission of Orsted).

The basic attractions of DC over AC are smaller capital costs and power losses in cables as the transmission distance increases. And in the case of *submarine* transmission, there is a very important additional factor—cable capacitance.

In our discussion of AC electricity in Section 3.4.1, we saw how resistors dissipate real power, whereas inductors and capacitors demand reactive power. Long submarine cables have much higher capacitance than equivalent onshore cables strung between pylons because their conductors are separated by thin layers of insulation and surrounded by a metal sheath. The capacitance appears in parallel with the load and demands reactive power that must flow through the cable's resistance, causing power losses. There are also *dielectric losses* in the cable's insulation materials. As cable length increases, reactive power flow takes up more and more of the cable's current-carrying capacity which, of course, should be primarily devoted to transporting real power from wind turbines to the shore. The cable could be increased in size and weight to compensate, but beyond a certain length this is no longer economic.

The capacitance problem is solved by HVDC transmission. Cable capacitance is charged only once when the system is first energized; thereafter, the entire current-carrying capacity can be used to transmit useful power. HVDC has several further advantages:

- It needs fewer conductors than a three-phase AC system.
- For a given peak (rated) voltage, each conductor can carry more power. DC operates constantly at peak voltage, but AC power depends on the rms voltage, which is only 71% $(1/\sqrt{2})$ of the peak.
- HVDC isolates the wind farm's AC system from that of the grid. The offshore and onshore systems need not be synchronized, offering advantages in stability and the independent control of real and reactive power.

The main disadvantage of HVDC transmission is the need for expensive power converters at the sending and receiving ends of the link. Fortunately, electronic power converters—rectifiers and inverters based on high-power transistors and other solid-state devices[2]—have made great strides in the past 30 years. The underlying principles and technology have many parallels with the power converters used for variable-speed wind turbines (see Section 3.4.2). However, we are now talking about the much greater power levels of whole wind farms, rather than that of individual turbines.

High-power submarine cables, whether transmitting AC or DC, must work for many years in a hostile environment with extremely high levels of electrical stress. For example, a HVDC link rated at 1 GW and operating at 200 kV must carry up to 5000 A. The insulation must withstand 200 kV without breakdown, and the conductors must carry 5000 A without overheating or wasting too much precious power. Present interest in HVDC for offshore wind is giving a big boost to the development of new cable designs and materials,[2] including fiber optics for data transmission. In land, HVDC transmission operating at 800 kV is in operation today, and the industry is developing 1.1 MV technology; see Chapter 7, Grid integration, for further discussion on HVDC.

Appendix 4.A: Specifications of turbines

Generic turbines are research-based turbines and details are provided for four such turbines in Table 4.A.1. They are NREL 5 MW, LW 8 MW, and DTU 10 MW and IEA 15

Table 4.A.1 Characteristics of wind turbines.

Turbine	NREL	LW	DTU	IEA	Unit
Power rating	5	8	10	15	MW
Rotor diameter	126	164	178.3	240	m
Hub height	90	110	119	150	m
Rotor speed range	6.9–12.1	6.3–10.5	6–9.6	5-7.56	rpm
Cut-in, rated	3, 11.4	4, 12.5	4, 11.4	3,10.59	m/s
Cut-out wind speed	25	25	25	25	m/s
Nacelle mass	296.78	375	551.56	991	tons
Blade mass	17.74	35	41.72	65	tons
Tower mass	347.46	558	605	860	tons
Tower height	87.6	106.3	115.6	135	m
Tower top diameter	3.87	5	5.5	7.94	m
Tower bottom diameter	6	7.7	8	10	m

NREL: National Renewable Energy Laboratory

LW: LEANWIND Project funded by European Commission

DTU: Denmark Technical University

IEA: International Energy Agency

Other sources of information on commercial turbines:

1. Section 4.5 (Offshore Wind Farm Case Studies)
2. Chapter 1 of Bhattacharya (2019)
3. https://www.ge.com/research/newsroom/ge-researchers-unveil-12-mw-floating-wind-turbine-concept
4. https://www.ge.com/renewableenergy/wind-energy/offshore-wind
5. https://www.siemensgamesa.com/en-int/products-and-services
6. https://www.vestas.com
7. https://www.4coffshore.com/offshorewind/

Appendix 4.B: Details of offshore wind farms on different types of foundations

This appendix provides details of various offshore wind farms. Table 4.B.1 provides data on UK offshore wind farms showing the year of commissioning, turbine used, mean sea depth, and types of foundation. Table 4.B.3 lists where gravity-based foundation (GBS) is used.

V90-3MW [Vestas 3 MW Rated Power Turbine with 90 m Rotor Diameter]

SWT-3.6-120 [Siemens Wind Turbine 3.6 MW Rated Power with 120 m rotor diameter] Tables 4.B.2 and 4.B.4.

Table 4.B.1 UK wind farms.

Name	Commissioned	Turbine manufacturer	Turbine model	Distance from shore (km)	Mean sea depth (m)	Foundation types
Robin Rigg	2009	Vestas	V90-3.0 MW	11	6.5	Monopile
Blyth Offshore	2000	Vestas	V66-2 MW	1	8.5	Monopile
North Hoyle	2003	Vestas	V80-2 MW	7	7.5	Monopile
Scroby Sands	2004	Vestas	V80-2 MW	2.3	5	Monopile
Kentish Flats	2005	Vestas	V90-3.0 MW	8.5	3.5	Monopile
Barrow	2006	Vestas	V90-3.0 MW	7	17.5	Monopile
Burbo Bank	2007	Siemens	SWT-3.6-120	7	4	Monopile
Beatrice	2007	REPower	5 MW	13	42.5	Monopile
Lynn and Inner Dowsing	2009	Siemens	SWT-3.6-120	5	5.5	Monopile
Rhyl Flats	2009	Siemens	SWT-3.6-120	8	8	Monopile
Thanet	2010	Vestas	V90-3.0 MW	10	22.5	Monopile
Gunfleet Sands 1	2010	Siemens	SWT-3.6-120	7	9	Monopile
Gunfleet Sands 2	2010	Siemens	SWT-3.6-120	7	9	Monopile
Walney	2010	Siemens	SWT-3.6-120	15	24.5	Monopile
Greater Gabbard	2012	Siemens	SWT-3.6-120	22	26	Monopile
Sheringham Shoal	2012	Siemens	SWT-3.6-120	16	16	Monopile
Ormonde	2012	REPower	5 MW	10	18.5	Jacket
Teesside	2013	Siemens	SWT-2.3	1.5	12	Monopile
Lincs	2013	Siemens	SWT-3.6-120	6	13	Monopile
London Array	2013	Siemens	SWT-3.6-120	20	12.5	Monopile
Gunfleet Sands 3	2013	Siemens	SWT-6.0-154	7	9	Monopile
Methill Demonstrator	2013	Samsung	S7.0-171	0.05	5	Jacket
West of Duddon Sands	2014	Siemens	SWT-3.6-120	14	21	Monopile
Humber Gateway	2015	Vestas	V112-3.0MW	8	13	Monopile
Gwynt y Mor	2015	Siemens	SWT-3.6-120	13	24	Monopile
Westermost Rough	2015	Siemens	SWT-6.0-154	25		Monopile
Dudgeon	2017	Siemens	SWT-6.0-154	32	21.5	Monopile
Hywind Scotland	2017	Siemens	SWT-6.0-154	25	105	Floating
Blyth Offshore Demonstrator	2017	Vestas	V164-8 MW	6.5	52.5	Monopile
Burbo Bank Extension	2017	Vestas	V164-8 MW	7	11.5	Monopile
Rampion	2018	Vestas	V112-3.45 MW	13	20	Monopile
Race Bank	2018	Siemens	SWT-6.0-154	17	20.5	Monopile
Galloper	2018	Siemens	SWT-6.0-154	27	28	Monopile
Walney Extension Part 1	2018	Siemens	SWT-7.0-154	19	24.5	Monopile
Walney Extension Part 2	2018	Vestas	V164-8.25MW	19	24.5	Monopile
European Offshore Wind Deployment Centre	2018	Vestas	V164-8MW	3	27.5	Suction Bucket
Beatrice Extension	2019	Siemens	SWT-7.0-154	13	53	Tetrapod
East Anglia One	2020	Siemens	SWT-7.0-154	43	5	Jacket
Hornsea One	2020	Siemens	SWT-7.0-154	120	28.5	Monopile

Fourth Column Turbine Model: Each model has letter/letters and two numbers. Letter is related to the manufacturer of the turbines. The smaller number is the rated capacity, and the larger number is the rotor diameter.

Table 4.B.2 The diameter of monopiles for various UK projects.

Name	Turbine model	Foundati on types	Pile diameter (m)
Robin Rigg	V90-3.0 MW	Monopile	4
North Hoyle	V80-2 MW	Monopile	4
Scroby Sands	V80-2 MW	Monopile	4
Kentish Flats	V90-3.0 MW	Monopile	4
Barrow	V90-3.0 MW	Monopile	5
Burbo Bank	SWT-3.6-120	Monopile	5
Lynn and Inner Dowsing	SWT-3.6-120	Monopile	5
Rhyl Flats	SWT-3.6-120	Monopile	5
Thanet	V90-3.0 MW	Monopile	5
Gunfleet Sands 1	SWT-3.6-120	Monopile	5
Gunfleet Sands 2	SWT-3.6-120	Monopile	5
Walney	SWT-3.6-120	Monopile	6
Greater Gabbard	SWT-3.6-120	Monopile	6
Sheringham Shoal	SWT-3.6-120	Monopile	6
Teesside	SWT-2.3	Monopile	5
Lincs	SWT-3.6-120	Monopile	5
London Array	SWT-3.6-120	Monopile	5
Gwynt y Mor	SWT-3.6-120	Monopile	5
Westermost Rough	SWT-6.0-154	Monopile	6.25
Walney Extension Part 1	SWT-7.0-154	Monopile	7
Walney Extension Part 2	V164-8.25 MW	Monopile	7

Table 4.B.3 Details of offshore wind farm supported on GBS either operating or planned.

Rated power (MW)	Country	Project	Water depth (m)	Hub height (m)	Distance to shore (km)
0.45	Denmark	Vindeby	2–4	35	2.5
0.5	Denmark	Tuno Knob	3–7	45	5.5
2	Denmark	Middelgrunden	3–6	-	-
2.3	Denmark	Nysted-I (Rodsand I)	6–9	69	10.7
2.3	Denmark	Rodsand II (Nysted II)	6–12	68.5	9
2.3	Sweden	Lillgrund	4–13	68	10
3	Germany	Breiitling	0.5	80	0.3
3	Sweden	Karehamn	21	80	7
3	Sweden	Vindpark Vanern	10	90	10.1
3	Denmark	Sprogo	6–16	70	10.6
3.6	Denmark	Avedore Holme	2	-	0.5
5	Belgium	Thornton Bank Phase 1	25	94	27
8	United Kingdom	Blyth Demonstrator	38	-	-
5	Netherlands	Tromp Binnen	25.5	89	65
5	Brazil	Asa Branca 22	14–21	-	35.7
5	Brazil	Asa Branca 23	12–24	-	36.9
3	Finland	Kemi Ajos I	1–7	88	2.6
3	Finland	Kemi Ajos II	1–7	88	2.6
2	Finland	Pori I	16–19	80	6.6
5	France	Calvados	19–22	-	11
6	Ireland	Oriel Wind Farm	17–27	-	7.8
5.7	France	Poweo	19–22	90	11

Table 4.B.4 Table of some key floating wind turbine projects (operating on planned).

#	Name	Location	Type of foundation	Total capacity (MW)	Type of turbine	Water depth (m)	Installed
1	Wind Float Atlantic	Portugal	Semi-submersible	25	MHI V164-8.4 MW	100	2020
2	Kincardine	UK	Semi-submersible	50	5× (V164-9.5 MW) 1× (V80-2 MW)	60–80	2021
3	Hywind Tampen	Norway	Spar-buoy	88	SG 8.0–167 DD	260–300	2022
4	Hywind Scotland	UK	Spar-buoy	30	SWT-6.0–154	95–120	2017
5	KF wind	Korea	Semi-submersible	1200	IEA 15-theoretical 16 MW	>200	-
6	Redwood coast offshore wind project	USA	Semi-submersible	120–150	Not known	610–1100	2024
7	Hibiki wind farm	Japan	Barge	3	Demonstration project	55	2030

Self-assessment questions

Q4.1 What are the main components of offshore wind farm?

Q4.2 Why are foundations important?

Q4.3 Why does the tower height increase with turbine-rated power?

Q4.4 Discuss the items that are to be considered in the design of offshore wind turbine foundations?

Q4.5 What are the current technologies to store wind energy?

Q4.6 What are the merits and demerits of large offshore wind turbines?

Q4.7 Are wind farms built close to population centers?

Q4.8 What are the advantages of floating systems?

Problems

4.1 A wind farm site comprises of 50 wind turbines of 8 MW capacity each. The average wind speed is 7 m/s at 10 m height. Find the average power output assuming a density of air of 1.223 kg/m^3. Make assumption on the blade diameter and other data you may need. Assume turbine efficiency and ignore wake effects.

4.2 When the wind farm will be decommissioned for the site in Problem 4.1, it is planned to install 40 wind turbines of 10 MW machines. Find the average power output assuming a density of air of $1.223\,\text{kg/m}^3$. Make assumption on the blade diameter and other data you may need. Use same assumptions as problem 4.1 and compare the results.

4.3 For offshore wind turbine ranging between 2.3 and 14 MW, plot the rotor diameter versus the turbine rated power (MW). Making an assumption that the blade is half the rotor diameter, plot the blade length versus the turbine rated power. What are the manufacturing and installation challenges as the blades get longer?

4.4 For the three turbines in Appendix 4.A, find the blade tip speed for the following conditions: (a) cut-in condition; (b) rated power condition. What are the implications if the tip speed is too high.

4.5 For offshore wind turbine ranging between 2.3 and 14 MW, plot the hub height versus the turbine rated power (MW). You may take 30 m as the average water depth and monopile foundation. How does the stiffness of the tower change with turbine size? How will the turbine size affect the design foundation loads? Make assumptions for data required.

4.6 For offshore wind turbine ranging between 2.3 and 14 MW, plot the rotor Nacelle assembly (RNA) mass versus the turbine rated power (MW). How does this affect the design. Combining the data on RNA mass together with the stiffness of the tower (based on Question 4.5), comment on the vibration period of the system.

4.7 Using the data in Appendix 4.B, comment on the historic growth of the offshore wind farm industry in terms of turbine size, system (grounded or floating) used, and distance from the sea. Can this be used as a guidance for countries willing to develop offshore wind farm.

4.8 Plot the existing data of wind turbine installation of existing offshore wind farms in terms of turbine capacity, water depth, and foundation chosen and comment on the graph. Apart from using the data in the Appendix 4.B.2, you can get more data from the literature. Can this graph be used as a preliminary guidance for choosing foundations?

Answers to questions

Q4.1 The main components are (a) Turbines and its support structure; (b) offshore substations which collect all the power from the individual turbines; (c) array cable which transmits powers from turbines to the substation; (d) export cable which transfers power from offshore substation ultimately to the grid.

Q4.2 Foundations support the loads and provide the stability of the whole system. For example, the thrust of the blades is applied at the hub level and is taken by the springs.

Q4.3 As the turbine rated power increases, the blade length increases. It is important that the blades does not experience the wave action. There must a considerable air gap between the tip of the blade and the surface of the water.

Q4.4 The main considerations for choosing foundations are turbine size, seabed condition, ground profile, water depth, available fabrication, and installation expertise.

Q4.5 Batteries and Hydrogen.

Q4.6 Pros: Can produce large energy per unit use of materials, less use of seabed space, and can be installed faster and less offshore time to install Cons: Large sizes of components (longer blades, larger diameter tower, and foundation) need large fabrication facilities, need specialized vessels (heavy lift vessels with large crane), and expertise to install.

Q4.7 Yes, they are close to population centers (see Figure 4.8 for example) Examples: (a) South Fork (United States): Location: Proposed offshore wind farm to be located off the coast of Long Island (New York) and close to the New York metropolitan area. (b) London Array (UK): Located in the Thames Estuary, off the coast of Kent and Essex in the United Kingdom. Proximity to greater London, one of the largest metropolitan areas in Europe.

Q4.8 There are many advantages: (a) energy can be harvested from deeper water and further away from offshore without disrupting the shipping channels. (b) Foundation requirements needs are lot simpler, and the sizes are lot smaller than bottom-fixed turbines. (c) Easy to install and occupy less sea-bed space. (d) Major repair and maintenance of bottom fixed are done in situ and are expensive. However, floating systems can be towed to the dock and repaired.

References

1. Gustavson. Limits to wind power extraction. *Science*, 204(4388), 13–17 (1979).
2. L. Arany *et al.* Closed form solution of Eigen frequency of monopile supported offshore wind turbines in deeper waters incorporating stiffness of substructure and SSI. *Soil Dynamics and Earthquake Engineering*, 83(83), 18–32 (2016). doi: 10.1016/j.soildyn.2015.12.011.
3. S. Bhattacharya. *Design of Foundations for Offshore Wind Turbines*, 1st edition. Wiley: Chichester (2019). doi: 10.1002/9781119128137.
4. E. Gaertner *et al.* IEA Wind TCP Task 37-Definition of the IEA Wind 15-Megawatt Offshore Reference Wind Turbine-Technical Report. Denver. doi: NREL/TP-5000-75698 (2020).
5. F. Manzano-Agugliaro *et al.* Wind turbines offshore foundations and connections to grid. *Inventions*, 5(1), 8 (2020). doi: 10.3390/inventions5010008.
6. E. Osmanbasic The Future of Wind Turbines: Comparing Direct Drive and Gearbox, Engineering .com. https://www.engineering.com/story/the-future-of-wind-turbines-comparing-direct-drive-and-gearbox (Accessed on October 11, 2021) (2020).
7. Power Engineering Direct Drive vs. Gearbox: Progress on Both Fronts. https://www.power-eng.com/2011/03/01/direct-drive-vs-gearbox-progress-on-both-fronts/#gref (Accessed on October 11, 2021) (2011).
8. Letcher. *A Handbook for onshore and offshore Wind Turbines*, 2nd edition (2023).
9. Nikitas, G. A study on soil-structure interaction of offshore wind turbine foundations (Doctoral dissertation, University of Surrey) (2020).

5 Offshore wind farm engineering

Following the considerations of site wind resources and wind turbine technologies described in Chapter 4, this chapter describes what it takes to develop a wind farm project. The basic elements include concept feasibility assessment, site investigation, engineering design for foundations, array design (i.e., layout of the wind farm), project commissioning, operation and maintenance plan, and end-of-life decommissioning plan. Of course, project economics are also paramount, and this will be discussed in Chapter 8.

The concept feasibility assessment examines the viability of the project. It entails a preliminary investigation of proposed project along the lines of site resources, type of turbine and system (grounded or floating), and electrical infrastructure (cable and substation type, cable route, and grid connection). The feasibility assessment is followed with a site-specific survey so that detailed design can be carried out. One of the important aspects is a reliable ground survey, which is required not only for designing but also for planning of installation. The project plan development follows with detailed design and commissioning of wind farm (i.e., installation of turbines on the support structure), operation and maintenance (OPEX) for power production, and finally a plan for the eventual end-of-life decommissioning.

The key aspects in offshore wind farm engineering are as follows:

(a) Wind farm design (i.e., layout of the wind power generation assets, i.e., turbines, transmission assets, i.e., substation and cables and in some cases planning the energy storage assets). The associated engineering decisions are spacing of the turbines, location of the substations, and finally the cable (both interarray and export cable) layout. This step involves region-specific techno-economic analysis and requires optimization. The essential data required for such analysis are turbine type, wind and wave data, geology of the seabed, local skills, availability of port, and construction logistics including vessel availabilities.

(b) To enable efficient and economic wind farm design, detailed site investigations are required, which involves collection of water depth information (bathymetry), wind field, wave information, and ground profile data. Wind data are needed to estimate the energy yield from the wind farm as well as thrust loads on the structure and

Onshore and Offshore Wind Energy: Evolution, Grid Integration, and Impact, Second Edition.
Vasilis Fthenakis, Subhamoy Bhattacharya, and Paul A. Lynn.
© 2025 John Wiley & Sons Ltd. Published 2025 by John Wiley & Sons Ltd.
Companion website: www.wiley.com/go/fthenakis/windenergy2e

foundation. Wave data are used to estimate the hydrodynamic loads on the structure and the foundations. Ground data on geology and soil conditions is necessary to choose the appropriate foundations, cable sizes, and scour protection design.

(c) Choice of the wind turbine system: The main decision is to choose grounded system (i.e., anchored to the seabed) or floating system for a given water depth and the ground condition. For example, within the grounded system—jacket or monopile and examples of floating system are barge, spar, semi-sub, tension leg platform (TLP), see Figures 4.24 and 4.25.

(d) Choosing the appropriate foundation system, associated design, and the method of installation.

(e) Choice of substation system (AC or DC) and capacity needed.

(f) Methods of installation of foundation and assembly of the turbine and grid connection.

(g) Planning operations and maintenance of these assets.

(h) Decommissioning and end-of-life aspects.

5.1   Conceptual design of wind arms

In the concept design stages, the main drivers are as follows:

(a) Annual energy yield (AED), that is, revenue generated by selling electricity over a year. This is dependent on the wind resource at the location and the turbine-rated power. The readers are referred to the World Bank wind resource map (Figure 4.15 in Chapter 4).

(b) CAPEX and OPEX. This is mainly dependent on the geology of the site, water depth, turbine type, technology employed (floating or bottom-fixed system), and the installation vessel requirements and availability.

Concept design is multiparameter optimization process, and most methods use the levelized cost of energy (LCOE) ($/MWh) as the optimization goal. It is important to highlight the variables and most important ones are (i) turbine availability; (ii) wake (interactions among the individual turbines can have significant impact on the overall power generation); (iii) cable lengths and grid connection costs; (iv) CAPEX (cost models for the turbines, support structures, cables); (v) operation expenditure (OPEX), installation and decommissioning.

In terms of a given site, the key factors that influence CAPEX are: (i) turbine size, (ii) water depth, (iii) distance to shore, (iv) geology of the site (soil types), and (v) installation and commissioning costs.

5.1.1   ASIDE 1: spacing of turbines and changes in geology and water depth

Larger spacing between turbines will yield higher power output and AED. However, as the turbines are spaced, there will be increase in CAPEX cost due to longer cable lengths, change in water depth, and/or change in geology. Figure 5.1 shows recommended typical spacing

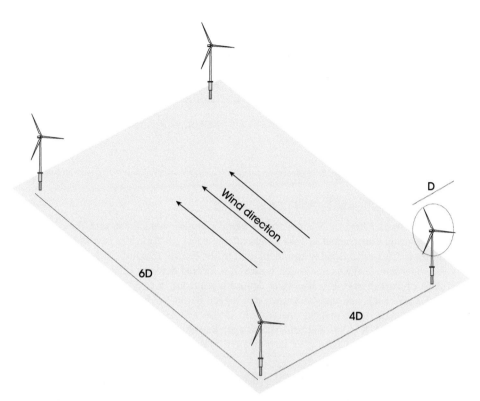

Figure 5.1 Typical spacing of turbines (Source: Bhattacharya[1]/with permission of John Wiley & Sons).

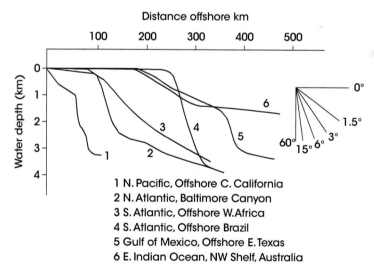

Figure 5.2 Water depth variation with distance from the shore (Source: Bhattacharya[1]/with permission of John Wiley & Sons).

of turbines for preliminary design purposes; spacing of 6D (D is rotor diameter) is recommended along the main wind direction. For a typical 8 MW turbine, the rotor diameter is 164 m, and therefore, the spacing will be approximately 1 km. Figure 5.2 shows the change in water depth for various oceans. It is clear from the figure that offshore California has a steep continental slope, and therefore, many of the developments are more likely to be floating. It is prudent to highlight examples of some wind farms showing the variability in layout and other aspects.

5.2   Cases studies

5.2.1   Dogger bank offshore wind farm

Dogger Bank Offshore Wind Farm is world's largest wind farm with 3.6 GW capacity and is being constructed in UK waters off the coast of Yorkshire; see Figure 5.3 for site location. This is an isolated sandbank within the central to southern North Sea spanning UK, German, Danish, and Dutch waters. It is located between 125 and 290 km off the east coast of Yorkshire (UK). The water depth ranges from 18 to 63 m. There are three phases of 1.2 GW each and will be known as Dogger Banks A, B, and C. The first turbine at Dogger Bank A has started operating in October 2023 and is producing electricity that is fed to the UK's national grid via Dogger Bank's high-voltage direct current (HVDC) transmission system. The wind farm has GE Haliade-X 13 MW turbines with 107 m long

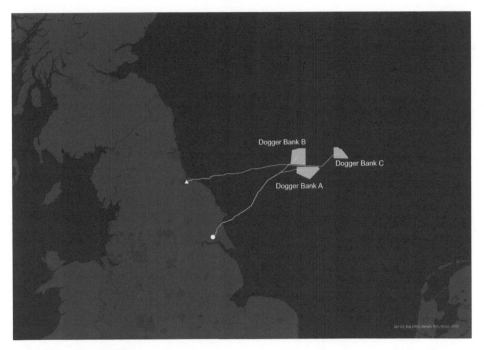

Figure 5.3 Dogger Bank Wind Farm.

Figure 5.4 Dogger Bank Wind Farm.

blades (Figure 5.4). Dogger Bank has the potential to improve resilience of electricity in Europe by linking coordinated network to Europe through interconnectors. The large wind farm can also support large-scale hydrogen production and connect to a wider hydrogen network that will help decarbonization of energy-intensive industries.

5.2.2 Westermost rough wind farm

Figure 5.5 shows the location of Westermost Rough Wind Farm also in the UK. The nameplate capacity is 210 MW with 35 turbines of 6-MW-rated power. Figure 5.6 shows the vertical ground profile of Westermost Rough Wind Farm where the changes in geology (defined by BC, Bo, Sw, and Ch) and water depth may be noted. The water depth at the site varied between 16 and 26 m.

If the geology of the site changes, it is likely that the theoretical optimized foundation dimensions will also change as different ground offers different level of resistance. Furthermore, it is expensive and operationally challenging to have many types of foundations for the same wind farm. The challenge is from the point of fabrication, transportation, and installation. It will be uneconomic to design 35 foundations individually having different dimensions, which may have a low material content but a higher fabrication cost as economy of scales is lost. Therefore, overall optimization is necessary.

Another important consideration is the possible impact of wind farm on shipping, fishing, and other activities. Figure 5.7 shows the 28 days maritime traffic survey of the Westermost Rough site, carried out using a combination of shore-based radar, AIS, and visual observations taken from the nontechnical summary required for obtaining consent/permissions outside the main traffic. This shows that Westermost Rough wind farm is in a fishing area.

5.2.3 Karehamn offshore wind farm

The Karehamn Offshore Wind Farm, the layout of which is shown in Figure 5.8, is located in the Swedish part of the Baltic Sea, approximately 7 km away from the coastal town of

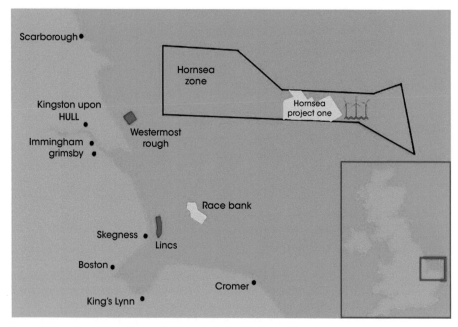

Figure 5.5 Location of four offshore wind farms including Westermost Rough.

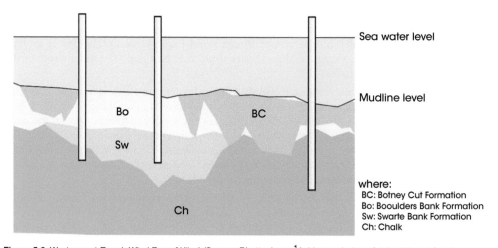

Figure 5.6 Westermost Rough Wind Farm (Wiley) (Source: Bhattacharya[1]/with permission of John Wiley & Sons).

Kårehamn. This is a 48 MW wind farm, which was connected to the grid in 2013. Sixteen Vestas V112 turbines with a hub height of 80 m and a capacity of 3 MW each are installed on gravity-based foundations in water depths of up to 20.5 m. The layout of the farm is a single line of 16 turbines (Figure 5.8).

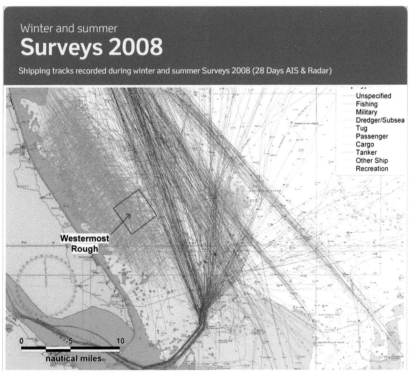

Winter and summer
Surveys 2008
Shipping tracks recorded during winter and summer Surveys 2008 (28 Days AIS & Radar)

Unspecified
Fishing
Military
Dredger/Subsea
Tug
Passenger
Cargo
Tanker
Other Ship
Recreation

Westermost
Rough

0   5   10
nautical miles

Figure 5.7 Navigation channels around the Westermost Rough Offshore Wind Farm https://www.eib.org/attachments/registers/53897501.pdf.

5.2.4 West of Duddon sands

The West of Duddon Sands Offshore Wind Farm comprises 108 wind turbines with a total installed capacity of 389 MW located in the East Irish Sea approximately 14 km from the nearest coast on Walney Island, Cumbria. It is situated near the wind farms of Barrow, Walney, and Ormonde. As shown in Figure 5.9, there is an open space in the middle of the farm because of igneous rock shape intrusion forming dikes. There are uncertainties associated with the drilling through the rocks, and therefore, it was prudent to avoid that region.

5.2.5 Development of offshore wind in the USA

The US East Coast, and particularly the Atlantic Outer Continental Shelf (OCS), is developing commercial offshore wind due favorable wind resource, close to major population centers as well as shallow water depths (<100 m). BOEM subdivides the Atlantic OCS into four planning areas (i) North Atlantic, (ii) Mid-Atlantic, (iii) South Atlantic, and (iv) Straits of Florida, see Figure 5.10. Figure 5.11 provides tentative details of the wind farms in these regions, and Figure 5.12 shows the corresponding wind speed categories.

Figure 5.8 Karehamn Wind Farm Array. It may be noted that this is a single curved line of 16 turbines (Source: RWE/https://se.rwe.com/en/projects-and-locations/wind-farm-karehamn//last accessed February 14, 2024).

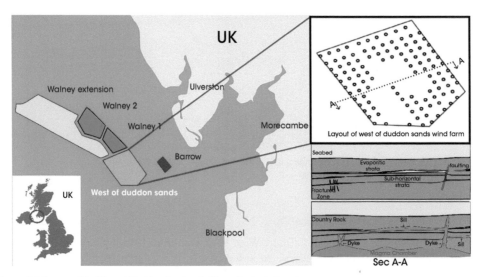

Figure 5.9 Layout of the West of Duddon Sands wind farm (Author's figure).

5.2.6 Wind farm development in China

Figure 5.13 shows offshore wind farm development off the east coast of China. Preliminary estimates show that there is 200 GW of potential wind power within water depth of 25 m and an additional 300 GW available in water depth between 25 and 50 m. Figure 5.14 shows the bathymetry around the Chinese seas with water depths. It is worth noting that the electricity demand is concentrated in the developed eastern coastal regions; thus, offshore wind development appears to be an ideal solution to meet a growing demand.

In the concept design stage, the main aim is to develop a project design envelop by narrowing down the options on number of turbines and the farm layout (i.e., array design), type of support structures, substation locations, inter-array cables, and the export cable. The concept design will enable the site investigation schedule and is discussed in the next section.

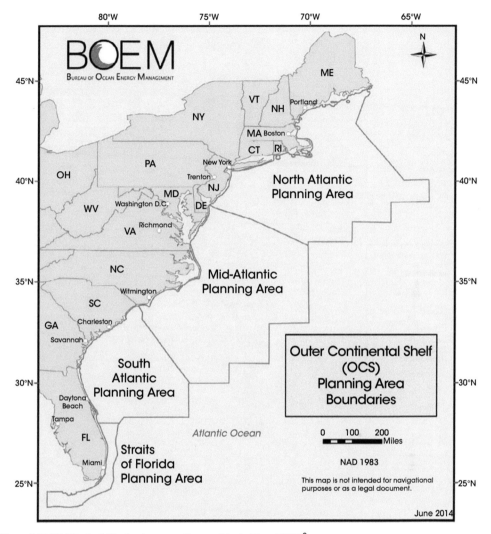

Figure 5.10 US Atlantic OCS planning areas (Source: Adapted from BOEM[2]).

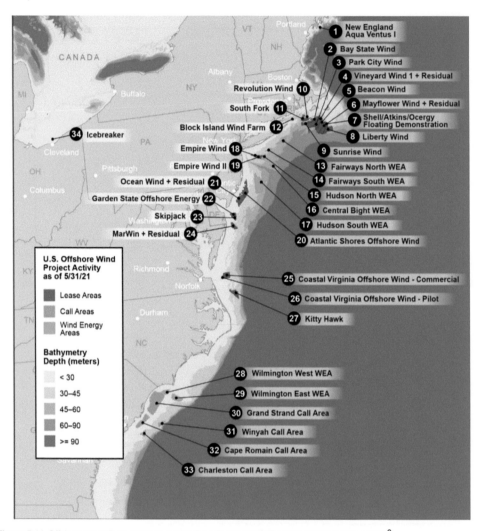

Figure 5.11 Offshore Wind Farm development in East Coast of US (Source: Adapted from BOEM[2]).

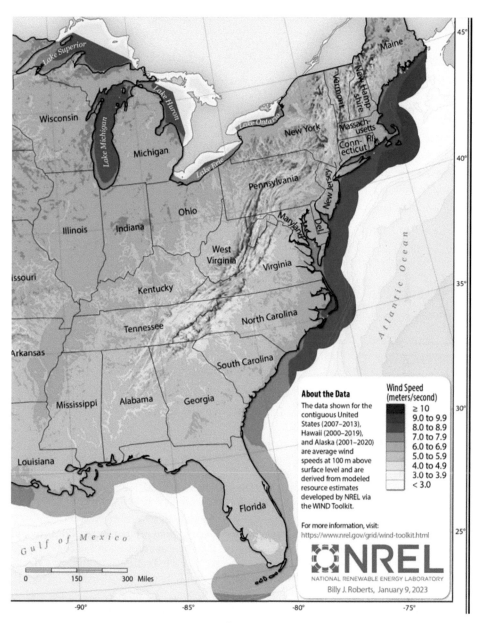

Figure 5.12 Wind speeds (Source: NREL Wind Exchange[3]).

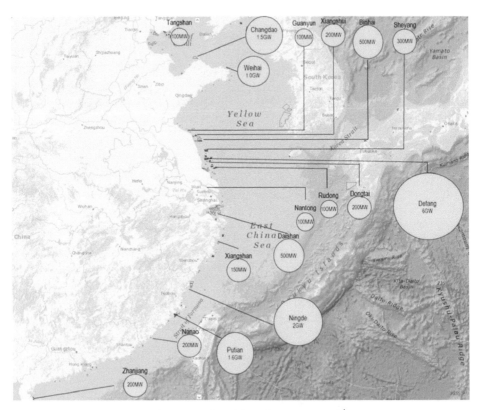

Figure 5.13 Wind farm development in China (author figure). From Bhattacharya et al.[4].

5.3   Detailed site investigation

Site investigation follows the conceptual design phase; it typically comprises of the following steps:

(1) *Control system design and revenue generation calculations.* Wind, wave, and current time series data for the location are acquired so that the scour protection and control systems for the turbine can be designed. The time series of wind can also support the estimation of the wind energy that can be produced in the lifetime.

(2) *Foundation design for the turbine support structure and offshore substation structure.* There can be wide variations in geology across wind farms as well as at a particular site. Figure 5.15 shows geological profile for 12 European wind farms where turbines of different capacity are installed. The foundations for these wind farms are monopiles with different dimensions (i.e., length, diameter, and wall thickness). Furthermore, the foundations are also embedded in different depths. The figure shows that ground is variable and highlights the importance of good understanding of the site.

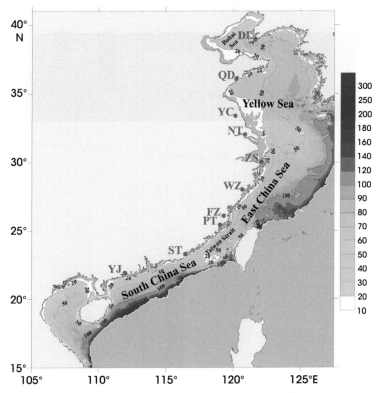

Figure 5.14 Bathymetry (m) of the waters adjacent to China following Lian *et al.*[5]. The regions (water depth < 10 m and >300 m) are covered by gray. The red dots are cities with OWF deployment plans. From south to north, the cities are Yangjiang (YJ), Shantou (ST), Putian (PT), Fuzhou (FZ), Wenzhou (WZ), Zhoushan (ZS), Nantong (NT), Yancheng (YC), Qingdao (QD), and Dalian (DL). (Source: Zhang lian et al./MDPI/CC BY 4.0/https://www.mdpi.com/2077-1312/10/12/1872).

(3) *Cable route assessment.* There are two types of cables (interarray and export cables) to transmit the electricity generated to the onshore grid. Investigation is necessary along the cable route for design purposes and installation planning. The major considerations are scour prediction, cable protection, and turbidity changes in the water. An example of scour-related early decommissioning of wind farm is discussed in Aside (2) for Robin Rigg Wind Farm.

(4) *Planning and analysis of installation of foundations.* Jack-up rigs are often used to install foundation, and detailed ground profile is necessary for jack-up rig leg penetration analysis. If the soil is soft, such systems are not suitable. Another example is whether suction can be used to install foundations.

(5) *Geo-hazard assessment.* Slope stability (submarine slides) or presence of gassy soils. Seismic analysis would require prediction of liquefaction susceptibility of the ground and site-response analysis, unexploded ordnance (UXO).

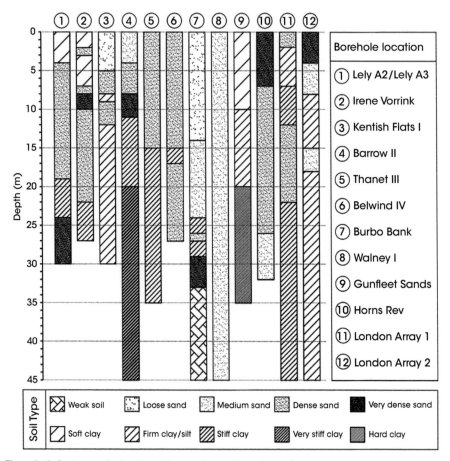

Figure 5.15 Geology profile for 12 wind farms in Europe (Author's own figure).

5.3.1 ASIDE 2: observed scour in Robin Rigg Wind Farm which had to be decommissioned mid-way in the lifecycle

The Robin Rigg wind farm, commissioned in 2009, is located 13 km Northwest of the Port of Workington (North England, UK) and was composed of 58 Vestas V90 wind turbines, each 3 MW. The seabed consists of different sandy deposits (17–21 m in depth) located across the estuary; these are interbedded with silts and clays and lie above glacial till. During these years, scour around the foundations (monopiles) developed as shown in Figures 5.16 and 5.17. The considerable reduction of the embedment length of the foundation resulted in the drop in natural frequency of the system which implied a significant structural risk, which led to the decommission of the turbine.

5.3.2 Variability in ground profiles

Understanding the ground conditions is fundamental in designing of foundations. Good quality ground significantly de-risks the foundation construction and performance and is

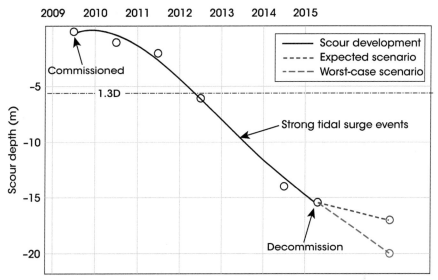

Figure 5.16 Development of the scour in the Robin Rigg Wind Farm between 2009 and 2015 following Menéndez-Vicente et al.[6]. The reference level is 2009, and the depths are given in m. (Source: Menéndez-Vicente et al.[6]/with permission of Elsevier/https://doi.org/10.1016/j.soildyn.2023.107803).

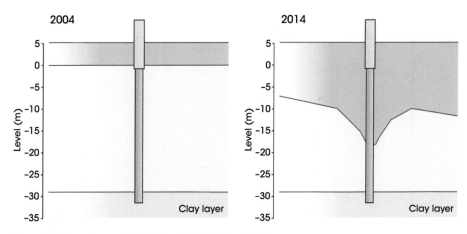

Figure 5.17 Scour issues in Robin Rigg Wind Farm (own figure).

conducive to the development of offshore energy. Ground is variable and depends on the formation. The geology of Northern Europe is generally very good due to the glacial age while the ground in Chinese seas is very poor. Figure 5.15 shows geology and ground profile of 12 wind farms in Europe, and Figures 5.18 and 5.19 show the ground profile for Chinese Seas.

Figure 5.18 shows the ground profile of Bohai Sea, which is formed by sediment deposition brought from mountains and Yellow River. Therefore, the ground is soft silty clay or clayey silt. Figure 5.19 shows the ground profile of the Taiwan strait.

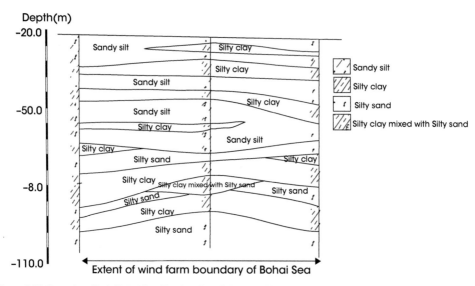

Figure 5.18 Ground profile in Bohai Sea (Northen East China, see Figure 5.12 for location) (Source: Bhattacharya et al.[4]/ with permission of Elsevier).

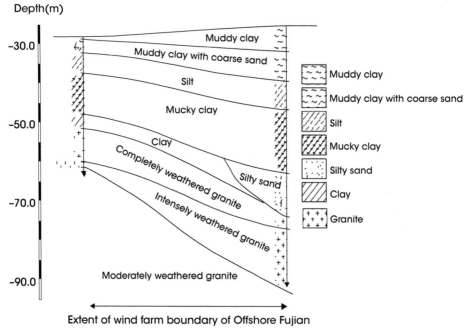

Figure 5.19 Ground profile in offshore Fujian Sea or Taiwan Strait (Source: Bhattacharya et al.[4]/with permission of Elsevier).

Table 5.1 Wind farm specifications (tentative) in the East Coast of USA.

Wind farm	Distance from shore (km)	Planned total capacity (MW)	Hub heights (m)	Turbines	Type of foundation
CVOW Commercial Project (Number 25 in Figure 5.11)	43.5	2640	134–154	176 Siemens Gamesa SG 14–222 DD Turbines	Monopile
Revolution Wind (Number 10 in Figure 5.11)	24	704–880	115–156	~80 × 11 MW turbine	Monopile
Ocean wind	24	1100	156	~98 × Haliade-X 12 MW	Monopile
Vineyard wind	24	804	109–144	~62 × GE Haliade-X 13.6 MW	Monopile/Jacket
Empire wind	24	2100	160	~138 × Vestas V236-15 MW	Monopile
Sunrise	48	880		~110 × SG 8–167	Monopile
South fork	26	132		~11 × SG 11–167	Monopile

On the other hand, the geological history of US Atlantic OCS (i.e., East Coast of USA; Table 5.1) region is highly variable with significant layering and spatial variations due to diverse sediment origin and post-deposition history (Figure 5.10). There is boulder in the northern part and calcareous soils in the southern part of the coast. A new type of soil called "green sand" is encountered in three of the wind farm development (Empire Wind, Sunrise, and

Beacon Wind) off coast of Massachusetts and one south of Long Island (New York) shown in Figure 5.20. Green sand is coarse-grained materials having Glauconite mineral (which is an iron potassium mica) and is characteristically green. Green sand has the tendency to transform from a stiff, high permeability coarse-grained material to a weak, low permeability fine-grained material due to particle crushing. In terms of wind farm construction, this material can pose significant risks during installation of foundations for grounded systems (i.e., bottom-fixed turbines). Further details on the characteristics can be found in Westgate et al.[7].

5.3.3 Stages of investigation for ground model

Site investigation consists of geological, geophysical, and geotechnical, and they are described in this section:

5.3.3.1 Geological study

A study based on geological history of formation can be the basis of planning site investigation and concept study and is used to create the first ground model. Ground-related hazards depend on the location of the wind farm. This can be classified as

(a) Man-made hazards: examples include existing infrastructure such as oil and gas pipelines either on the seabed or buried, communication cables, ship wrecks, jack-up footprints, unexploded ordnance, etc.

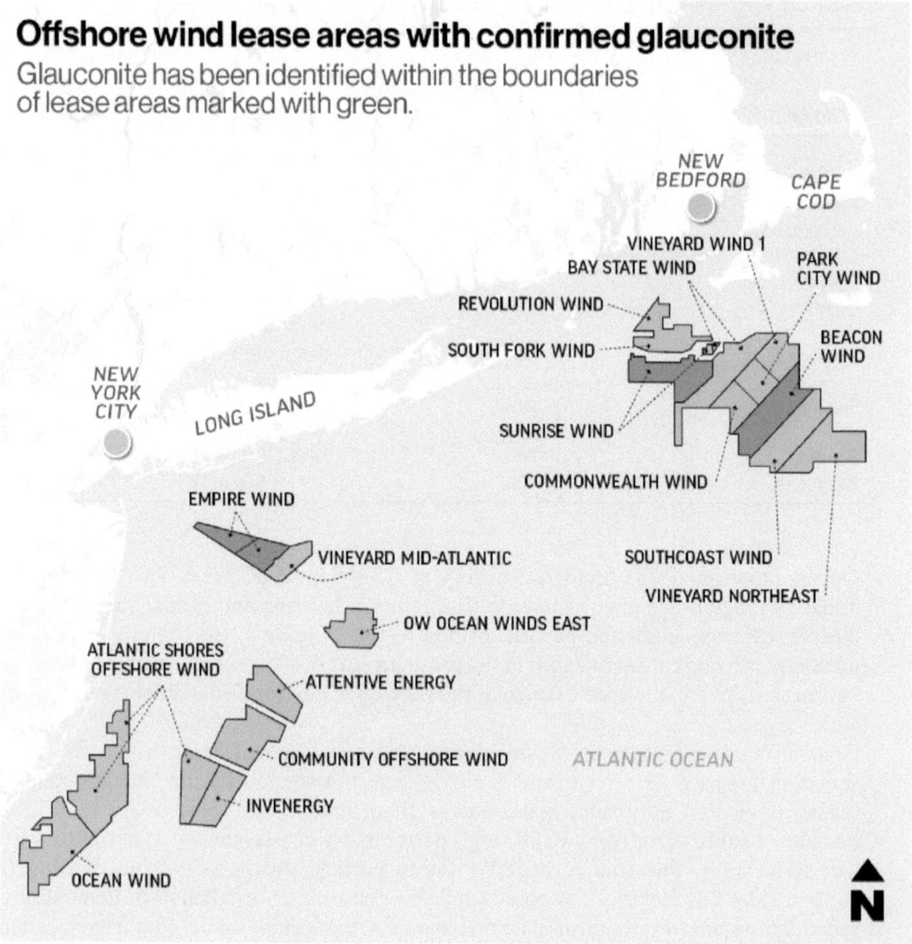

Offshore wind lease areas with confirmed glauconite
Glauconite has been identified within the boundaries
of lease areas marked with green.

Figure 5.20 Identified location on "green sand" or Glauconite in offshore locations in East Coast of USA (Source: The New Bedford Light/https://newbedfordlight.org/a-tricky-sticky-mineral-thats-challenging-offshore-wind-developers// CC BY 4.0).

(b) Natural seabed hazards: this includes rock outcrops, seabed topography, seabed channels, unstable and steep slopes, etc. Example is the West Duddon Sand project (see Section 5.2.4).

(c) Subsurface geological hazards: examples include rapidly changing stratigraphy, buried infilled channels, igneous intrusions near seabed, tectonic faults, gas hydrates zone, presence of gassy soils, and so on.

5.3.3.2 Geophysical survey

This involves mobilizing specialist vessels to map the seabed surface as well as subsea floor, and the data generated are used to update the ground model. This is often carried out before the final business decision is taken.

5.3.3.3 Geotechnical investigation

This is detailed and is carried out when the final investment decision is taken. Typical cost for geotechnical site investigation in Europe is circa \$15 M (2024 costs) for a 1 GW wind farm. Geotechnical investigation involves construction of boreholes to get soil samples for laboratory testing. There are other types of soil testing carried out such as cone penetration test (CPT). For details, the readers are referred to Chapter 4 of Bhattacharya[1] for further details.

5.4   Offshore construction

5.4.1   Introduction

The offshore construction phase involves installation of foundations, cables, and commissioning of turbines and connection to the grid. Figure 5.21 shows schematic of transportation and installation of gravity base system (GBS), whereby the foundation is placed on the seabed or shallowly embedded inspired by Karehamn Offshore Wind Farm (see Figure 5.8 for the layout). The main steps are seabed preparation, which includes leveling the seabed, transportation of the foundations in a barge, installation of the foundations, ballasting of the foundation, scour protection around the foundation, and finally installation of the tower and the turbine. Figure 5.22 shows the transportation and installation of Jacket type of system based on Seagreen Offshore Wind Farm.

Different types of vessels are required for installing different wind turbine parts; main ones are Jack-up rigs or floating heavy lift vessels (HLV), support vessels, tug/anchor handlers, cable lay vessels, survey vessels, seabed preparation vessels, crew transfer vessels, scour protection installation vessels, and cable protection installation vessels. Helicopters may also be used during the construction phase for equipment and personnel transfer.

Seabed preparation	Transportation of foundation	Installation of foundation
Ballasting of foundation	Scour protection	Wind turbine installation

Figure 5.21 Installation of GBS system (Source: Courtesy of Jan De Nul).

Figure 5.22 Installation of Seagreen Offshore Wind Farm (Scotland) Jacket system (Source: Seagreen Wind Energy Limited/https://www.seagreenwindenergy.com/post/john-hill-seagreen-project-director-provides-an-update-on-progress/last accessed February 14, 2024).

5.4.2 Choosing foundation systems

The choice of wind turbine foundation depends on the water depth and the geology of the bottom. For waters deeper than 50–60 m, floating turbines would be used regardless of the geology of the bottom. Figure 5.23 shows the collation of worldwide data on the type of foundation used based on the turbine-rated power and water depth. Figure 5.24 schematically shows the design envelope of floating systems. There are two choices in terms of mooring systems: (i) Taut mooring and (ii) catenary mooring. The anchoring options are also shown in the figure. Table 5.2 provides some examples where the different foundation options have been used. Figure 5.25 shows a particular type of foundation system known as Suction Bucket Jacket (SBJ).

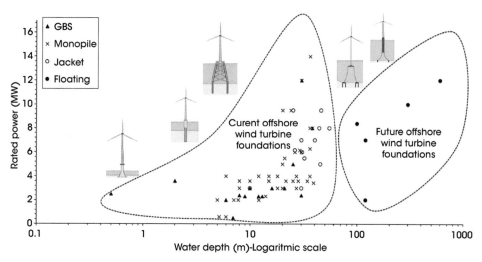

Figure 5.23 Choosing foundation based on geology and water depth (Author's own figure).

Table 5.2 Examples of wind farm systems (operating or planned).

Foundation options	Water depth	Geology	Example wind farm
Gravity base, see Figure 5.21	Approximately 30 m	Unsuitable for soft clay at the surface due to settlement/ deformation. Preferred in sites with sand, medium to stiff clay and rocky sites.	Karehamn Wind farm
Suction bucket jacket (SBJ), see Figure 5.25	Approximately 25–45 m	Preferred in sites with fine to medium sand and stiff clay. Unsuitable in sites with cobbles, boulders, and soft clay.	Seagreen project (Scotland, see Figure 5.22) Aberdeen Project
Jacket on piles; see Figure 4.25	30–60 m	Preferred in soft soil sites, seismic zones with liquefiable soils.	Formosa Wind Farm (Taiwan)
Monopiles	Approximately 35 m	This can be used in a wide range of soils such as loose to medium dense sandy deposits, dense sandy deposits, stiff clay, layered soils with soft deposits in the top layers.	Walney, London Array, CVOW (Coastal Virginia Offshore Wind)
Floating (spar buoy, semi-submersible); see Figure 5.24	60–500 m	Most type of geology is suitable, and the type of anchor will depend on the ground profile, see Figure 5.24	Hywind Scotland, Hywind Tampen

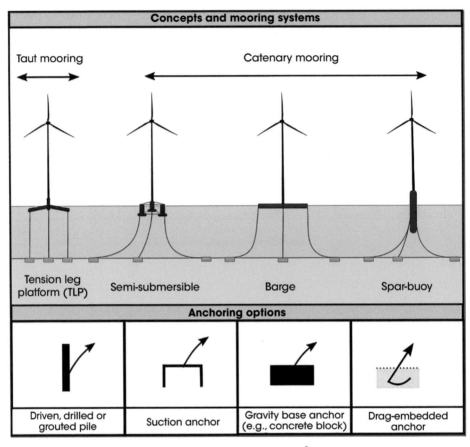

Figure 5.24 Different types of FOWTs and anchoring systems (Source: Amani[8]/with permission of University of Surrey).

Figure 5.25 Example of suction bucket jacket (Source: Jalbi et al.[9]/with permission of Elsevier).

5.5 Engineering models of wind turbine systems

This section provides salient features of these systems in terms of load transfer mechanism of loads from structure to the ground and modes of vibration of the overall structure.

5.5.1 Load transfer mechanism from the structure to the ground

One of the main aims of the overall design and the foundation is to transfer of loads from the wind turbine structure (i.e., aerodynamic thrust acting on the blades and hydrodynamic loads acting in the tower) to the supporting ground without the soil yielding/failing. The load on the foundations depends on the chosen foundation system and three distinct cases are presented:

(a) Single foundation for a grounded system, see Figure 5.26(a) where a monopile is shown. This system is very similar to mono-caisson or a GBS.

(b) Multiple foundation for a grounded system shown in Figure 5.26(b) where a jacket on multiple piles is shown.

(c) Foundation for a spar-type floating system shown in Figure 5.27.

The difference between the load transfer processes of single foundations and multiple foundations is explained in Figure 5.26(a) by taking the example of single large diameter monopile and multiple piles supporting a jacket. In the case of monopile-supported wind

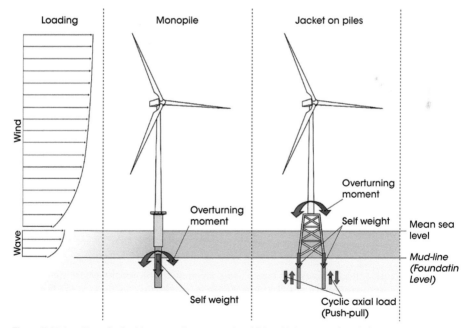

Figure 5.26 Load transfer for (a) a monopile supported and (b) multiple support foundations (Source: Bhattacharya[1]/with permission of John Wiley & Sons).

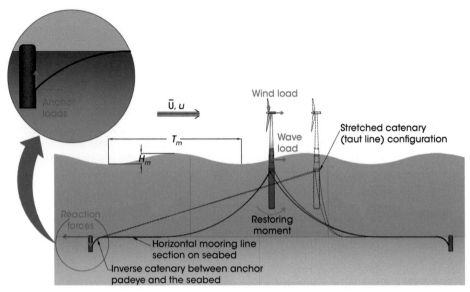

Figure 5.27 Load transfer for floating offshore wind turbine system (Source: Bhattacharya[1]/with permission of John Wiley & Sons).

turbine structures or for that matter any single foundation, the load transfer is mainly through overturning moments where the monopile/foundation transfers loads to the surrounding soil, and therefore, it is lateral foundation soil interaction. On the other hand, for multiple support structure, the load transfer is mainly through push–pull action, that is, axial load as illustrated in Figure 5.26(b).

Figure 5.27 shows a particular type of floating system known as spar-supported floating offshore wind turbines with catenary mooring and suction caisson or pile anchors. This is effectively the example of the Hywind concept, the first floating offshore wind farm. For foundation design, it is necessary to estimate an upper bound for the ultimate load on the anchor. This can be obtained by taking the configuration where the mooring line is completely stretched and there is no part of it lying on the seabed. This is similar to the configuration of a single taut mooring line. In this case, the load is transferred directly to the anchor without the effect of soil friction on a horizontal section of the mooring line, and the angle of the mooring line at the seabed is maximal. Figures 5.28 and 5.29 are examples of models of floating systems for seismic fault rupture of seabed i.e. large movement of the supporting foundations. Figure 5.29 shows the model of a TLP system.

5.5.2 Modes of vibration

Offshore wind turbines are located in a dynamic environment where there are plenty of cyclic and dynamic loads. Modes of vibration are also an important design consideration. Vibration characteristics play a significant role in choosing a particular structural system to support wind turbine generator (WTG), that is, four-legged pile or four-legged suction caisson. There are mainly two types of vibrations for grounded wind turbines: (i) Sway-bending (see Figure 5.30) and (ii) Rocking (see Figure 5.31). If the foundation is

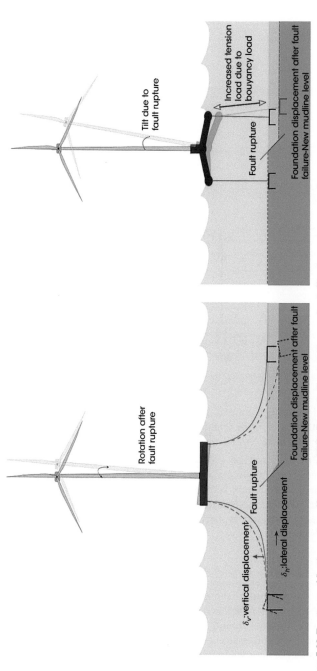

Figure 5.28 Two types of floating systems (Catenary & Tension Leg Platform). Failure mechanism is shown due to seismic fault rupture (Source: Amani[8]/with permission of University of Surrey).

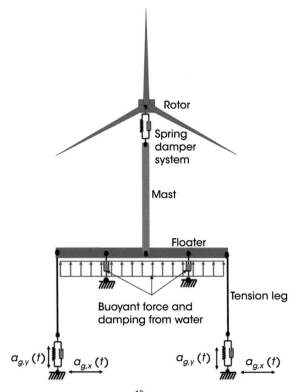

Figure 5.29 TLP system (Source: Bhattacharya *et al.*[10]/MDPI/CC BY 4.0).

stiff/rigid vertically compared to the flexibility of the tower, sway-bending mode is expected. On the other hand, if the foundation is not sufficiently rigid, rocking modes combined with flexible modes of tower may occur.

Rocking modes must be avoided at any cost for offshore wind turbine structures as low-frequency rocking mode may interact with the rotor frequency. There is, therefore, a requirement of minimum vertical stiffness of suction caissons so that rocking mode does not occur. Readers are referred to Chapter 3 in Bhattacharya[1] for further discussion on the modes of vibration. Appendix 5.A provides modes of vibrations for different types of structural systems.

5.5.3 Mechanical model of the whole system

Most of the existing offshore wind turbine structure is supported on monopiles, and therefore, monopile system is considered for discussion. Monopiles are large steel column inserted deep into the ground. Essentially, there are two steel columns (Tower and Foundation) joined by another steel column known as transition piece (TP). The tower supports the rotor, nacelle, landing platforms, electrical, and communication cables. Nacelle houses the machineries, gearbox, control system and the generators. Typical hub height, Rotor Diameter, Rotor RPM range, Nacelle Mass, Blade Mass, and Tower Mass for a range of

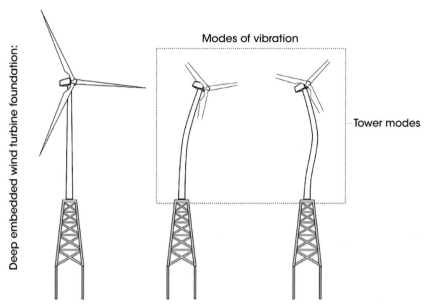

Figure 5.30 Sway-bending modes of vibration for offshore wind turbines supported on deep foundations (Source: Bhattacharya[1]/with permission of John Wiley & Sons).

Table 5.3 Typical turbine specifications.

Turbine rated power (MW)	Hub height (m)	Rotor diameter (m)	Rotor speed range (RPM)	Nacelle mass (MT)	Blade mass (MT)	Tower mass (MT)
5	90	126	7–12	297	18	347
8	110	164	6–11	375	35	558
10	119	178	6–10	551	42	605
12	135	218	7.8	600	55	860

turbines are listed in Table 5.3. Figures 5.32 and 5.33 shows a mechanical model of a wind turbine structure supported on monopile system.

Few points to note:

(1) The wind turbine structure passes through three media: air, water, and ground, and therefore, any vibrations arising are damped by them. These are termed as aero-dynamic, hydro-dynamic, and soil damping.

(2) The foundation (the part of the structure under water) is represented by a set of four uncoupled springs (K_L, K_R, K_{LR}, and K_V) representing the foundation–soil interaction. For example, K_V is the vertical spring and mainly supports the mass of the structure. These are also known as "macro-elements" or "linearized foundation stiffness". Turbine engineers need these stiffness values while carrying out simulations to obtain loads.

Appendix 5.B provides definitions of the stiffness terms and reference to closed-form solutions.

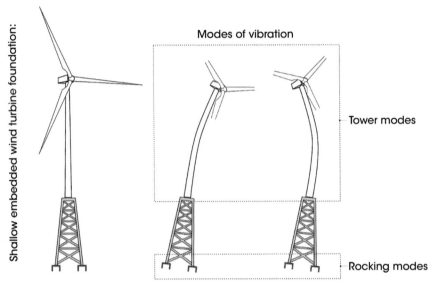

Figure 5.31 Rocking modes of vibration for offshore wind turbines supported on deep foundations (Source: Bhattacharya[1]/with permission of John Wiley & Sons).

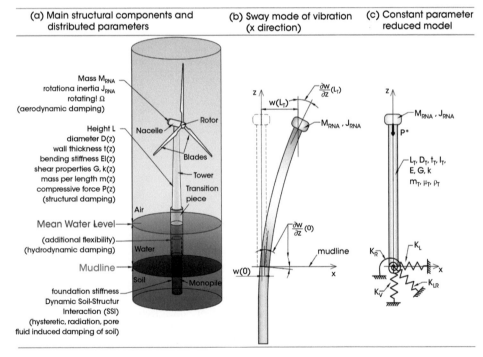

Figure 5.32 Mechanical model of a monopile system.

5.5.3.1 Why are foundations important?

(1) Foundations support the loads and provide the stability of the whole system. For example, the thrust of the blades is applied at the hub level and is ultimately taken by the springs. Proper analysis and design ensure that whole system will not fail under the extreme loads. Examples include 1 in 100-year event.

(2) If the springs (K_L, K_R or K_{LR}) are too soft, they may deform excessively hampering the operations of the turbine.

(3) Foundation stiffness is also necessary to estimate the natural frequency of the whole system. Figure 5.33 shows a mechanical model from Arany *et al.*[11] where the different variables are shown. Interested readers are referred to Arany *et al.*[11] and Bhattacharya[1] for mathematical formulations of natural frequency of monopile supported wind turbines.

5.6   Loads on a wind turbine structure

Figure 5.34 shows a schematic diagram of a bottom-fixed wind turbine support structure with a monopile foundation. The loads exercised on the structure, which are ultimately be carried to the foundation. There are four main loads apart from the weight of the whole system and they are wind, wave, and loads due to rotation of the blades (known as 1P load) and loads on the tower (also known as 2P/3P loads) loads. The loads are described later:

Wind load on the hub. The load produced by the thrust of the wind on the blades and tower. The cyclic component of the load depends on the turbulence of the wind at that location (wind speed variation over a period) and turbine operating characteristics.

Wave load. The load caused by waves crashing against the substructure (i.e., the part of the structure exposed to the wave), the magnitude of which depends on the wave height and wave period.

1P load. Load caused by the vibration at the hub level due to the mass and aerodynamic imbalances of the rotor. This load has a frequency equal to the rotational frequency of the rotor (referred to as 1P loading). Since most of the industrial wind turbines are variable speed machines, 1P is not a single frequency but a frequency band between the frequencies associated with the lowest and the highest revolutions per minute (RPM). The lowest is also known as "cut-in" speed, and the maximum is known as "rated RPM". Typical 1P frequencies for different wind turbines are provided by the turbine manufacturer and depends on the turbine rated power. The range is usually between 5 and 12 RPM.

Blade passing load (2P/3P). Loads in the tower due to the vibrations caused by blade shadowing effects (referred to as 2P/3P). As the blades of the turbine pass through the front of the tower, the temporary shadowing effect reduces the thrust on the tower. Figure 5.34(b) shows the shadowing effect due to two configurations of the blades, and the differential load on the tower is noted. This is a dynamic load having frequency equal to three times the rotational frequency of the turbine (3P) for three-bladed wind turbines. For a 2-bladed turbine, this will be two times (2P) the rotational frequency

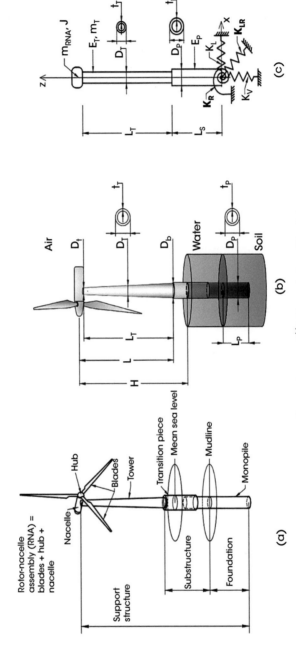

Figure 5.33 Mechanical model of a wind turbine structure (Source: Arany et al.[11]/with permission of Elsevier).

of the turbine (1P). The 2P/3P loading is also a frequency band like 1P and is simply obtained by multiplying the limits of the 1P band by the number of the turbine blades.

5.6.1 Load cases for foundation design

Codes of Practices or Best Practice guide describe hundreds of load cases that need to be analyzed to ensure the safe operation of wind turbines throughout their lifetime of 20–30 years. However, in terms of foundation design, not all these cases are significant or relevant. All design load cases are built as a combination of four wind and four sea states. The wind conditions are as follows:

(U-1) Normal turbulence scenario: the mean wind speed is the rated wind speed (U_R) where the highest thrust force is expected, and the wind turbulence is modeled by the normal turbulence model.

(U-2) Extreme turbulence scenario: the mean wind speed is the rated wind speed (U_R), and the wind turbulence is very high, the extreme turbulence model (ETM) is used.

(U-3) Extreme gust at rated wind speed scenario: the mean wind speed is the rated wind speed (U_R), and the 50-year extreme operating gust (EOG) calculated at U_R hits the rotor. The EOG is a sudden change in the wind speed and is assumed to be so fast that the pitch control of the wind turbine has no time to alleviate the loading. This assumption is very conservative and is used for simplified foundation design.

(U-4) Extreme gust at cut-out scenario: the mean wind speed is slightly below the cut-out wind speed of the turbine (U_{out}), and the 50-year EOG hits the rotor. Due to the sudden change in wind speed, the turbine cannot shut down.

The wave conditions are as follows:

(W-1) 1-year extreme sea state (ESS): a wave with height equal to the 1-year significant wave height $H_{S,1}$ acts on the substructure.

(W-2) 1-year extreme wave height (EWH): a wave with height equal to the 1-year maximum wave height $H_{m,1}$ acts on the substructure.

(W-3) 50-year ESS: a wave with height equal to the 50-year significant wave height $H_{S,50}$ acts on the substructure.

(W-4) 50-year EWH: a wave with height equal to the 50-year maximum wave height $H_{m,50}$ acts on the substructure.

The 1-year ESS and EWH are used as a conservative estimation of the normal wave height (NWH). It is important to note here in relation to the ESS that the significant wave height and the maximum wave height have different meanings. The significant wave height H_S is the average of the highest one-third of all waves in the 3-hour sea state, while the maximum wave height H_m is the single highest wave in the same 3-hour sea state.

According to the relevant standards (IEC[12, 13]; DNV[14]), in the probability envelope of environmental states, the most severe states with a 50-year return period have to be considered. Indeed, extreme waves and high wind speeds tend to occur at the same time; however, the highest load due to wind is not expected when the highest wind speeds occur. This is partly because the pitch control alleviates the loading above the rated wind speed, but also because turbines shut down at high wind speeds for safety reasons. Idle or shut-down turbines, as well as turbines operating close to the cut-out wind speed, have a significantly reduced thrust force acting on them compared with the thrust force at the rated wind speed.

The highest wind load is expected to be caused by scenario (U-3), and the highest wave load is due to scenario (W-4). In practice, the 50-year extreme wind load and the 50-year extreme wave load have a negligible probability to occur at the same time, and the DNV[14] code also does not require these extreme load cases to be evaluated together. The designer has to find the most severe event with a 50-year return period based on the joint probability of wind and wave loading. Therefore, for the ultimate limit state (ULS) analysis, the following combinations of wind and wave energies are suggested by Arany *et al.*[15]:

(1) the ETM wind load at rated wind speed combined with the 50-year EWH—the combination of wind scenario (U-2) and wave scenario (W-4). This will provide higher loads in deeper water with higher waves.

(2) the 50-year EOG wind load combined with the 1-year maximum wave height. This will provide higher loads in shallow water in sheltered locations where wind load dominates.

Five load cases important for simplified foundation design are identified and described in Table 5.4.

These scenarios are somewhat more conservative than those required by standards and can be adopted for simplified analysis. From the point of view of serviceability limit state (SLS) and fatigue limit state (FLS), the single largest loading on the foundation is not representative because the structure is expected to experience this level of loading only once throughout the lifetime.

The load cases in Table 5.4 are representative of typical foundation loads in a conservative manner and may serve as the basis for conceptual design of foundations. However, detailed analysis for design optimization and the final design may require addressing other load cases

Table 5.4 Load combination.

#	Name and description	Wind model	Wave model	Alignment
E-1	Normal operational conditions *Wind and wave act in the same direction (no misalignment).*	NTM at U_R (U-1)	1-year ESS (W-1)	Collinear
E-2	Extreme wave load scenario *Wind and wave act in the same direction (no misalignment).*	ETM at U_R (U-2)	50-year EWH (W-4)	Collinear
E-3	Extreme wind load scenario *Wind and wave act in the same direction (no misalignment).*	EOG at U_R (U-3)	1-year EWH (W-2)	Collinear
E-4	Cutout wind speed and extreme operating gust scenario *wind and wave act in the same direction (no misalignment).*	EOG at U_{out} (U-4)	50-year EWH (W-4)	Collinear
E-5	Wind-wave misalignment scenario *Same as E-2, except the wind and wave are misaligned at an angle of $\phi = 90°$. The dynamic amplification is higher in the cross-wind direction due to low aerodynamic damping.*	ETM at U_R (U-2)	50-year EWH (W-4)	Misaligned at $\phi = 90°$

as well. These analyses require detailed data about the site (wind, wave, current, geological, geotechnical, bathymetry data, etc.) and also the turbine (blade profiles, twist and chord distributions, lift and drag coefficient distributions, control parameters and algorithms, drive train characteristics, generator characteristics, tower geometry, etc.).

Figure 5.34 shows the time history (wave form) of the main four loads. The salient points of the loads are discussed here:

(a) Each of these loads has unique characteristics in terms of magnitude, frequency, and number of cycles applied to the foundation.

(b) The loads imposed by the wind and the wave are random in both space (spatial) and time (temporal), and therefore, they are often described statistically.

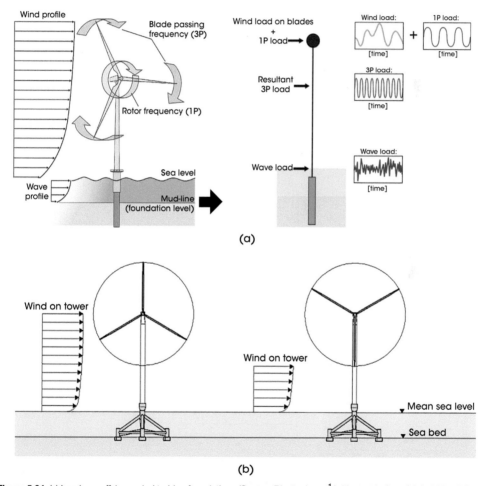

(a)

(b)

Figure 5.34 (a) Loads on offshore wind turbine foundations (Source: Bhattacharya[1]/with permission of John Wiley & Sons). (b) Explanation of 3P load (Source: Bhattacharya[1]/with permission of John Wiley & Sons).

(c) Apart from the random nature, these two loads may also act in two different directions for a wind turbine structure. The blowing wind above the sea causes waves, and therefore, wind and wave are colinear. However, in order to have a steady power output, the control system often performs yaw action causing wind-wave misalignment.

(d) 1P loading is caused by mass and aerodynamic imbalances of the rotor and the forcing frequency equals the rotational frequency of the rotor.

(e) On the other hand, 2P/3P loading is caused by the blade shadowing effect, wind shear (i.e., the change in wind speed with height above the ground), and rotational sampling of turbulence. Its frequency is simply two or three times the 1P frequency.

The readers are referred to Bhattacharya[1] for methods to calculate the loads.

Figure 5.35 shows a simplified representation of the four loads on monopile supported structure for 1 minute duration. Loads for wind and wave are in Mega Newton (MN) range, and the loads for 1P and 3P are in kN (kilo Newton) range. Combination of wind and wave provides the magnitude of the design load and is shown in Figure 5.36. It may be noted that wind and wave have different time periods. Wave frequency depends on the sea and in the simplified version the predominant wave frequency can be considered. In the example chosen, the wind has a time period of 100 seconds and wave 10 seconds. Therefore, the combined wind and wave load is cyclic in nature with minimum (M_{min}), maximum (M_{max}), and mean values (M_{mean}).

The next two figures show examples of loads on the foundation. Figure 5.36b shows a plot of overturning moment for a 10 MW turbine in 10 m of water depth. Figure 5.36c shows a 3.6 MW turbine in 21.5 m water depths. The load cases used for estimating the moments are

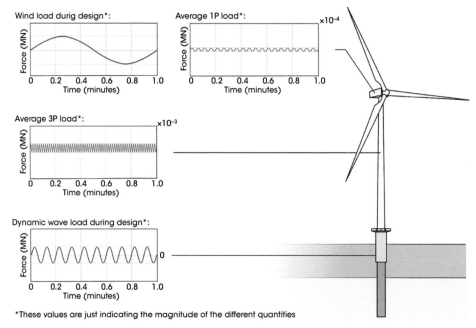

*These values are just indicating the magnitude of the different quantities

Figure 5.35 Schematic diagram showing the loads on a wind turbine structure.

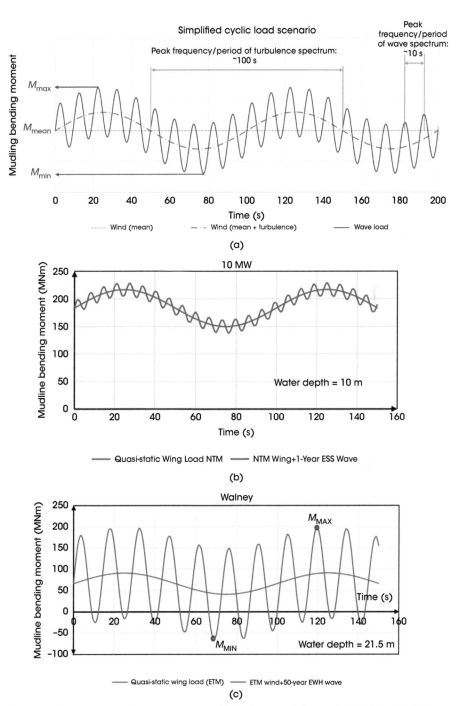

Figure 5.36 (a) Combination of wind and wave load. (b) Loading scenario for a typical 10 MW turbine in 10 m water depths. (c) Loading scenario for 3.6 MW turbine in 21.5 m water depth.

provided in the caption. The figure shows that the overturning moment in the foundation (i.e. mudline bending moment) can be one-way cyclic (as in Figure 5.35b) or asyymmetric two-way cyclic (as in Figure 5.35c) and depends on the turbine rated power and water depth. Larger turbines will have longer blades (by extension taller towers) and therefore higher thrust load when other factors remains constant. Therefore, 10MW turbine will have a higher thrust than 3.6MW turbine which will lead to higher mudline moment. On the other hand, wind farm locations will have a good wind speed and wave loads are proportional to water depth. This explains why 21.5m water depth in Figure 5.35c is different from load pattern in 10m water depth. The readers are referred to Jalbi et al.[16] for a simplified method to predict the nature of loading on the foundation i.e. whether it is one-way or two cyclic.

5.7 Frequencies of loads acting on wind turbine structure

Figure 5.36 shows a schematic diagram of the main frequencies of the main loads (wind, wave, 1P, and 3P) so that the dynamic design constraints can be visualized. Each of the loads have a frequency range and they are shown in the figure. Wind and wave loads are expressed through normalised PSD (Power Spectral Density). The formulation for developing wind and wave spectrum is provided in codes of practices, and an example spectrum is provided in Appendix 5.C. It is advisable to have the global natural frequency of the whole wind turbine system apart of the forcing frequencies so that amplification of responses (also known as dynamic amplification (loosely speaking resonance) can be avoided. Many codes of practices suggest to have a safety margin on either side of the loading. It is worth noting that resonance under operational condition has been reported in the German North Sea projects[17].

Wind energy is concentrated at low frequencies (typically 100 second time period), and the wave frequency depends on the sea. Typically, the predominant frequency of wave is

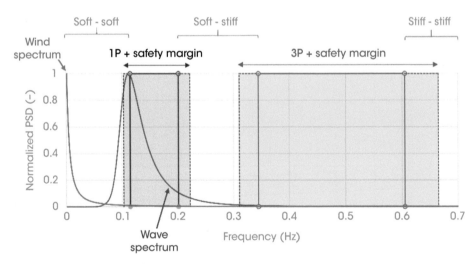

Figure 5.37 Frequency range of the four main loads acting on the wind turbine structure.

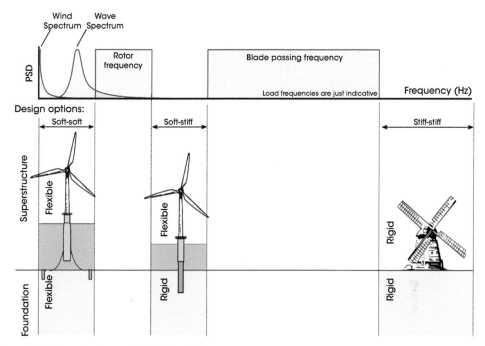

Figure 5.38 Schematic diagram of the design choices.

0.1–0.2 Hz. 1P and 3P depend on the operating range of the turbine. These are visualized in Figure 5.37. An example is shown in the next ASIDE 3 section.

Three design space may be noted in Figure 5.36: soft–soft, soft–stiff, and stiff–stiff. These terms are essentially concerned based on relative flexibility of the tower with respect to the foundation and explained schematically in Figures 5.37 and 5.38.

Few points may be noted from the design choices available:

(1) In the "soft–stiff" design, the natural frequency or the resonant frequency is very close to the upper end of 1P (i.e., frequency corresponding to the rated power of the turbine) and lower bound of the 3P (i.e., cut-in speed of the turbine). This will inevitably cause vibration of the whole system as the ratio of forcing to natural frequency is very close to 1.

(2) For a soft–stiff 3 MW WTG system, 1P and 3P loading can be considered as dynamic (i.e., ratio of the loading frequency to the system frequency very close to 1). Most of the energy in wind turbulence is in lower frequency variations (typically around 100 second peak period), which can be considered as cyclic. On the other hand, 1P and 3P dynamic loads change quickly in comparison to the natural frequency of the WTG system, and therefore, the ability of the WTG to respond depends on the characteristics, and dynamic analysis is therefore required.

(3) As a rule of thumb, if the natural frequency of the WTG structure is more than five times the forcing frequency—the loading can be considered cyclic, and inertia of the system may be ignored. For example, for a 3 MW wind turbine having a natural frequency

221

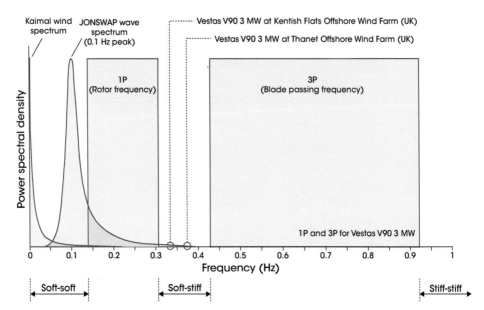

Figure 5.39 Frequency range of the loads along with natural frequency of the turbines for 3 MW turbines.

of 0.3 Hz, any load having frequency more than 0.06 Hz is dynamic. Therefore, wave loading of 0.1 Hz is dynamic.

ASIDE 3: Target frequency for 3 MW turbine in North Sea

The operating range of Vestas V90 3 MW turbine is 8.6–18.4 RPM, and they are shown in Figure 5.39. Wind load spectrum is based on Kaimal spectrum, and wave load spectrum is potted using JONSWAP spectrum. Current design aims to place the natural frequency of the whole system in between the upper bound of 1P and lower bound of 3P in the so-called "soft–stiff" design. In the plot, the natural frequency of two Vestas V90 3 MW wind turbines from two wind farms (Kentish Flats and Thanet) are also plotted. Though the turbines are same, the variation in the natural frequency is due to the different ground and site conditions.

5.8 Design requirements

It is important to highlight the main calculations that are necessary:

(a) Verification of *safe load transfer* from the superstructure to the supporting ground in the case of extreme load events. In limit state design philosophy, this is termed as ultimate limit state (ULS). Essentially satisfying the ULS criteria ensures that the supporting foundation or the ground does not collapse or the material of the tower/monopile does not yield.

(b) Predicting the *modes of vibration* of the adopted structural system, that is, rigid rocking modes or flexible modes or a combination. Modes of vibration dictate the dynamic response and therefore the deformations and stresses in the materials such as tower or monopiles. Furthermore, modes of vibration also influence the rotor Nacelle assembly (RNA) acceleration. In limit state design philosophy, this satisfies fatigue limit state

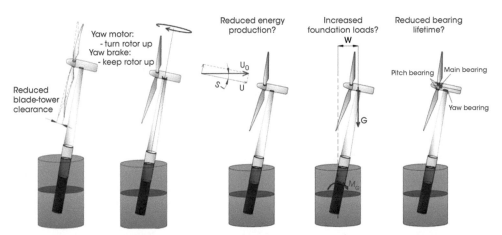

Figure 5.40 Aspects governing the SLS requirements for monopile foundation (Source: Bhattacharya[1]/with permission of John Wiley & Sons).

(FLS) as well as serviceability limit state (SLS). A strict limit of rotation is imposed on monopile supported wind turbines. Figure 5.40 provides plausible reasons for such stricter rotational limits, which includes increased foundation loads, and wear and tear of bearings. Wind turbine structures are subjected to millions of cycles of repeated loading (due to wind, wave, 1P, and 2P/3P), and there are many welded joints in the whole structure. Welded joint arises during the fabrication of tower or monopile from a steel sheet. Therefore, fatigue of the material is a problem, and cracks initiation must be under control and which is ensured through FLS criteria. Fatigue damage at a crack initiation location is affected by potential flaws such as initial defects, stress concentration, incomplete weld root penetration (weld defect), number of wave-induced stress range cycles, corrosion, and performance of coatings. This also shows the importance of operation and maintenance (O & M) of wind turbine systems.

(c) Long-term deformation so that *SLS* requirements (example rotation of the foundations) are not violated under the combined actions of wind and wave. Table 5.5 shows some typical criteria for design of wind turbine structure.

(d) Long-term *change in dynamic characteristics* (i.e., change in natural frequency and damping) of the whole system due to millions of cycles of loading. This change can be driven by various changes such as scour depth, change in ground properties supporting the foundations due to cycles of loading, and corrosion of the materials. This also satisfies SLS and FLS.

Table 5.5 ULS and SLS criteria.

Limit state	Typical criteria
ULS	(i) Ground failure (soil failure) around the foundation causing foundation collapse. (ii) Foundation should remain elastic.
SLS	(i) Permanent tilt at pile head <0.75° (these are typical for grounded systems). (ii) Rotor Nacelle acceleration (RNA) acceleration <0.2–0.4 g. (iii) Acceptable pile head deformation.
FLS	(i) Wind + wave loading imposes a large number of cycles during the operational life of the turbines Fatigue limit must be checked.

5.9   Summary of offshore wind turbine engineering

Offshore environments present diverse met ocean (wind and the wave climate) and geological conditions. Examples of geological conditions are stiff clay and sand in European seas, soft clay layered deposits in Chinese seas, chalk in the English Channel, and carbonate deposits. Although existing offshore wind farms have been constructed mainly in soft clay, stiff clays, and sand deposits, future projects are likely to be deployed in more challenging geological conditions (soft rocks in shallow seabed, carbonate soils, and liquefiable soils) for which prior experience is limited.

This chapter shows the layout of various offshore wind farm around the world explaining the reason behind such layout. The different constraints while designing these are also discussed together will example layouts. This chapter also provides an overview of the range of data required for carrying out design. The different types of loads that the offshore wind turbine structure experience are described together with the salient features. Engineering models that are needed for carrying out various calculations are discussed together with the design criteria.

Appendix 5.A: Modes of vibrations

Figure 5.A.1 shows the modes of vibration of wind turbine tower assuming fixed base, that is, the tower is rock-socketed. The formula can be obtained from standard dynamics textbook. A list of vibration formulas can be found in Varghese et al.[18].

5.A.1   Wind turbine systems on deep foundation

This can be two types: monopile and jacket on pin piles and the modes of vibration are shown in Figure 5.A.2. Further details of vibrations can be found in Arany et al.[11]. As the foundation is rigid compared to the tower, one may note the flexible modes of tower.

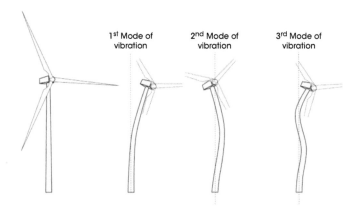

Figure 5.A.1 Modes of vibration and the formulation can be obtained.

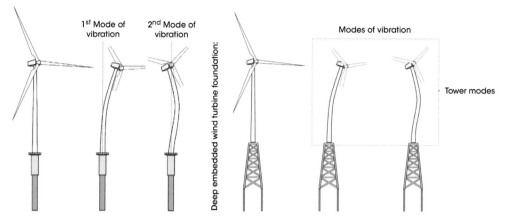

Figure 5.A.2 Modes of vibration for monopiles and jacket on piles.

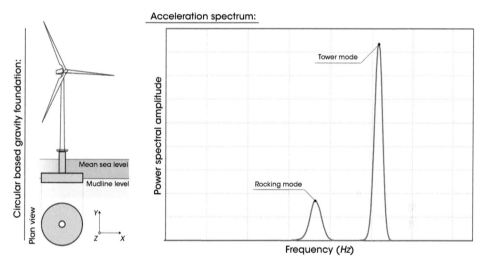

Figure 5.A.3 Modes of vibration for a GBS structure.

5.A.1.1 Modes of vibration of GBS structure

Figure 5.A.3 shows the modes of vibration of GBS and two modes are expected: Rocking mode of the foundation and flexible modes of the tower.

Appendix 5.B: Foundation stiffness

In the mechanical model shown in Figure 5.32, the foundation is represented by a set of four springs:

K_L (Lateral stiffness), that is, lateral load needed for unit deflection having units of MN/m.

K_R (Rotational stiffness), that is, moment needed for unit rotation having units of MNm/rad.
K_{LR} (Cross-coupling spring), that is, moment required for unit deflection having unit of MN.
K_V (Vertical spring), that is, vertical load needed for unit settlement having units of MN/m.
The formulas for different types of foundations are provided in Tables 5.B.1 to 5.B.4.

5.B.1 Definition of terms (K_L, K_R, and K_{LR}) through the example of a cantilever beam

The stiffness terms are defined though the example of a cantilever beam. Two cases are considered here: (a) concentrated load at the free end, see Figure 5.B.1, and (b) moment at the free end, see Figure 5.B.2.

K_L = Lateral load required for unit deflection, that is, when $\delta_{max} = 1$, P is the lateral stiffness, see Figure 5.B.1.

$$K_L = \frac{3EI}{L^3}$$

$K_{LR} = $ Lateral load required for unit rotation (when $\theta = 1$, P is the cross-coupling stiffness), see Figure 5.B.1

$$K_{LR} = \frac{2EI}{L^2}$$

K_R = Moment required for unit rotation: $K_R = \dfrac{EI}{L}$

K_{RL} = Moment required for unit deflection = $K_{LR} = \dfrac{2EI}{L^2}$

The above example is for a cantilever beam fixed at the bottom. However, foundations are surrounded by soil and the bottom of the foundation may not be fixed to the ground. Therefore, obtaining stiffness is complex and is dependent of soil properties as well. Many closed form formulations are available for obtaining stiffness of different types of foundations. A comprehensive review is available in Varghese et al.[18] and Bhattacharya[1]. Arany et al.[15] provide formulation for monopiles. Jalbi et al.[19] provide stiffness for suction caissons. Salem et al.[20] provide formulation for vertical stiffness for suction caissons.

This section of the appendix provides few examples.

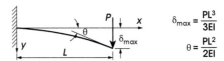

Figure 5.B.1 Concentrated load at the free end.

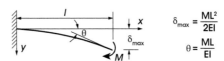

Figure 5.B.2 Moment at free end.

Table 5.B.1 Formulae for stiffness for slender piles applicable to Jacket on pin piles based on Shadlou and Bhattacharya[22].

Lateral stiffness K_L	Cross-coupling stiffness K_{LR}	Rotational stiffness K_R
Shadlou and Bhattacharya[22], slender pile, homogeneous soil (over-consolidated soils)		
$\dfrac{1.45 E_{S0} D_P}{f_{(v_s)}} \left(\dfrac{E_{eq}}{E_{s0}}\right)^{0.186}$	$-\dfrac{0.30 E_{S0} D_P^2}{f_{(v_s)}} \left(\dfrac{E_{eq}}{E_{s0}}\right)^{0.50}$	$\dfrac{0.18 E_{S0} D_P^3}{f_{(v_s)}} \left(\dfrac{E_{eq}}{E_{S0}}\right)^{0.73}$
Shadlou and Bhattacharya[22], slender pile, linear inhomogeneous soil (normally consolidated soils)		
$\dfrac{0.79 E_{S0} D_P}{f_{(v_s)}} \left(\dfrac{E_{eq}}{E_{s0}}\right)^{0.34}$	$-\dfrac{0.26 E_{S0} D_P^2}{f_{(v_s)}} \left(\dfrac{E_{eq}}{E_{S0}}\right)^{0.567}$	$\dfrac{0.17 E_{S0} D_P^3}{f_{(v_s)}} \left(\dfrac{E_{eq}}{E_{S0}}\right)^{0.78}$
Shadlou and Bhattacharya[22], slender pile, parabolic inhomogeneous soil (typical of sandy soils)		
$\dfrac{1.02 E_{S0} D_P}{f_{(v_s)}} \left(\dfrac{E_{eq}}{E_{S0}}\right)^{0.27}$	$-\dfrac{0.29 E_{S0} D_P^2}{f_{(v_s)}} \left(\dfrac{E_{eq}}{E_{S0}}\right)^{0.52}$	$\dfrac{0.17 E_{S0} D_P^3}{f_{(v_s)}} \left(\dfrac{E_{eq}}{E_{S0}}\right)^{0.76}$

Parameter definitions:

$$E_{eq} = \frac{E_P I_P}{\dfrac{D_P^4 \pi}{64}};$$

D_p = pile diameter,

E_{S0} = Young's modulus of the soil at $1 D_P$ below the ground

$f_{(v_s)} = 1 + |v_s - 0.25|$ for Shadlou and Bhattacharya[22]

Table 5.B.2 Stiffness of rigid piles based on Shadlou and Bhattacharya[22].

Lateral stiffness K_L	Cross-coupling stiffness K_{LR}	Rotational stiffness K_R
Shadlou and Bhattacharya[22], rigid pile, homogeneous soil		
$\dfrac{3.2 E_{S0} D_P}{f_{(v_s)}} \left(\dfrac{L_P}{D_P}\right)^{0.62}$	$-\dfrac{1.7 E_{S0} D_P^2}{f_{(v_s)}} \left(\dfrac{L_P}{D_P}\right)^{1.56}$	$\dfrac{1.65 E_{S0} D_P^3}{f_{(v_s)}} \left(\dfrac{L_P}{D_P}\right)^{2.5}$
Shadlou and Bhattacharya[22], rigid pile, linear inhomogeneous soil		
$\dfrac{2.35 E_{S0} D_P}{f_{(v_s)}} \left(\dfrac{L_P}{D_P}\right)^{1.53}$	$-\dfrac{1.775 E_{S0} D_P^2}{f_{(v_s)}} \left(\dfrac{L_P}{D_P}\right)^{2.5}$	$\dfrac{1.58 E_{S0} D_P^3}{f_{(v_s)}} \left(\dfrac{L_P}{D_P}\right)^{3.45}$
Shadlou and Bhattacharya[22], rigid pile, parabolic inhomogeneous soil		
$\dfrac{2.66 E_{S0} D_P}{f_{(v_s)}} \left(\dfrac{L_P}{D_P}\right)^{1.07}$	$-\dfrac{1.8 E_{S0} D_P^2}{f_{(v_s)}} \left(\dfrac{L_P}{D_P}\right)^{2.0}$	$\dfrac{1.63 E_{S0} D_P^3}{f_{(v_s)}} \left(\dfrac{L_P}{D_P}\right)^{3.0}$

Parameter definitions:

$f_{(v_s)} = 1 + |v_s - 0.25|$

D_p = pile diameter,

E_{S0} = Young's modulus of the soil at $1 D_P$ below the ground

Table 5.B.3 Vertical stiffness for shallow skirted foundations exhibiting rigid behavior based on Salem et al. (2021).[20]

Ground profile	$\dfrac{K_V}{DE_{SO}f(v_s)}$
Homogeneous	$2.31\left(\dfrac{L}{D}\right)^{0.52}$
Parabolic	$2.16\left(\dfrac{L}{D}\right)^{0.96}$
Linear	$2.37\left(\dfrac{L}{D}\right)^{1.28}$

$$f\left(v_s\right) = \left[\left(10v_s^3 - 5.88v_s^2\right)\left(-0.34\ln\tfrac{L}{D} + 0.77\right)\right] + 0.91v_s\left(-0.57\ln\tfrac{L}{D} + 0.6\right) + 1$$

Table 5.B.4 Stiffness for shallow skirted foundations exhibiting rigid behavior $0.5 < L/D < 2$ following the work of Jalbi et al.[19].

Ground profile	$\dfrac{K_L}{DE_{SO}f(v_s)}$	$\dfrac{K_{LR}}{D^2E_{SO}f(v_s)}$	$\dfrac{K_R}{D^3E_{SO}f(v_s)}$
Homogeneous	$2.91\left(\dfrac{L}{D}\right)^{0.56}$	$-1.87\left(\dfrac{L}{D}\right)^{1.47}$	$2.7\left(\dfrac{L}{D}\right)^{1.92}$
Parabolic	$2.7\left(\dfrac{L}{D}\right)^{0.96}$	$-1.99\left(\dfrac{L}{D}\right)^{1.89}$	$2.54\left(\dfrac{L}{D}\right)^{2.44}$
Linear	$2.53\left(\dfrac{L}{D}\right)^{1.33}$	$-2.02\left(\dfrac{L}{D}\right)^{2.29}$	$2.46\left(\dfrac{L}{D}\right)^{2.9}$

L, embedded length.
D, diameter of the caisson
v_s, Poissons ratio of soil.
$$f(v_s) = 1.1 \times \left(0.096\left(\tfrac{L}{D}\right) + 0.6\right)v_s^2 - 0.7v_s + 1.06.$$

Appendix 5.C: Construction of wind and wave spectrum

5.C.1 Construction of wind spectrum

Turbulence of wind is usually estimated as a fluctuating wind speed component (u) superimposed on the mean wind speed $\overline{U}$, and therefore, the total wind speed can be written as $U = \overline{U} + u$. The degree of turbulence is usually characterized by the turbulence intensity I, given by Equation (C.1) later

$$I = \sigma_U/\overline{U} \tag{C.1}$$

where σ_U is the standard deviation of wind speed around the mean $\overline{U}$ (which is usually taken over 10 minutes). The turbulence intensity varies with mean wind speed, with site location, and with surface roughness and is also modified by the turbine itself.

Taylor[21] states in his frozen turbulence hypothesis that the characteristics of eddies can be considered constant (frozen) in time, and vortices travel with the mean horizontal wind speed, see Figure 5.C.1. This assumption is found to be acceptable for wind turbine applications, and the turbulence is usually analyzed in the frequency domain by a power spectral density (PSD) function, which describes the contribution of different frequencies to the total variance of the wind speed.

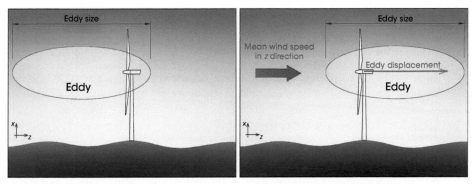

Figure 5.C.1 Taylor's frozen turbulence hypothesis: an eddy travels with the mean wind speed while its size and characteristic parameters remain constant.

The frequency of turbulence is connected to the size of eddies. A larger eddy means low-frequency variation in wind speed, while smaller vortices induce short, high-frequency wind-speed variations. If the characteristic size of an eddy is d(m) and it travels with $\overline{U}$(m/s) speed, the travel time through the rotor in $\tau = d/\overline{U}$ time. The frequency connected to this time period is $f = 1/\tau$ (Hz).

Through this, the length scale and time scale of turbulence can be connected. The typical length scales of high energy-containing large turbulent eddies are in the range of several kilometers. The large eddies tend to decay to smaller and smaller eddies with higher frequencies as turbulent energy dissipates to heat. Kolmogorov's law describing this process states that the asymptotic limit of the spectrum is $f^{-5/3}$ at high frequency.

There are two families of spectra commonly used in wind energy applications: the von Kármán and the Kaimal spectra. The main difference is that the Kaimal spectrum is somewhat less peaked, and the energy is contained in a bit wider frequency range. Kaimal spectrum is more suitable for modeling the atmospheric boundary layer, and the von Kármán spectrum is better for wind tunnel modeling. DNV suggests the Kaimal spectrum.

5.10 Kaimal spectrum

The theoretical Kaimal spectrum for a fixed reference point in space in neutral stratification of the atmosphere $S_{uu}(f)$ as suggested by DNV can be written as

$$S_{uu}(f) = \frac{\sigma_U^2 \left(\dfrac{4L_k}{\overline{U}}\right)}{\left(1 + \dfrac{6fL_k}{\overline{U}}\right)^{\frac{5}{3}}} \tag{C.2}$$

where L_k is the integral length scale (formula available in the DNV code). Based on the DNV code, $L_k = 5.67z$ for $z < 60$ m and $L_k = 340.2$ m for $z \geq 60$ m where z denotes the height above sea level.

f is the frequency,

$\overline{U}$ is the mean wind speed (from site measurements),

and σ_U is the standard deviation of wind speed.

These can be estimated from measurements or calculated using Equation (C.2), and turbulence intensity values may be obtained from standards IEC 61400-1 and IEC 61400-3.

5.11 Construction of wave spectrum

The wind blowing over the sea generates wind waves because of the increased pressure on the free surface of water. First, small waves are produced with high frequency and low wave height, and the energy is gradually transferred toward the higher amplitude waves with lower frequency and longer wavelength. The developing sea state depends on many factors, including but not limited to the water depth, the shape of the sea bottom, the mean wind speed, and the fetch. The latter is the typical leeward distance to shore considering the prevailing wind direction. The dependence on the water depth is apparent from the dispersion relation:

$$\omega^2 = gk \tanh(kS)$$

where ω[rad/s] is the angular frequency, $k = 2\pi/\lambda$[1/m] is the wave number with λ(m) being the wavelength, and S(m) is the mean sea depth.

Any offshore location consists of a large number of waves with various frequencies and wavelengths. The importance of each frequency is characterized by the power associated with it, which can be represented by the PSD function. The PSD can be produced from site measurements of the wave height using discrete Fourier transform, or alternatively the JONSWAP spectrum $S_{ww}(f)$ suggested by many codes such as DNV:

$$S_{ww}(f) = \frac{\alpha g^2}{(2\pi)^4 f^5} e^{-\frac{5}{4}\left(\frac{f}{f_p}\right)^4} \gamma^r \quad \gamma = 3.3$$

$$r = e^{-\frac{(f-f_p)^2}{2\sigma^2 f_p^2}} \quad \alpha = 0.076\left(\frac{\overline{U}_{10}^2}{Fg}\right)^{0.22} \quad f_p = \frac{22}{2\pi}\left(\frac{g^2}{\overline{U}_{10}F}\right)^{\frac{1}{3}}$$

$$\sigma = \begin{cases} 0.07 & f \leq f_p \\ 0.09 & f > f_p \end{cases}$$

where

f is the frequency,

α is the intensity of the spectrum,

F is the fetch,

f_p is the peak frequency,

γ is the peak enhancement factor,

g is the gravitational constant,

$\overline{U}_{10}$ is the mean wind speed at 10 m height above sea level.

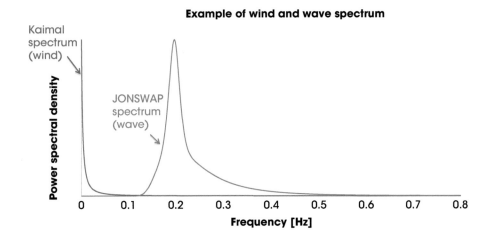

Example of wind and wave spectrum

Self-assessment questions

Q5.1 What are the pros and cons of turbines getting bigger? Hint: increasing the size of the turbine decreases the cost per unit of electricity generated.

Q5.2 Compare the offshore and onshore wind farm construction.

Q5.3 What are the steps in construction of a wind farm based on monopile?

Q5.4 What are the main loads acting on foundations of offshore wind turbine? Comment on the nature of the loads (cyclic/dynamic/monotonic).

Q5.5 What are the typical criteria for design of offshore wind turbine foundations?

Q5.6 Briefly describe the main geotechnical work that is necessary for the overall development of a wind farm.

Frequency diagram of OWTs

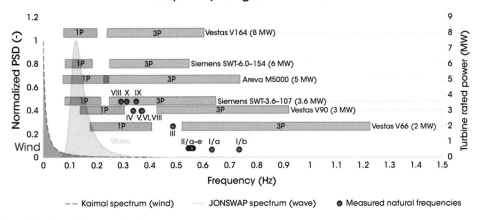

Figure 5.41 1P and 3P range of turbines.

Q5.7 How would you decide on the monopile thickness?

Q5.8 Discuss the typical allowable rotation and deflection for wind turbines for bottom fixed and floating foundations.

Q5.9 Figure 5.41 shows the 1P range for many turbines. Are there any issues with large turbines?

Q5.10 Figure 5.42 below shows the bathymetry data of the USA. Compare the development in the East Coast and the West Coast.

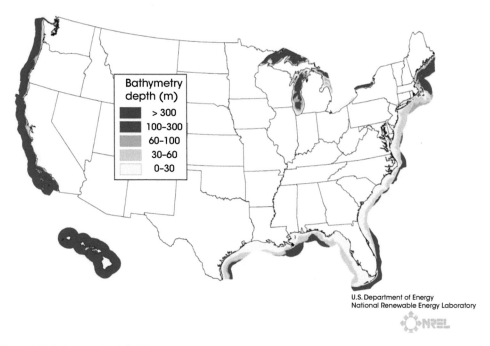

Figure 5.42 Bathymetry depth for US waters.

Problems

5.1 Using the data in Chapter 4, provide a view on the future of wind turbine in term of constructability. Is there a limit on the size of large monopiles that can be fabricated? What data and information you will need to carry out assessment.

Hint: Plot the monopile diameter and assess if the monopile can be fabricated. For the given steel plate, what is the minimum radius you can bend.

5.2 Continuing Problem 5.1, is there any limitation on lifting large pieces of items offshore?

Plot the blade length, blade weight, Nacelle weight, and monopile weight with capacity. All these pieces need to be lifted to install. Can they be installed in wind conditions or in high wave condition? How can this scenario be modeled?

5.3 One 10 MW turbine gives the same power output as two 5 MW turbines. Develop a layout for a 120 MW wind farm in a location where monopile is the foundation and the center of the wind farm is 40 km offshore with two options 5 and 10 MW. Compare the amount of steel and cables and time to construct. You need to simplify the problem and make assumptions by studying the existing operating wind farms.

5.4 A developer wishes to develop a wind farm with Vestas V164-8.0MW turbines capable of producing 8.0 MW, and the rotor diameter is 164 m. The site is North Sea with peak wave period of 10 seconds. The developer wishes to operate the turbine is 5–12 RPM, the site is moderately turbulent (18% turbulent intensity), and the average wind speed is 8 m/s. Sketch the range of frequencies acting on the system. Suggest a target natural frequency of the whole system.

5.5 A developer wishes to support a 8 MW turbine on monopiles and ground immediately below the mudline is medium dense sand and extends to about 30 m. The height of the tower is 100 m above water level, and the water depth is 25 m. Approximate calculation shows that the highest load at the hub level due to the wind load corresponds to 50 year extreme operational gust loading and has been estimated to be 2.5 MN. The wave load corresponding to EWH is estimated to be 2.8 MN and can be assumed to act at mean sea level. The certifier of the wind farm imposes 0.5° maximum rotation at the mudline level. What is the minimum rotational foundation stiffness required?

5.6 For the given offshore wind, the ground is uniform medium dense sand having the following properties: [Young's modulus of 40 MPa and Φ' (friction angle = 33 (degrees) and saturated density (γ') of 8 (KN/m³). Using any simplified method, estimate the stiffness of a 6 m diameter monopile with an embedded length of 30 m. The wall thickness of the monopile can be assumed to be 40 mm.

5.7 The developer in Problems 5.4, 5.5, and 5.6 wishes to install the 8 MW turbine in a seismic location. 6 m diameter monopile foundation is used with an embedment depth of 30 m. It is estimated that 5 m of soil will liquefy. Predict the change in natural frequency of the system. The ground is same as Problem 6 [Young's modulus of 40 MPa and Φ' (friction angle = 33 (degrees) and saturated density (γ') of 8 (KN/m³)]. Make assumptions for tower dimensions and RNA mass for 8 MW turbine.

Hint: As the ground is liquefied 5 m, the tower is now elongated by 5 m. The monopile is now embedded 25 m instead of 30 m.

5.8 A 5 MW turbine supported on a symmetrical four-legged jacket in deep waters, as shown in figure later. It is proposed to use 4 m diameter and 4 m length caisson as a foundation. The ground underneath is homogeneous stratum over bedrock with Young's modulus of 40 MPa. Details about the turbine specification and jacket dimensions are given in Table 5.6. Details on the foundation are given in Table 5.7.

a) Estimate the vertical stiffness of the caisson.
b) Make a prediction on whether the system will have rocking mode of vibration or sway bending mode of vibration.

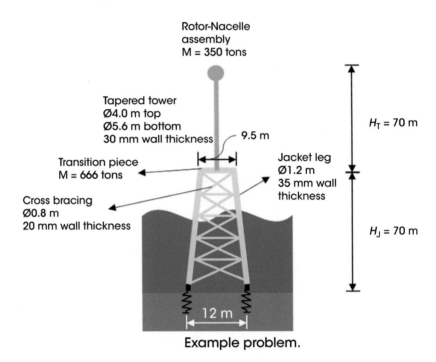

Rotor-Nacelle
assembly
M = 350 tons

Tapered tower
Ø4.0 m top
Ø5.6 m bottom
30 mm wall thickness

9.5 m

H_T = 70 m

Transition piece
M = 666 tons

Jacket leg
Ø1.2 m
35 mm wall
thickness

Cross bracing
Ø0.8 m
20 mm wall thickness

H_J = 70 m

12 m

Example problem.

Table 5.6 Details of the jacket structure.

Parameter	Value	Unit
Height of the jacket (hJ)	70	m
Jacket bottom width (L_{bottom})	12	m
Jacket top width (L_{top})	9.5	m
Area of jacket leg (AC)	0.1281	m^2
Distributed mass of the jacket including diagonals (mJ)	8150	kg/m
Tower height (h_T)	70	m
Bottom diameter of the tower (D_{bottom})	5.6	m
Top diameter of the tower (D_{top})	4.0	m
Distributed mass of the tower (mT)	3730	kg/m
Mass of rotor-nacelle assembly (MRNA)	350	tons
Mass of transition piece (MTP)	666	tons

Table 5.7 Foundation details for the example problem.

Parameter	Value	Unit
Foundation depth (L)	4	m
Foundation diameter (D)	4	m
Depth to bed rock (H)	50	m
Soil Young's modulus (Es)	40	MPa
Soil Poisson's ratio (vs)	0.28	Nondimensional

Answers to questions

Q5.1 Increasing the size of the turbine decreases the cost per unit of electricity generated. The reasons are

(c) Power output is proportional to the square of blade length, that is, a small increase in blade length leads to a large increase in power output.

(d) Longer blades require a higher hub height, which provides more consistent wind speed.

(e) Installation costs are relatively lower due to a smaller number of locations.

However, designing big turbines is challenging. Figure 5.41 compares 1P and 3P frequency range of a range of turbines. They are compared with wind and wave frequency. However, installation gets challenging and requires specialist vessels, that is, heavy Lift.

Q5.2 Onshore turbines are limited in size due to lack of transport infrastructure (roads and vehicles for transportation) and installation cranes. Offshore turbine components can be directly transported to the location using vessels.

Q5.3 The steps are shown in Figure 5.43.

Q5.4 The loads are wind, wave, 1P, and 3P. Wind is mostly cyclic loading causing fatigue types of issues. 1P, 3P, and wave are dynamic.

Q5.5 This depends on the bottom fixed or floating; see the text for answers.

Q5.6 Survey the locations of foundation for turbines and substation and the cable laying locations.

Q5.7 Apart from the requirement of ULS, that is, material does not yield under loading, there is also need to look into the driving stresses. How they will be installed?

Q5.8 Bottom fixed have more stringent rotation and deflection criteria.

1. Seabed preparation   2. Transportation of piles   3. Suspension of piles

4. Pile installation   5. Survey tool installation   6. Transportation of transition piece

7. Installation of transition piece   8. Bolting and grouting   9. Transportation of transition pices

Figure 5.43 Major steps in the installation of wind turbines.

Q5.9   As the turbines gets larger, the target frequency is lower and will impact the wave frequency. Therefore, wave loads get dynamic.

Q5.10   East Coast water depth is shallow and therefore bottom fixed. On the other hand, West Coast is deeper water and therefore floating.

References

1. S. Bhattacharya. *Design of Foundations for Offshore Wind Turbines*. John Wiley & Sons (2019).
2. BOEM (2023). https://www.boem.gov/renewable-energy
3. NREL Wind Exchange (2023). https://windexchange.energy.gov/
4. S. Bhattacharya *et al.* Civil Engineering Challenges Associated with Design of Offshore Wind Turbines with Special Reference to China, in *Wind Energy Engineering*. Academic Press, pp. 243–273 (2017).
5. Z. Lian *et al.* Potential influence of offshore wind farms on the marine stratification in the waters adjacent to China. *Journal of Marine Science and Engineering*, 10, 1872 (2022). doi: 10.3390/jmse10121872.
6. C. Menéndez-Vicente *et al.* Numerical study on the effects of scour on monopile foundations for Offshore Wind Turbines: the case of Robin Rigg wind farm. *Soil Dynamics and Earthquake Engineering*, 167, 107803 (2023).
7. Z. Westgate *et al.* Glauconite sand challenges for US offshore wind development. *Proceedings of the ASME 2022 4th International Offshore Wind Technical Conference. ASME 2022 4th International Offshore Wind Technical Conference*. Boston, Massachusetts, USA. December 7–8, 2022. V001T02A002. ASME. https://doi.org/10.1115/IOWTC2022-98666.
8. S. Amani. Seismic analysis and design of offshore wind turbines. Diss. University of Surrey (2023).
9. S. Jalbi *et al.* Dynamic design considerations for offshore wind turbine jackets supported on multiple foundations. *Marine Structures*, 67, 102631 (2019a).
10. S. Bhattacharya *et al.* Seismic design of offshore wind turbines: good, bad and unknowns. *Energies*, 14(12), 3496 (2021).
11. L. Arany *et al.* Closed form solution of Eigen frequency of monopile supported offshore wind turbines in deeper waters incorporating stiffness of substructure and SSI. *Soil Dynamics and Earthquake Engineering*, 83, 18–32 (2016).
12. IEC. 61400-1: Wind turbines part 1: Design requirements. *International Electrotechnical Commission*, 177 (2005).
13. IEC (2009). International Standard IEC 61400-3 wind turbines – Part 3: design requirements for offshore wind turbines, 1.0 ed. International Electrotechnical Commission (IEC).
14. DNV (2014). Offshore Standard DNV-OS-J101 Design of offshore wind turbine structures. Høvik, Norway.
15. L. Arany *et al.* Design of monopiles for offshore wind turbines in 10 steps. *Soil Dynamics and Earthquake Engineering*, 92, 126–152 (2017).
16. S. Jalbi *et al.* A method to predict the cyclic loading profiles (one-way or two-way) for monopile supported offshore wind turbines. *Marine Structures*, 63, 65–83 (2019b).
17. W.-H. Hui *et al.* Resonance Phenomenon in a Wind Turbine System Under Operational Conditions, in A. Cunha *et al.* (eds.). *Proceedings of the 9th International Conference on Structural Dynamics, EURODYN 2014*. Porto: Portugal, 30 June–2 July (2014).
18. R. Varghese *et al.* A compendium of formulae for natural frequencies of offshore wind turbine structures. *Energies*, 15(8), 2967 (2022).
19. S. Jalbi *et al.* Impedance functions for rigid skirted caissons supporting offshore wind turbines. *Ocean Engineering*, 150, 21–35 (2018).
20. A.R. Salem *et al.* Vertical stiffness functions of rigid skirted caissons supporting offshore wind turbines. *Journal of Marine Science and Engineering*, 9(6), 573 (2021).
21. G.I. Taylor. The spectrum of turbulence. *Proceedings of the Royal Society A: Mathematical, Physical and Engineering Sciences*, 164, 476–490 (1938).
22. M. Shadlou and S. Bhattacharya. Dynamic stiffness of monopiles supporting offshore wind turbine generators. *Soil Dynamics and Earthquake Engineering*, 88, 15–32 (2016).

6 Operations and maintenance (O&M)

6.1   Introduction

Wind turbines (WT) must be reliable for wind energy to become a major constituent in a much-needed transition from fossil fuels to renewable and sustainable energy systems. Operation and maintenance (O&M) systems and procedures are of paramount importance in ensuring the reliability of WT operation, while O&M data analysis can guide design and construction improvements. The first step in improving wind turbine reliability is to accurately determine the criticalities of WT and define factors affecting reliability and associated level of impact. Compilation and analysis of historical O&M data are required as a basis of reliability and availability analyses that combines prediction of potential failures with preventive measures for minimizing failures. To the extent that failures of WT are unpredictable, the uncertainty poses great risks associated with loss of energy production and loss of asset. When criticalities and factors affecting WT reliabilities are quantitatively assessed, then design improvements, operations scheduling, condition monitoring, and spare parts management can be done accordingly, and uncertainty can be reduced. O&M of offshore installations carry additional risks due to accessibility and operations dependence on weather conditions.

The advancement of the sector requires a strategic O&M implementation that will enhance availability, optimize profit, and minimize downtime. The O&M costs consist of failed components prices, personnel costs, crew and components transfer, and fixed cost such as fees and insurance. Turbine O&M costs represent the single largest component of overall OPEX and a major constituent (e.g., up to 30%) of the levelized cost of electricity (LCOE).[1] In addition, the cost of lost energy production during maintenance and repair downtime further increases the LCOE. Therefore, wind energy professionals need tools to predict time to failure, time to repair and cost for failures for WT. This chapter offers a review of the literature on wind turbine failures, a statistical analysis of failure data to understand the mechanisms and components associated with expected failures, and a review of methodologies for studying and improving reliability of wind turbine operations.

Onshore and Offshore Wind Energy: Evolution, Grid Integration, and Impact, Second Edition.
Vasilis Fthenakis, Subhamoy Bhattacharya, and Paul A. Lynn.
© 2025 John Wiley & Sons Ltd. Published 2025 by John Wiley & Sons Ltd.
Companion website: www.wiley.com/go/fthenakis/windenergy2e

6.2   Maintenance practices and tools

Maintenance practices for WT can be characterized as *preventive* and *corrective*. Preventive maintenance can be defined as scheduled maintenance and condition-based maintenance. Scheduled maintenance is done upon recommended intervals by wind turbine manufacturers, and it is generally done on 6-month intervals for onshore WT. Condition-based maintenance requires a condition monitoring system (CMS) involving an extra investment cost, but it is more likely to prevent unexpected high failure costs, as compared to just scheduled maintenance. Corrective maintenance is, on the other hand, done upon a failure which requires repair or replacement.

There are several condition monitoring techniques that can be used in scheduled maintenance as well as in condition-monitoring-based maintenance; these are listed below:

- Performance monitoring is used as wind turbine condition monitoring (e.g., monitoring of power in association to wind speed and direction, rotor blade angle and rotor speed).
- Visual inspection.
- Ultrasonic testing techniques (UTTs) are applied on wind turbine towers and rotor blades to determine surface and subsurface structural deterioration.
- Acoustic emission employing transducers and optic fiber displacement sensors to detect faults in gearboxes, bearings, shafts, and blades.
- Thermography used for detecting hot spots in electrical and electronic equipment as well as rotor blades by using infrared cameras.
- Radiographic inspection using X-ray imaging for revealing tight delamination or cracks in a wind turbine component.

The main detection methods for wind turbine condition examination are:

- Vibration analysis (VA) applied to WT components such as shafts, bearings, and rotor blades and subsystems such as gearbox.
- Oil analysis (OA) applied to determine the quality of oil inside a wind turbine gearbox whether there is debris contaminant because of a damage in bearings and gearings in a gearbox.
- Strain measurement (SM) applied mostly to wind turbine blades by using strain gauges to quantify stress level in situ and forecast lifetime.
- Electrical effects applied to electrical equipment such as motors, generators, and accumulators in a wind turbine.
- Shock pulse method (SPM) used to detect damages in a bearing using transducers by reading the signals.

6.3   Wind turbine O&M data

In wind farm O&M practice, historical failure events can be recorded from CMSs, supervisory control and data acquisition (SCADA) systems, failure reports, and maintenance logs.

Historical events are commonly summarized for reliability studies and presented in terms of *failure rate*, which is the failure frequency given by the number of failures per turbine per unit time, and *downtime per failure*, which is given as the time during which a turbine does not produce power output due to a failure. Thus, the *failure rate* λ_i of subassembly i, estimated as the number f failures per turbine per year, can be determined by the below equations [2]:

$$\lambda_i = \frac{\sum_{p=1}^{P} n_{i,p}}{\sum_{p=1}^{P} N_p(T_p/8760)}, \tag{6.1}$$

where $n_{i,p}$ is the number of failures of subassembly i in period p, N_p is the number of WTs considered in period p, and T_p is the time duration of period p in hours. The failure rate of the entire wind farm is calculated as the sum of the failure rates of all its subassemblies:

$$\lambda = \frac{\sum_{i=1}^{I} \sum_{p=1}^{P} n_{i,p}}{\sum_{p=1}^{P} N_p(T_p/8760)} \tag{6.2}$$

where I is the total number of subassemblies in the turbine. The numerator of Equation (6.2) represents the total number of failures of the turbine, and the denominator represents the total number of turbine years within the survey.

Similarly to the failure statistics, the downtime estimates are either average calculated downtime per failure for each subassembly D_i, or total downtime in a certain interval. The calculation of D_i is shown in Equation (6.3).

$$\overline{D}_i = \frac{\sum_{p=1}^{P} D_{i,p}}{\sum_{p=1}^{P} n_{i,p}}, \tag{6.3}$$

where $D_{I,p}$ is the total downtime due to failures of subassembly I in period p. The numerator of Equation (6.3) is the total downtime caused by failures of subassembly i, and the denominator is the total number of failures in the entire duration of a database.

The reliability data, including failure rate and downtime, are critical for the design, operation, maintenance, and performance assessment of WTs but are often difficult to obtain; that is why we devote a good fraction of this chapter in the analysis of such data.

6.3.1 Onshore WT O&M data

The earliest comprehensive O&M survey is the "Wissenschaftliches Mess- und Evaluierungsprogramm" (WMEP), which included wind turbine operations in Germany between 1989 and 2008. This survey included all scheduled and unscheduled maintenance logs for 1500 onshore turbines with 64,000 individual events. Figure 6.1 shows the failure rate and downtime values for a subset of 575 turbines from the WMEP database for which complete log data were available.[3] For a schematic of the associated components see Figure 1.13. As shown, electric and control systems had by far higher failure rates than the rest of the subsystems whereas hub, control system, and rotor blades have the highest downtime values (Table 6.1).

Pfaffel et al.[4] added data on the WMEP analysis by reviewing WT O&M surveys around the world. They showed failure rate and downtime results from the CIRCE database with 13,000

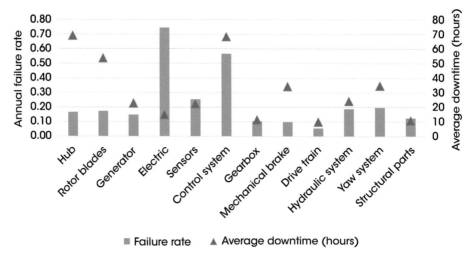

Figure 6.1 Average annual failure rate and downtime values of WT subsystems for 575 wind turbines in the WMEP database (*Source:* Ozturk/MDPI/CC BY 4.0).

Table 6.1 Components of a wind turbine.

Systems	Components
Brake system	Brake disk, spring, Motor
Cables	
Gearbox	Toothed gear wheels, Pump, oil heater/cooler, Hoses, Low-speed shaft, High-speed shaft,
Generator	Bearings, rotor, stator, Coil
Main shaft	Low speed shaft, high Speed shaft, bearings, Couplings
Nacelle housing	Nacelle
Pitch system	Pitch motor, gears
Power converter	Power electronic switch, cable, DC bus
Rotor blades	Blades
Rotor hub	Hub, air brake
Screws	
Tower	Tower, Foundation
Transformer	Controllers
Yaw system	Yaw drive, yaw motor

operational years data from Spain, the Elforsk/Vindstat database with 3100 operational years data from Sweden, the Huadian database with 547 operational years and the University Nanjing databases with 330 operational years both from China, the LWK with 6000 operational years and WMEP with 15,357 operational years databases from Germany, and the VTT Finish database. It is noted that although every initiative has its own data collection, there is always the possibility of overlapping data sets, especially among the databases that correspond to the same geographical region. As shown in Figure 6.2, except for University Nanjing database, the rest of the databases show similar results for both failure rate and downtime of WT. The rotor system is usually the most frequently failed subsystem followed by the transmission and control systems. For the downtime, the highest values are mostly for drive train system followed by rotor and transmission systems.[4]

The high failure rate (47 failures per WT and year) shown by the University Nanjing data is more than 100 times as high as the lowest failure rate (0.4) shown by the Elforsk/Vindstat data. On the other hand the University Nanjing data show the lowest mean down time per failure (0.18 days per failure) which is just 3.3% of the Elforsk/Vindstat mean down time (5.4 days per failure). These differences are mainly due to the different sources and different consideration of events of these initiatives. Comparing the total annual downtime as

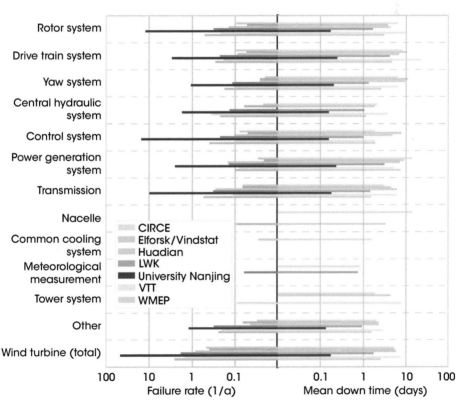

Figure 6.2 Wind turbine failure rate and downtime results from various databases around the world (*Source:* Pfaffel et al.[4]/MDPI/CC BY 4.0).

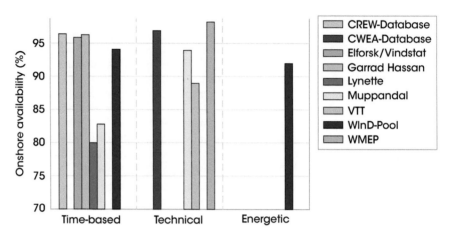

Figure 6.3 Availability of onshore wind turbines based on various databases (*Source:* Pfaffel et al.[4]/MDPI/CC BY 4.0).

the product of failure rate and mean down time, downtime numbers vary between 2.2 and 10.6 days per year.

Accordingly, availability is estimated to be as high as 97–99% in the inland WT best cases (Figure 6.3).

Sheng also reports availability of about 98% and mean time between failures (MTBFs) of more than 7000 hours, implying a failure rate of a little over 1 failure(s)/turbine per year for most modern land-based WT.[5]

It is noted that the Lynette database which shows a low *time-based availability* of 80% is 35 years old. Another low estimate (83%) is derived by the Muppandal database, but the same database shows a *technical availability* of 94%; this shows the significance of availability definitions. Let us examine these definitions:

Time-based availability provides information on the share of time where a WT is operating or able to operate in comparison to the total time. Various definitions for the calculation of time based availability exist in the wind sector. The definition used in these results shown in this figure follows the "System Operational Availability" of the IEC 61400-25-1 Standard where all times of full and partial performance and low wind conditions are considered as available.

On the other hand, *technical availability* is a variation of the time-based availability which gives information on the share of time where a WT is available from a local perspective. For this purpose, down time with external causes like grid failures or lightning is considered as available and scheduled maintenance and associated forced shutdown are excluded from the calculation.

A third definition of WT availability, *energetic availability*, also known as production-based availability, gives an indication about the turbine's energy yield compared to the potential output and thereby includes long down time during high wind speed phases and derated operation. Thus, all differences between potential and actual production are assumed to be losses. The estimation of energetic availability requires wind speed data.

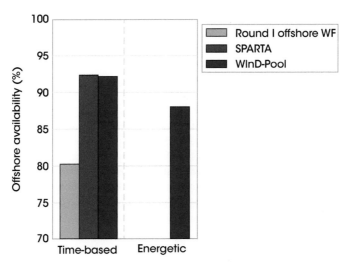

Figure 6.4 Availability of offshore wind turbines based on various databases (*Source:* Pfaffel et al.[4]/MDPI/CC BY 4.0).

Due to harsher environmental conditions and more difficult accessibility, offshore WT tends to have lower availability than onshore ones; an 80–92% availability is shown in Figure 6.4. The *time-based availability* based on the SPARTA and WInD-Pool data are almost the same. The low availability of the offshore WF is caused by major technical problems during the first years of this first offshore tender in the UK. It can be assumed that these problems have now been overcome and the results of WInD-Pool and Sparta, showing ~92% availability, are more representative of the current situation.

6.3.2 Offshore WT O&M data

Data on off-shore WT failure are scarce, but offshore wind turbine life costing analyses show that O&M costs can represent up to more than 30% of the total costs of WT throughout their life.[6] Furthermore, turbine blades along with the gearbox and electrical generators have been identified as the components with the greater failure rates of the turbines. In contrast with gearboxes, electrical generators, and other components, where the use of condition monitoring is much more mature and is currently implemented in some turbines, the blades are rarely instrumented, and their inspections and maintenance are usually calendar based, which creates an opportunity for improvement on their reliability and life extension by switching to condition-based types of maintenance.

6.3.3 Onshore and offshore O&M data analysis

A recent wind turbine reliability data review[2] included data from 18 databases with 18,000 WTs, corresponding to over 90,000 turbine-years of which four of the databases a totaling 1551 WTs correspond to 3300 turbine-years of offshore operation in European coasts. This section is based entirely on this source. Their findings associated to onshore WT are like those based on the WMP database that were shown earlier. Thus, *subassemblies exhibiting higher average failure rates are the electrical, control system, pitch system, blades, and hub,* whereas subassemblies like blakes, shafts and bearings, nacelle, and structure exhibit low average failure rates. A comparison between onshore and offshore WT reliability, although is based on a few data offshore WT, is instructive of the risks experienced in the emerging offshore wind market.

Two populations of onshore and offshore WTs are analyzed simultaneously to identify the similarities and differences. The top five critical subassemblies in onshore WTs in order of significance are electrical, control system, blades and hub, pitch, and generator (Figure 6.5). These subassemblies are also critical in offshore wind turbines (OWTs), but the order of criticality is slightly different. The pitch system is the most critical subassembly in offshore data sources, appearing in three of the four offshore sources, while the generator is critical in two of the four sources. The electric system, control system, blades and hub, and yaw system appear to be critical in terms of failure rates for both onshore and offshore WTs.

The critical subassemblies in terms of downtime are presented in Figure 6.6. Although there are some discrepancies in the average downtime per failure between sources, the criticality ranking of subassemblies in terms of downtime is quite consistent for both onshore and

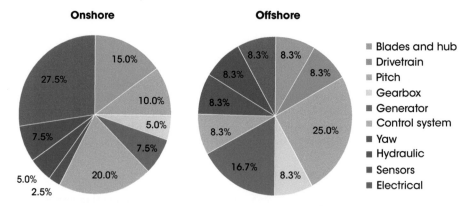

Figure 6.5 Critical subassemblies in terms of failure rate (*Source:* Dao et al.[2]/John Wiley & Sons/CC BY 4.0).

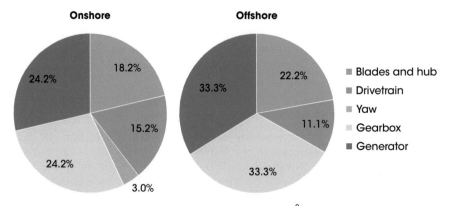

Figure 6.6 Critical components in terms of downtime (*Source:* Dao et al.[2]/John Wiley & Sons/CC BY 4.0).

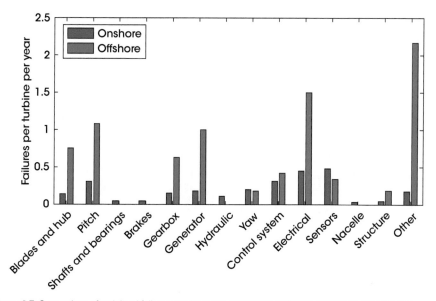

Figure 6.7 Comparison of weighted failure rates between onshore and offshore databases (*Source:* Dao et al.[2]/John Wiley & Sons/CC BY 4.0).

offshore populations. Failures of the *gearbox, generator, and blades* and *hub* result in the largest *downtimes per failure* for both onshore and offshore WTs.

Figure 6.7 shows onshore and offshore failure rates; the onshore failure rates are determined from 12 onshore data sources, whereas the offshore failure rates are from a single data source. As shown in this figure, the failure rates for offshore WTs are generally higher than those for onshore WTs, and this applies to almost all subassemblies. *Electrical, control system, generator, blades and hub, and pitch systems all experience high failure rates for both populations, and their average failure rates are higher offshore than onshore.* Structure and gearbox subassemblies follow the same pattern, but their failure rates are outside the top five highest failure rate subassemblies. The two interesting and irregular subassemblies are the yaw system and sensors which have higher failure rates onshore than offshore.

When individual subassembly data are combined, the offshore WT failure rate is roughly three times the weighted average onshore WT failure rate. The difference between onshore and offshore failure rates can partly be explained by the offshore severe operating conditions, such as higher mean wind speed and corrosive saltwater. Under the effects of the marine environment, offshore WT structure is subjected to a larger loading variation in high wind speeds. Thus, the failure rates of many major subassemblies such as blades and hub, gearbox, generator, structure, and electrical components of offshore WTs are higher than those of onshore WTs. In addition, larger WTs tend to experience more failures compared with small ones, so it should be noted that part of the difference may be due to the larger size/power rating of the offshore WTs than the onshore average in the databases.

The onshore stop rates shown in Figure 6.8 are calculated from two databases, CREW (USA) and East China, and the offshore stop rates are from the Noordzee Wind database. We observe that the subassembly stop rates are much higher than their failure rates from

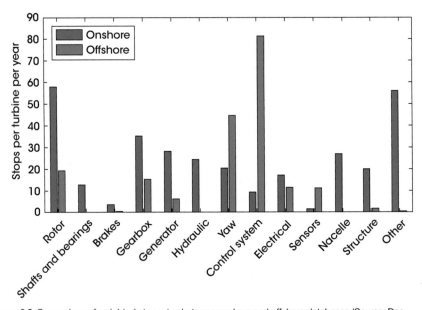

Figure 6.8 Comparison of weighted stop rates between onshore and offshore databases (*Source:* Dao et al.[2]/John Wiley & Sons/CC BY 4.0).

other sources and, although the failure rates for offshore subassemblies are higher than those for onshore subassemblies, the stop rate comparison shows a different trend. Almost all subassemblies, except control and yaw systems, experience higher stop rates in the onshore population. Control system, yaw, and rotor are the high stop-rate subassemblies offshore, while rotor, gearbox, generator, and hydraulic are the high stop rate subassemblies onshore.

Figure 6.9 compares the stop time per event (the duration that a turbine does not generate power due to its stops) based on onshore and offshore data sources which report alarm event rates. For a given subassembly, the average stop time per event for offshore WTs is generally higher than that for onshore WTs. Across all subassemblies, on average, stop events last 3.9 hours onshore while offshore events last 7.7 hours. Accessibility is likely to be a reason behind this difference. Also, there are two particular subassemblies, gearbox and generator, which are associated with very high stop times in offshore WTs.

The offshore WT population in this comparison is on the Egmond aan Zee Wind Farm, where there was a manufacturer's fault that caused gearbox failure in almost all the turbines, thereby skewing the data. This gearbox failure propagated to the generator and caused the generator to fail, which is shown by the spikes in Figure 6.9. However, even without this unusual gearbox failure, the stop rates of major subassemblies, such as the electrical and rotor, are higher offshore than onshore. However, it is noted that there is only one offshore data source in the downtime/stop time comparison and that the population of offshore WTs is much smaller than that of onshore WT.

Thus, the offshore wind industry needs to further improve component reliability and to develop and apply O&M strategies to reduce the LCOE that is produced from WT.

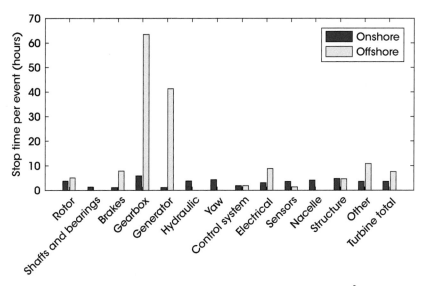

Figure 6.9 Stop time per event for onshore and offshore wind turbines (*Source:* Dao et al.[2]/John Wiley & Sons/CC BY 4.0).

6.3.4 The impact of O&M on the LCOE

As detailed in Chapter 8, the LCOE is the net present value (NPV) of the cost to produce electricity over a WT's lifetime based on the expected power generated. It includes all the life-cycle costs of the turbine that can be classified into two groups: capital investment expenditure (CAPEX) and operation and maintenance expenditure (OPEX). CAPEX includes the cost of the turbine, infrastructure, installation, and other upfront costs for wind farm development, deployment, and commissioning. OPEX covers all the costs to operate and maintain the turbine such as leases, insurance, administration, management, scheduled maintenance, unscheduled maintenance, spare parts, and other day-to-day operation costs.

6.3.4.1 Land-based WT

Lawrence Berkeley National Laboratory (LBNL) has compiled O&M cost data for 202 installed wind power projects, totaling 22,393 MW with commercial operation dates from 1982 to 2020.[7] This section of the book is entirely based on this LBNL report. It is noted that a full time series of O&M cost data, by year, is available for only a small number of projects, whereas in most cases, O&M data are available for just a subset of years of project operations. In most cases, the reported values include the costs of wages and materials associated with operating and maintaining the wind project, as well as rent. Other ongoing expenses, including general and administrative expenses, taxes, property insurance, depreciation, and workers' compensation insurance are generally not included. As such, the O&M cost data shown in Figures 6.10 and 6.11 may not represent the total operating expenses for wind power projects.

The data shown in these figures indicate that O&M costs are not uniform across projects. However, as shown in Figure 6.10, projects installed in the past decade have, on average, incurred lower O&M costs than those installed earlier. Specifically, capacity-weighted

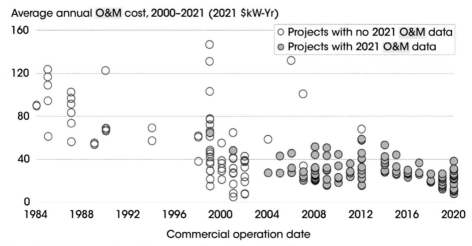

Source: Berkeley Lab; some data points suppressed to protect confidentiality

Figure 6.10 Onshore wind turbines average O&M costs, by commercial operation date. The chart highlights the 112 projects, totaling 18,330 MW, for which 2021 O&M cost data were available (*Source:* Wiser et al.[7]/U.S. Department of Energy Office of Scientific and Technical Information/Public Domain).

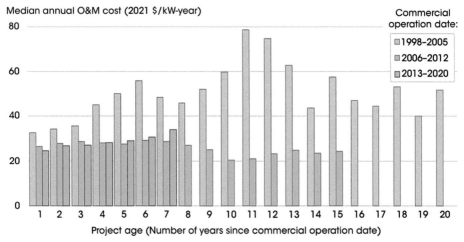

Source: Berkeley Lab; medians shown only for groups of two or more projects, and only projects >5 MW are included

Figure 6.11 Onshore wind turbines. Median annual O&M costs by project age and commercial operation date (*Source:* Wiser et al.[7] Land based Wind Market Report: 2022/U.S. Department of Energy Office of Scientific and Technical Information/Public Domain).

average 2000–2021 O&M costs for the 24 projects in the sample constructed in the 1980s equal $79/kW-year, dropping to $64/kW-year for the 37 projects installed in the 1990s, to $29/kW-year for the 65 projects installed in the 2000s, and $21/kW-year for the 76 projects installed since 2010. This decline may be due to at least two factors: (i) O&M costs generally increase as turbines age and component failures become more common and (ii) projects installed more recently, with larger and more mature turbines and more sophisticated O&M practices, may experience lower overall O&M costs.[6]

Limitations in the underlying data do not permit the influence of these two factors to be clearly distinguished. Nonetheless, to help illustrate key trends, Figure 6.11 shows median annual O&M costs over time, based on project age (i.e., the number of years since the commercial operation date) and segmented into three project-vintage groupings. Though sample size is limited, the data show a general upward trend in project-level O&M costs as projects age, at least among the oldest projects in the sample. Figure 6.11 also shows that projects installed over the past 15 years have had, in general, lower O&M costs than those installed in the earlier years of 1998–2005, at least for the first 15 years of operation.

6.3.4.2 Offshore wind turbines

Compared with an onshore WT, an offshore WT requires a higher initial investment and expectantly higher O&M costs due the harsh environment where they operate. During the construction phase, unfavorable weather can result in the delay of the commissioning date due to lack of accessibility of installation vessels. During the operation phase, in the occurrence of a failure, weather-related risks can increase the total downtime of the wind farm by impeding the access of the support vessels dispatched to perform maintenance activities, leading to revenue losses.[8]

However, offshore WT are typically larger than the land-based ones, and this reduces balance of system and O&M costs on a per-kilowatt basis. Also, the higher wind speeds that offshore access result in higher electricity production and may offset the higher O&M and installation costs associated with distance to shore and harsh meteorological conditions.

The cost of O&M adds to the LCOE and unexpected O&M components can make wind energy less competitive. The frequency and intensity of unpredictable failures can be reduced by building more robust and condition resilient WT at areas with well-established failure causes.

Although the LCOE in offshore wind generation has decreased recently, it is still much higher compared with onshore wind generation. As reported by Lazard (2023), the LCOE of off-shore WT is $86/MWh versus $26-$54/MWh for land-based WT.

However, this is a big improvement from 2015 prices when off-shore wind was three times more expensive than land-based wind ($181/MWh versus $61.MWh). Reducing LCOE is critical in making the offshore wind industry more competitive. Together with the improvement of turbine design and technology to reduce CAPEX, optimizing O&M strategy is essential to OPEX reduction and ultimately to LCOE optimization.

To reduce O&M costs associated with WT and drive down LCOE, the use of different strategies is important. For example, O&M of offshore WT can be inhibited by accessing difficulties resulting in cost increases. Traditionally, boats are used in carrying out crews for inspection, repair, and maintenance of OWTs. However, the use of helicopters can be beneficial because it reduces the travel time and the associated downtime (Figure 6.12).

Figure 6.12 Offshore helicopter service for accelerated crew transfer (*Source:* Edward et al. (2012) *Offshore Wind Cost Reduction Pathways Study*. https://www.thecrownestate.co.uk/media/1770/ei-offshore-wind-cost-reduction-pathways-study.pdf).

6.4   Statistical tools of reliability analysis

Statistical tools can be used to assess wind turbine reliability, availability, and failure costs considering environmental and operational factors. Such tools can assist wind turbine manufacturers for design decisions, wind farm operators for O&M activities and also inform other stakeholders such as insurance companies, government officials, and wind turbine researchers. Specifically, statistical tools such as failure modes, effect, and criticality analysis (FMECA), clustering and survival analyses, and Bayesian updating can reveal failure and downtime behaviors of WT, subsystems, and components depending on the operational and environmental conditions, and, thus, inform the wind energy community about expected wind turbine reliabilities and availabilities. Machine learning applications such as logistic regression (LR) and artificial neural networks (ANNs) can build data-driven models to be used as decision support tools. FMECA will be applied to evaluate criticalities of wind turbine subsystems and determine the failure and downtime behavior of components in critical subsystems, cluster and survival analyses will assess factors effecting wind turbine reliability, and LR and ANN models will model and predict wind turbine reliability. Modeling with LR and ANN methods can be done using statistical analysis software. Bayesian updating can be used to predict time to failure and time to repair probabilities, and ANNs can be used to develop a decision support tool to forecast failure cost of WT considering operational and environmental conditions.

6.4.1 Failure modes, effect, and criticality analysis (FMECA)

FMECA consists of four main components such as failure modes, failure causes, effects of the failures, and failure mode criticality numbers. Failure modes represent the type of failure occurring in every subsystem, whereas failure mechanisms are the causes that lead to failures. Effects of the failures are simply consequences, whereas failure mode criticality numbers are calculated as sum of expected cost from the failures and loss of energy production for every subsystem.

6.4.1.1 Failure modes

Failure modes can be classified as the failures which happen in the specific subsystem (i.e., blade failure, gearbox failure, and generator failure), or more specific ones such as fatigue and fractures in the toothed shaft of a gearbox, loss of function in the lubricant system depending on the data availability. The database which is used in this study does not include detailed data about the failure modes beyond the subsystem where the failure occurred.

6.4.1.2 Failure causes

Table 6.2 shows failure causes in the WMEP database. They are high wind, grid failure, lightning, icing, malfunction of control systems, component wear or failure, loosening of parts, other causes, and unknown causes. Grid failures in the WMEP database are assumed to occur only if there is a systematic grid failure and are not part of this investigation.

Hurricane and its category is detailed in Table 6.3.

Table 6.2 Locations, causes, and effects of the failure which are included in WMEP database.

Failure locations	Failure causes	Failure effects
Structures failures	High wind	Overspeed
Rotor blade failures	Grid failure	Overload
Mechanical brake failures	Lightning	Noise
Drive train failures	Icing	Vibration
Gearbox failures	Malfunction of control system	Reduced power
Generator failures	Component wear or failure	Causing follow-up damage
Yaw system failures	Loosening of parts	Plant stoppage
Sensor failures	Other causes	Other consequences
Hydraulic system failures	Cause unknown	–
Electrical system failures	NA[a]	NA
Control system failures	NA	NA
Hub failures	NA	NA

a) NA, data are not available.

Table 6.3 Definition of Hurricane.

Saffir-Simpson Hurricane Wind Scale
Climatology | Names | Wind Scale | Extremes | Models | Breakpoints

The Saffir-Simpson Hurricane Wind Scale is a 1–5 rating based only on a hurricane's maximum sustained wind speed. **This scale does not take into account other potentially deadly hazards such as storm surge, rainfall flooding, and tornadoes.**
The Saffir-Simpson Hurricane Wind Scale estimates potential property damage. While all hurricanes produce life-threatening winds, hurricanes rated Category 3 and higher are known as major hurricanes.[a] Major hurricanes can cause devastating to catastrophic wind damage and significant loss of life simply due to the strength of their winds. Hurricanes of all categories can produce deadly storm surge, rain-induced floods, and tornadoes. These hazards require people to take protective action, including evacuating from areas vulnerable to storm surge.

Category	Sustained winds	Types of damage due to hurricane winds
1	74–95 mph 64–82 kt 119–153 km/h	**Very dangerous winds will produce some damage:** well-constructed: frame homes could have damage to roof, shingles, vinyl siding, and gutters. Large branches of trees will snap and shallowly rooted trees may be toppled. Extensive damage to power lines and poles likely will result in power outages that could last a few to several days.
2	96–110 mph 83–95 kt 154–177 km/h	**Extremely dangerous winds will cause extensive damage:** well-constructed frame homes could sustain major roof and siding damage. Many shallowly rooted trees will be snapped or uprooted and block numerous roads. Near-total power loss is expected with outages that could last from several days to weeks.
3 (major)	111–129 mph 96–112 kt 178–208 km/h	**Devastating damage will occur:** well-built framed homes may incur major damage or removal of roof decking and gable ends. Many trees will be snapped or uprooted, blocking numerous roads. Electricity and water will be unavailable for several days to weeks after the storm passes.
4 (major)	130–156 mph 113–136 kt 209–251 km/h	**Catastrophic damage will occur:** well-built framed homes can sustain severe damage with loss of most of the roof structure and/or some exterior walls. Most trees will be snapped or uprooted and power poles downed. Fallen trees and power poles will isolate residential areas. Power outages will last weeks to possibly months. Most of the area will be uninhabitable for weeks or months.
5 (major)	157 mph or higher 137 kt or higher 252 km/h or higher	**Catastrophic damage will occur:** a high percentage of framed homes will be destroyed, with total roof failure and wall collapse. Fallen trees and power poles will isolate residential areas. power outages will last for weeks to possibly months. Most of the area will be uninhabitable for weeks or months.s

a) In the western North Pacific, the term "super typhoon" is used for tropical cyclones with sustained winds exceeding 150 mph.

Source: NOAA: https://www.nhc.noaa.gov/aboutsshws.php

6.4.1.3 Failure effects

Failure effects can be classified in the same way as failure causes which are given in Table 6.2. For example, we can consider the following effects of failures: overspeed, overload, noise, vibration, reduced power, causing follow-up damage, plant stoppage, and other consequences.

6.4.1.4   Criticality of failure modes

Criticality priority number (CPN) is the one of the most important outcomes of an FMECA application. CPN is estimated as follows:

$$Criticality\ Priority\ Number = Occurrence\ Severity * Consequence\ Severity$$
$$* Nondetection\ Severity \qquad (6.4)$$

Ozturk and Fthenakis applied FMECA on the WMEP data.[9] Their study has four methodological dimensions, namely: (i) obtaining reliability and availability metrics for an identical turbine model for the four Koppen–Geiger climatic regions in Germany:

- Cfa: Temperate-without dry season-hot summer
- Cfb: Temperate-without dry season-warm summer
- Dfb: Cold-without dry season-warm summer
- Dfc: Cold-without dry season-cold summer

Figure 6.13 shows the average failure rate and downtime values if the breakdown for climatic regions are not considered.

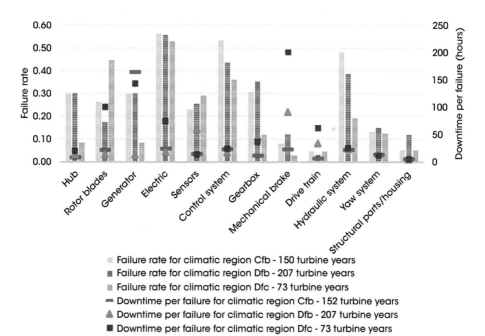

Failure rate for climatic region Cfb - 150 turbine years
Failure rate for climatic region Dfb - 207 turbine years
Failure rate for climatic region Dfc - 73 turbine years
Downtime per failure for climatic region Cfb - 152 turbine years
▲ Downtime per failure for climatic region Dfb - 207 turbine years
■ Downtime per failure for climatic region Dfc - 73 turbine years

Figure 6.13 Averaged failure rate and downtime per failure values for subsystems of a 500-kW geared-drive wind turbine model with 40 m tower height in different climatic regions (FR, failure rate; DT, downtime).

The most critical components in the generator subsystem are bearings; gearbox parts are the most affected components in every climatic region, followed by seals in gearbox failures. Wear is the dominant failure cause. Bearing failures have the noise and vibration effect following plant stoppage predominantly in the Dfc climatic region.

The highest failed component in the electric system is switches. The dominant cause is wear in all climatic regions. Rotor blade failures seem to be affected by the cold climatic region. Dfc climatic regions have an impact on the critical subsystems and cause failure in WT. This implies that the wind turbine operations and maintenance strategies for subsystems should be arranged taking into account of local climatic conditions of the turbines. For example, rotor blade downtimes and failure rates are impacted by colder climates, where longer downtime and higher failure rates are observed. Also, lightning became an important failure cause in cold climatic regions for rotor blade failures.

It is expected that most of the equipment failures would give malfunctioning signals about the potential for equipment malfunction. CMSs enable the detection of the failures in wind turbine subsystems such as gearbox, drive train, generator, and tower by the use of vibration, heat, and pressure sensors. In WT, however, failures often appear suddenly and cannot be detected. Visual inspection, checking the lubrication level in gearboxes, VA, and nondestructive testing methods, which include ultrasound and acoustic emissions in scheduled maintenance, are also other ways to detect potential anomalies in a wind turbine.

6.4.2 Logistic regression (LR)

LR is a machine learning method which examines the relationship between binary outcomes (dependent variable) such as whether a turbine is being frequently failing or nonfrequently failing and inputs (independent variables) such as factors affecting turbine failing frequency. LR uses maximum likelihood method to produce a best-fitting function to the data. The procedure for best fitting is applied by several iterations to maximize the probability of the observed data to be part of the correct classified outcome.

6.4.3 Artificial neural networks (ANNs)

ANN method is a machine learning technique to model complex problems and predict outcomes utilizing collected data and independent variables. The method mimics the learning and recalling behavior of human brain, which consists of massive number of neurons. Neurons in ANN are randomly connected with three different layers, namely input layer, hidden layer, and output layer.

Input layer has all information from the input pattern, whereas hidden layer connects the input layer to the output layer by an activation function, and the outcome layer transforms the hidden layer activation into the scale of desired output. The most widely used neural network is multilayer feed forward (MLF) with back propagation learning. The network algorithm utilizes gradient descent on the weights of the networks. The neural network algorithm repeats feed-forwarding and back propagating until it finds the optimum weights between the nodes which minimize the outcome error.[10]

6.5   Workforce education and training

An offshore wind energy industry is emerging in several countries, including the USA. This growth is supported by national- and state-level targets and commitments, including the Biden administration's national offshore wind target to reach 30 GW of installed offshore wind capacity by 2030 while developing a domestic workforce and state-level procurement commitments. As of 2022, there are two fully operational projects totaling 42 MW, whereas 19 projects in the USA offshore pipeline that have reached the permitting phase, and eight states have set offshore wind energy procurement goals totaling 39,322 MW by 2040.[11]

As detailed in a US-DOE/NREL report,[12] offshore wind energy projects are complex and require an extensive, varied, and well-trained workforce. It is expected that the offshore wind industry will create new jobs among engineering, management, land-based and offshore skilled trades, and professional services, requiring the development of unique skills, education, and experience to meet industry needs.

Specifically to O&M, this workforce includes long-term jobs for managing, overseeing, and conducting monitoring, testing, repair, and maintenance of WT and their components. O&M jobs also include marine crews who transport technicians to operate and maintain OWTs as well as onshore staff who provide management and engineering support.

O&M occupations can generally be categorized into two types: O&M crew and wind plant operation. The O&M crew includes wind turbine technicians and foundation and support structure maintenance engineers. The wind turbine technician will have training for repairs and troubleshooting using digital tools. The foundation and support structure maintenance engineers are structural specialists expected to have a civil or structural engineering degree and be trained in offshore-specific health safety and environmental protocols. The activities of these offshore O&M crews are supported by an onshore team that determines an operations plan and coordinates the execution of that plan for the offshore crew. This engineering, management, and oversight crew includes specialists with higher degrees, licenses, and field expertise, such as environmental science, computer science and data science, or business management. For example, data scientists are becoming increasingly valuable as smart asset management approaches for optimizing operations and maintenance are being implemented.

6.5.1   Safety training

Standardized safety training for workers who perform installation and operation activities at sea has been identified as one of the highest priority areas in the USA, to address to ensure that an adequately trained workforce is available to build projects. Currently, there is no official industry training standard; however, the Global Wind Organization (GWO) standards[13] are the most widely adopted. GWO is a nonprofit group of manufacturers and offshore wind energy project owners who join to "understand and reduce the risk associated with safety hazards in the wind industry" for land-based and offshore wind. GWO sets guidelines, based on risks encountered in the wind turbine generator environment, that are incorporated into training curricula by GWO-certified training providers.

As of 2022, in the USA, there are about 50 university and community colleges that offer some form of offshore wind energy courses as part of their curricula. College and vocational programs are typically designed to create a workforce entering the skilled trades in construction and manufacturing, and university programs are predominantly focused on creating a workforce by providing the education required for engineers, professional support roles, and scientists and researchers. The curricula offered vary widely and is difficult to assess the extent to which they are addressing workforce requirements. However, these programs do have the potential to create awareness and interest in the offshore wind industry and can potentially be expanded to further meet requirements as the industry develops. In addition, existing engineering programs, in particular offshore engineering or offshore energy, are well suited to address many of the requirements of the offshore wind industry and serve as a good foundation for an offshore-wind-specific supplement. Notable programs specific to offshore wind energy include those at the University of Maine and the University of Rhode Island both active in supporting east coast offshore wind farm development projects.

6.6 Summary

In the early years of wind turbine development in the 1970s and 1980s, the industry has been affected by reliability and availability issues, ranging from structural collapses to premature component failures in the early prototypes. Historical data show that operations and maintenance costs varied by project age and commercial operations date. Despite limited data availability, projects installed over the past 15 years have, on average, incurred lower O&M costs than older projects in their first years of operation. The data also suggest that O&M costs tend to increase as projects age, at least for the older projects in the databases.

O&M costs can be decreased by proactive maintenance by eliminating the level of effect of a potential failure as well as reducing downtime. Wind turbine reliability prediction and maintenance optimization may result in a substantial monetary benefit of preventive maintenance. Industry surveys indicate that corrective maintenance can save hundreds of thousands O&M costs during the operation life of a multi-MW wind turbine. O&M cost of onshore wind turbine can add 14 $/MWh or more to the cost of producing wind electricity. Furthermore, the cost for lost energy production during the maintenance activities further decreases the revenue.

As wind farm deployment is increasing and wind energy becomes a major constituent of electricity grids, around the world, reliability prediction for WT is becoming more important.

Therefore, the wind industry needs tools to help for their decision-making in their maintenance activities. Statistical models using machine learning applications, namely LR and ANN, can aid in decision-making of wind turbine maintenance activities and ultimately improve wind turbine reliability while minimizing O&M cost of wind energy. LR models provide odds ratios for risk factors, which can be interpreted to examine wind turbine reliability. ANN also provides complex relationship between risk factors and turbine reliability and provide prediction of turbine reliability.

Self-assessment questions

Q6.1   What is the expected availability of modern land-based and off-shore WT?

Q6.2 What is the reason for the difference in WT availability in land-based and off-shore installations?

Q6.3 What components are most likely to fail in land-based WT?

Q6.4 What components are most likely to fail in off-shore WT?

Q6.5 Give three availability definitions and explain their differences.

Problems

6.1 Calculate failure rates of the control system of WT from the data shown in Figure 6.1.

6.2 Calculate failure rates and downtime estimates of wind farms comprising 575 WT from the data shown in Figure 6.1.

6.3 (a) Calculate wind turbine failure rates and associated downtimes from the seven databases listed in Figure 6.2. (b) Compare the results obtained from the different databases. (c) Compare these results with the results obtained from the data of Figure 6.1.

6.4 Compare gear box failures and associated wind turbine downtimes for onshore and offshore WT and give likely reasons for the difference.

Hint: The data in Figures 6.5 and 6.6 show higher failures and longer downtimes in offshore turbines. This likely happened because of corrosion effects in and more difficult access to offshore WT.

6.5 Offshore wind maintenance following a hurricane. The students are advised to start working on this problems by reviewing the information is the US-DOE Wind Energy Program website: US DOE Wind Energy Program: https://www.energy.gov/eere/wind/wind-market-reports-2022-edition#offshore, and the NOAA website: https://oceanservice.noaa.gov/facts/cyclone.html

US Atlantic coast poses significant risk to OWTs from hurricanes. Figure A1 shows North Atlantic Tracking Chart for 2011, Figure A2 shows the wind farms in the east coast of USA, and Table 6.1 shows some details of the wind farms. Figure A3a shows a general pattern of major hurricane events affecting the US-NW. You should look for updates to these figures.

(a) The aim is to develop contingency planning in the case of hurricane. Failure rates for turbines are shown in Table 6.2. Consider a scenario where 2011 events are repeated. Estimate the number of turbines that will be unoperational in the assumed year. Make simplified assumptions such as the hurricane affects the whole wind farm. Use Geographical Information System (GIS) or Google Maps to locate the wind farms and follow the hurricane tracks. An example is done in Figure A3b, whereby it may be seen 1960s Hurricane Donna 5 passes through the edge of CVOW wind farm.

(b) Estimate the time and resources (vessel and cost) required to reinstate the wind farms. You may use the data in the appendix for your calculations.

(c) Estimate the downtime and the loss of revenue for the developers.

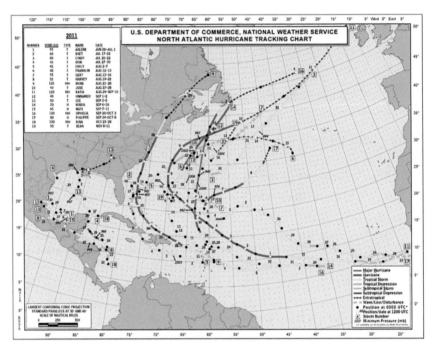

Figure A1 North Atlantic hurricane track.

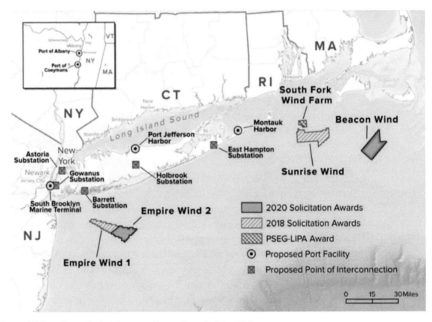

Figure A2 Locations of wind farms on the northeast of the USA.

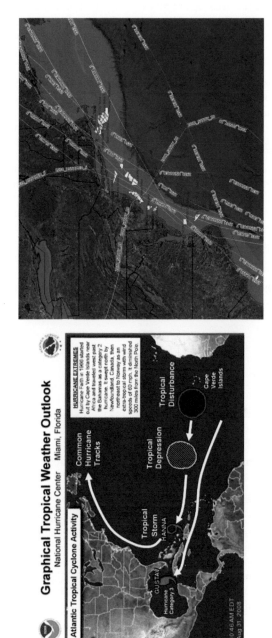

Figure A3 (a) General understanding of hurricane tracks; (b) hurricane tracks through some wind farms.

6.6 Maintenance following typhoon in Taiwan

Hurricanes (as it called in Problem 1) and typhoons are essentially the same weather phenomenon. They are all large tropical storm systems that revolve around an area of low pressure and produce heavy rain and wind speeds exceeding 74 mph (118 kph). Taiwan's location makes it vulnerable to the western North Pacific Ocean tropical cyclone that regularly produces damaging typhoons (Figures A4 and A5).

(a) Consider a scenario where a typhoon sweeps through the entire Taiwan coasts. The typhoon track is equivalent to hurricane category 3. Estimate the failures.

(b) Estimate the time and resources (vessel and cost) required to reinstate.

6.7 Maintenance following an earthquake

Offshore wind farms are also constructed in earthquake areas. Figure A6 shows the seismicity in Taiwan. It is estimated that that for a scenario earthquake subsurface liquefaction is expected, and this will require remedial work for scour protection of the foundation and cable protection systems. For a particular wind farm, there are 100 km of inter array cables (IAC) and 60 km of export cable.

Estimate the cost for remedial measures.

6.8 Scour protection design and maintenance

For a given location with seabed current of 1.2 m/s and foundation dimensions shown in Figure A7, what are the available designs for scour protection? Compare the materials needed

Use the example from Vineyard Project (United States)

6.9 For the conditions of Problem 6.4, what is the cost of remedial action if scour protection was not used? (Figures A8 and A9).

6.10 Cable damage during storms and maintenance

Following cables used in a wind farm with no-scour protection. Following a hurricane, during the survey, it was found the scour depth is 3 m and the cables need to be protected (Figures A10–A12).

(a) What are the possible options to protect the cables?

(b) In each case, compare the bending radius of the cable with the original.

6.11 Failure rates for different components in offshore wind turbines

Among the various classifications presented in academic literature, the reference designation system for power plants (RDS-PP) designation (VGB PowerTech, 2012) is widely adopted for data collection and aggregation from many wind farm developers and operators, offering sufficient and manageable details for system and subsystem identification and components' technical information. Table A2 presents a high-level classification that will be used later to present the findings from different sources (*Source:* Adopted from Pfaffel et al. (2017)).

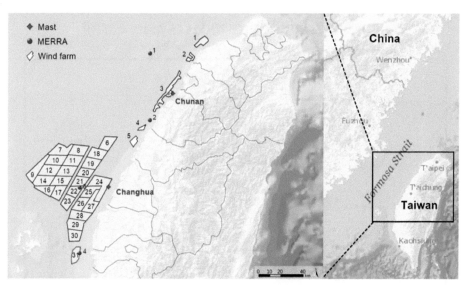

Figure A4 Wind farm in Taiwan (*Source:* European Chamber of Commerce Taiwan, 2022 Global Offshore Wind Summit Taiwan - 2022 Global Offshore Wind Summit Taiwan (ecct.com.tw)).

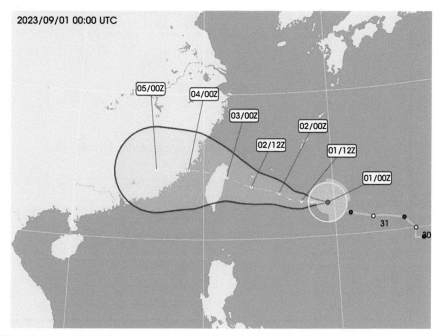

Figure A5 Typhoon tracks (*Source:* Taiwan English News / https://taiwanenglishnews.com/tag/weather//last accessed April 02, 2024).

The seismicity around Taiwan Strait

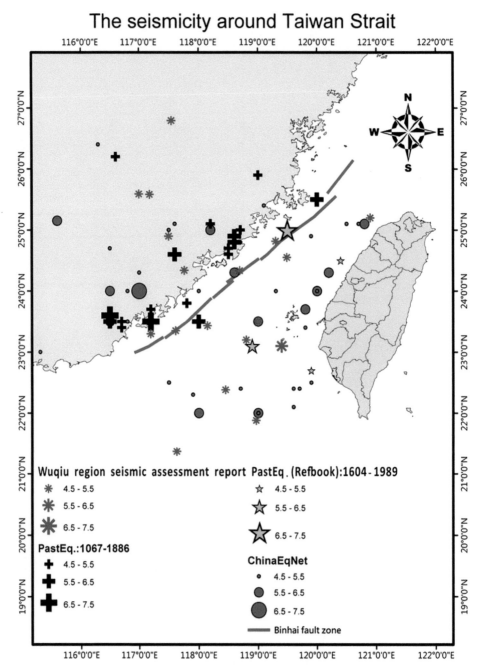

Figure A6 Seismicity in Taiwan (*Source:* Yu-Kai Wang/MDPI/CC BY 4.0/doi:10.3390/en9121036).

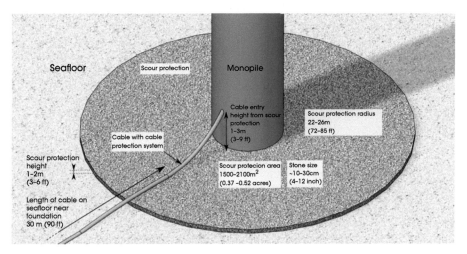

Figure A7 Schematic of a generic scour protection design.

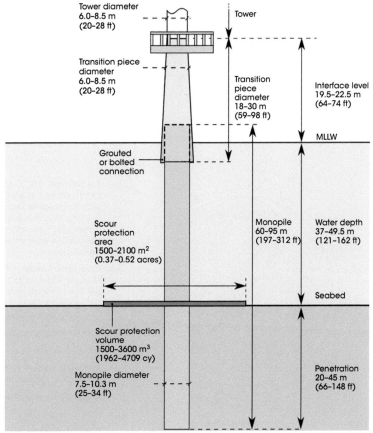

Vineyard Wing Project

Figure A8 Schematic of sample scour protection design for monopile foundations (*Source:* source vineyard wind).

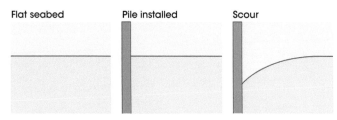

Figure A9 No scour protection and scour development.

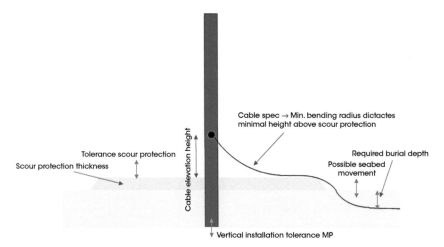

Figure A10 Sketch of major parameters influencing anchoring cable dimension.

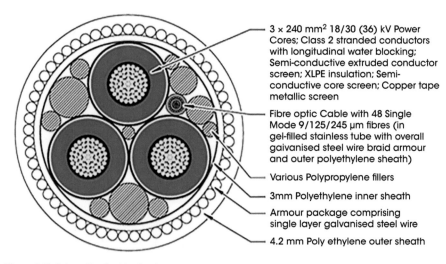

3 × 240 mm² 18/30 (36) kV Power Cores; Class 2 stranded conductors with longitudinal water blocking; Semi-conductive extruded conductor screen; XLPE insulation; Semi-conductive core screen; Copper tape metallic screen

Fibre optic Cable with 48 Single Mode 9/125/245 μm fibres (in gel-filled stainless tube with overall galvanised steel wire braid armour and outer polyethylene sheath)

Various Polypropylene fillers

3mm Polyethylene inner sheath

Armour package comprising single layer galvanised steel wire

4.2 mm Poly ethylene outer sheath

Figure A11 Schematic of cable structure.

Figure A12 Sketch of cable connection.

Table A2 Taxonomy of systems.

Labels

1. Rotor system	**6. Power generation system**
1a. Rotor blades	**7. Transmission**
1b. Rotor hub unit	7a. Converter system
1c. Rotor brake system	7b. Generator transformer system
1d. Pitch system	**8. Nacelle**
2. Drivetrain system	**9. Common cooling system**
2a. Speed conversion system	**10. Meteorological measurement**
2b. Brake system drivetrain	**11. Tower system**
3. Yaw system	11a. Tower
4. Central hydraulic system	11b. Foundation system
5. Control system	**12. Others**

	Strath offshore		OWEZ	
	Repair	replacement	Repair	replacement
1. Rotor system	0.227	0.003	0.704	
1a. Rotor blades	0.010	0.001		0.054
1b. Rotor hub unit	0.038	0.001		
1c. Rotor brake system				
1d. Pitch system	0.179	0.001	0.649	
2. Drive train system	0.038	0.154		0.498
2a. Speed conversion system	0.038	0.154		0.486
2b. Brake system drivetrain				0.012
3. Yaw system	0.006	0.001		
4. Central hydraulic system				
5. Control system	0.054	0.001		
6. Power generation system	0.321	0.095		0.206
7. Transmission	0.154	0.010		0.381
7a. Converter system	0.081	0.005		0.195
7b. Generator transformer system	0.003	0.001		
8. Nacelle				
9. Common cooling system	0.007			
10. Meteorological measurement				
11. Tower system	0.089		0.052	
11a. Tower			0.052	
11b. Foundation system	0.089			
12. Others	0.166	0.001		

Data required:

(1) Vessel rates and typical time taken for different operations of wind farms (Scour Protection Blade installation, Foundation Installation, Cable laying)
(2) Classification of hurricane
(3) Failure rate for different components

Compare the failure rates from typhoon and earthquakes

Answers to questions

Q6.1 ~98% and 92%

Q6.2 Severe weather creates more access and operation problems in off-shore installations than in land-based ones.

Q6.3 Electrical, Control, Blades+hum; see figures 6.1 and 6.5

Q6.4 Q6.4. Pitch, Generator; see figure 6.5

Q6.5 Q6.5. Time-based-, Technical-, Energetic -availabilities; see text for their differences

Appendix 6.A: Hurricanes, typhoons, and cyclones

Source of the figure:

https://www.geographyrealm.com/tropical-cyclones-hurricanes-typhoons-and-cyclones/

https://www.geographyrealm.com/copyright/

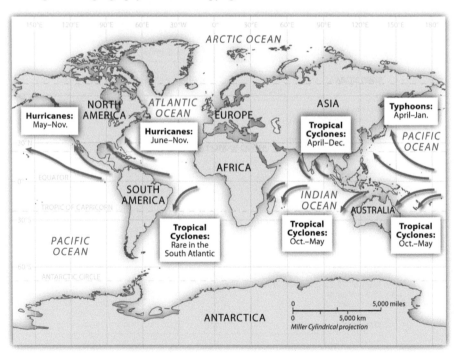

(*Source:* With permission of Lumen candela / https://courses.lumenlearning.com/atd-herkimer-worldgeography/
chapter/5-5-tropical-cyclones-hurricanes/).

Appendix 6.B: Price of OWF: Normalized cost: LCOE breakdown for European wind farms:

New turbines announced in the market by mid-twenties

Model	Company	Nameplate capacity (MW)	Serial production year	Height (m)	Blade length (m)	Rotor diameter (m)	W/m²
SG 14-220 DD	Siemens Gamesa	14 MW	2024	Site specific	108	220	368
SG 14-236 DD	Siemens Gamesa	14 MW	2024	Site specific	115	236	320
Haliade-X	General Electric	14 MW	2024	260	107	220	368
V236-15.0	Vestas	15 MW	2025	280	116	236	343
MySE 16.0-242	MingYang Smart Energy	16 MW	2026	264	118	242	348
TBD	TBD	20 MW	2030	300	136	270	350

For 1000 MW or more capacity wind farm:

LCOE cost breakdown for an offshore wind farm (2022 costs)

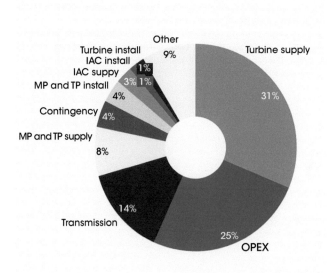

Total LCOE $100/MWh; MT; monopile; TP transition piece; IAC Inter Array Cable .(*Source*: Morten Kofoed Jensen/U.S. Department of Energy / https://www.nrel.gov/wind/assets/pdfs/engineering-wkshp2022-1-1-jensen.pdf. / Public Domain)

References

1. R. Wiser *et al*. *Assessing Wind Power Operating Costs in the United states: Results from a Survey of Wind Industry Experts*. Lawrence Berkeley National Laboratory (2019).
2. C. Dao *et al*. Wind turbine reliability data review and impacts on levelised cost of energy. *Wind Energy*, 22, 1848–1871 (2019).
3. S. Ozturk *et al*. Assessing the factors impacting on the reliability of wind turbines via survival analysis—a case study. *Energies*, 11, 3034 (2018).
4. S. Pfaffel *et al*. Performance and reliability of wind turbines: a review. *Energies*, 10(11), 1904 (2017).
5. S. Sheng and R. O'Connor. Reliability of Wind Turbines, Chapter 15, in T. Letcher (ed.). *Wind Energy Engineering*. Academic Press (2017).
6. J.C. Lopez and A. Kolios. Risk-based maintenance strategy selection for wind turbine composite Blades. *Energy Reports*, 8, 5541–5561 (2022).
7. R. Wiser *et al*. *Land-Based Wind Market Report*. Lawrence Berkeley National Laboratory (2022).
8. A. Ioannou *et al*. Informing parametric risk control policies for operational uncertainties of offshore wind energy assets. *Ocean Engineering*, 177, 1–11 (2019).
9. S. Ozturk *et al*. Failure modes, effects and criticality analysis for wind turbines considering climatic regions and comparing geared and direct drive wind turbines. *Energies*, 11, 2317 (2018).
10. S. Ozturk and V. Fthenakis. Predicting frequency, time-to-repair and costs of wind turbine failures. *Energies*, 13(5), 1149 (2020).
11. US DOE Wind Energy Program. https://www.energy.gov/eere/wind/wind-market-reports-2022-edition#offshore (2022).
12. US-DOE/NREL Offshore Wind Market Report. https://www.energy.gov/sites/default/files/2022-09/offshore-wind-market-report-2022-v2.pdf : (2022).
13. Global Wind Organization (GWO). https://www.globalwindsafety.org/.
14. H. Edward *et al*. *Offshore Wind Cost Reduction Pathways Study*. The Crown Estate. https://www.thecrownestate.co.uk/media/1770/ei-offshore-wind-cost-reduction-pathways-study.pdf (2012).

7 Grid integration

7.1 Setting the scene — national grids

The story of grid electricity goes back over a century. Edison's pioneering work in the late 1800s on the generation and distribution of direct current (DC) electricity was overtaken in the 20th century by the alternating current (AC) systems championed by Nikola Tesla (see Section 3.4.1), largely due to the ease with which AC voltage can be transformed – up for long-distance transmission, down for local distribution, and consumption. Early grids, often owned by private companies or municipalities, were generally small and local but were soon interconnected to form larger systems. Governments in industrial countries became keen to encourage the growth of electricity, realizing its importance for industrial innovation and efficiency and its key role in enhancing their citizens' quality of life. Today, it is impossible to imagine a modern economy and society without large amounts of electrical energy implemented via regional or national electricity grids.

The electricity grid is a complex network of power lines, designed to transport energy from suppliers to loads. In its normal configuration, power plants feed electricity into the high-voltage (>100 kV) transmission grid, which transports the energy to demand centers where the voltage is stepped down to deliver power to individual customers on the distribution grid. Both the European and the US electricity grids are more than a century old. The US electricity grid is a conglomerate of many smaller grid systems that were built in the late 19th century, each regulated by separate utility companies. The grid eventually grew into three major "Interconnects" on which all power plants operate synchronously. They are the Eastern Interconnection, Western Interconnection, and the Electric Reliability Council of Texas (ERCOT); there are very few links between the three interconnects. Figure 7.1(a) shows these interconnects, the North American Electric Reliability Corporation (NERC) interconnect subregions and about one hundred generator, load, and transmission balancing areas. Balancing authorities integrate power resources to meet demand within balancing areas; they are responsible for maintaining interconnection frequency and controlling the flow of power so that overloading of transmission lines is

Onshore and Offshore Wind Energy: Evolution, Grid Integration, and Impact, Second Edition.
Vasilis Fthenakis, Subhamoy Bhattacharya, and Paul A. Lynn.
© 2025 John Wiley & Sons Ltd. Published 2025 by John Wiley & Sons Ltd.
Companion website: www.wiley.com/go/fthenakis/windenergy2e

avoided. The USA power transmission grid consists of 300,000 km of lines operated by 500 companies.

The largest synchronous (by connected power) electrical grid is that of continental Europe connecting over 400 million customers in 24 countries. Although synchronous, some countries operate in a near-island mode, with low connectivity to other countries and there are plans to increase connectivity and host more renewable energy, while enhancing the reliability of the grid. To this effect, the European Electricity Grid Initiative (EEGI) has the objective to achieve a completely decarbonized electricity production by 2050 while integrating national networks into a market-based, truly pan-European network (Figure 7.1(b)).

A good example of an extensive system serving a developed economy is the UK's national grid,[1] originally developed in the 1930s to form Europe's largest synchronized AC electricity network. Operating at 50 Hz and up to 132 kV, it was upgraded in 1949 with some 275 kV links, followed by additional 400 kV links from 1965 onward. The highest voltages are used to transmit three-phase bulk electricity around the country, transformed down to progressively lower voltages for distribution to industrial, commercial, and domestic customers. There are also submarine links for bulk power transfer to and from France (2 GW) and The Netherlands (1 GW).

As grids expanded, gaining regional or even national coverage, the size of individual power plants increased dramatically. In the 1930s, the largest were typically rated at around 60 MW; half a century later they exceeded 1 GW. Currently, the largest UK power plant is the Drax coal-fired station in Yorkshire, producing up to 4 GW from six steam turbine generators each rated at 660 MW. The remarkable story of electricity supply in the 20th century is one of ever larger centralized power plants based on fossil fuel combustion and, in some countries, nuclear fission and hydroelectric generation.

The map of England, Wales, and Scotland in Figure 7.2 shows the current locations of power plants rated above 500 MW. These are principally coal-fired stations in traditional industrial areas; modern gas-fired units more widely dispersed; and nuclear plants, generally away from highly populated areas and close to the coast. Many smaller plants are not shown in the figure; nor is the vast distribution network ranging from high-power, high-voltage cables supported on huge pylons right down to the 230 V cables that bring single-phase AC electricity into individual households.

The UK's pattern of large power plants underlines a number of economic and operational criteria that apply equally well in other countries:

- Power plants are best located close to centers of population and industry, reducing the need for long transmission lines and minimizing transmission losses.
- Fuel transport costs are reduced by siting plants close to fuel sources (or coastal terminals if the fuel is imported).
- Power plants needing large quantities of cooling water are often placed beside shorelines, large lakes, or rivers.
- Public acceptability demands that nuclear plants are generally located well away from population centers.

Such considerations have largely determined the distribution of power plants shown in Figure 7.1. But what about renewable energy? Are the rules of the game being altered by the

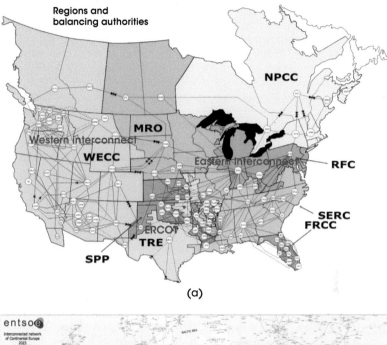

(a)

(b)

Figure 7.1 (a) The US electric grid: 3 major interconnects, 8 NERC subregions and about 100 balancing authorities (Source: Environmental and Energy Study Institute (EESI) / Public Domain).
(b) European electricity grid. Network elements are not located at their exact geographic location. The map shows transmission lines designed for 220 kV voltage and higher and generation stations with net generation capacity of more than 100 MW (Source: With permission of European Network of Transmission System Operators, 2023 / https://eepublicdownloads.entsoe.eu/clean-documents/Publications/maps/2023/230922/Map_Continental-Europe-2.500.000.pdf.)

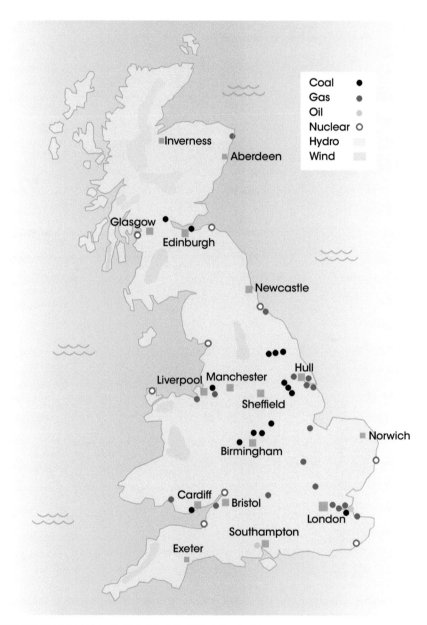

Figure 7.2 The large power plants that feed the UK's national grid.

rapid growth of wind power? Will grid networks need substantial modification, and must network designers and operators adopt new mindsets and learn new skills?

Nature does not generally cooperate by providing wind power where it is most needed, and turbines are often cited in areas that would not even be considered for traditional power plants. We are now in an era of increasing renewable generation, with grid penetration levels of 10%, 25%, or even higher on the horizon. Denmark, so often at the forefront of developments, aims to generate 50% of its electricity from the wind by 2050. Clearly, such plans have major implications for grid operators who must allow for increasing amounts of variable generation while maintaining security of supply. Big challenges undoubtedly lie ahead for electricity supply industries in many countries – probably the biggest since the arrival of nuclear power half a century ago.

You may already have noticed the colored tints on the map of Figure 7.1 – blue for hydroelectric and green for wind power – indicating regions favored for these two technologies. Needless to say, most do not meet the usual criteria for siting grid-connected power plants. Hydroelectric stations in the highlands of Scotland and a small mountainous area of North Wales are well away from centers of population and although Scotland looks set fair to become a major provider of renewable electricity from wind, wave, and tide, it will need heavy investment in new transmission from outlying regions. Many of the UK's onshore wind farms have been installed in the green-shaded areas, typically high uplands and coastal strips facing prevailing westerly winds, but it is neither cheap nor simple to connect them to the existing grid. And then, of course, there is a major new player – offshore wind, which needs submarine cable networks and associated substations.

Comparable challenges lie ahead for other countries. We have already mentioned Denmark. Referring to Figure 3.53 showing the liberal scattering of wind farms in northwest Germany, it is not hard to imagine the challenges posed by the safe and satisfactory delivery of their electricity. And how about the USA, with its vast tracts of windswept country far removed from big cities and major grid networks; or Spain, another leader, whose wind energy is often garnered from mountain territory? It is not simply a matter of grid *connection*, important though that is; rather it is one of *integration* – the successful blending of renewable and conventional energy sources into large grid networks to provide safe, reliable, and economic electricity on a national, or even international, scale. Grid integration is one of the big challenges facing the global wind and solar energy industries (Figure 7.4).

7.2 Electricity markets and types of power generators

A discussion of electricity markets is necessary to better understand the effects of variable generation on the grid.

Grid-friendly WT capabilities
Fault-ride through is the capability of electric generators to stay connected in short periods of voltage dips in electric distribution networks. It is needed to avoid a short circuit that could lead to a widespread loss of generation.

(Continued)

(Continued)

Automatic active power frequency control. The frequency in AC grids is kept constant within strict limits – typically at exactly 50 or 60 Hz. The frequency drops if more energy is consumed than the generators feed-in. The opposite occurs if there is an energy surplus – the grid frequency increases.

Static voltage support. In order to protect the connected loads, the voltage must be kept within defined limits – that applies to the distribution grid in particular. Inverters that have the ability to provide controlled inductive or capacitive reactive power can help guarantee the required voltage quality. Reactive power-compatible inverters can also be used to compensate for phase shifts that are caused by transformers, large motors, or long cable sections.

Dynamic grid support. Inverters with dynamic grid support functions act within milliseconds preventing a grid failure in spreading further. Dynamic grid support ensures that the inverter is ready to feed energy into the grid immediately after a drop in the grid voltage.

The price in energy markets is set by the dynamics of supply and demand under the geographical constraints imposed by the reach and limitations of the transmission grid. For this reason, large grid systems are split into smaller markets at major nodes or zones, each of them establishing a price for each hour of the day according to supply and demand. This is called *nodal pricing*, or locational-based marginal pricing (LBMP). Imbalances of supply and demand on single nodes can be overcome by importing electricity from other zones, if there are not transmission constraints.

The balancing of supply and demand on the grid is achieved by Independent Service Operators (ISO) through generator bidding using seasonal, week-ahead, day-ahead, and hour-ahead load forecasts. The ISO ranks all generator bids based on their bidding price and fills up the load forecast of each hour of the day starting from the cheapest bidder, considering transmission line limits. Any differences between the load following and actual load are settled, on real time, in the ancillary services market which consists of Regulation and Reserve resources. Regulation resources can quickly adjust their output to accommodate changes to the balance of supply and demand, upon receiving a signal from the ISO. To protect against the risk of a plant outage, the ISO also has in-service spinning reserves that can be on full capacity within 10 minutes, and nonspinning reserves that can respond within 30 minutes.

The grid operations outlined above are important when we consider the impacts of high penetration of variable generation (i.e., solar and wind) into the grid. Let us now discuss the different types of power plants according to the services they provide. Nuclear and large coal-fired and gas-fired power plants, which operate 24 hours a day serve a baseload, typically the demand that is required throughout the year (8760 hours). Peaker plant can be photovoltaics, at residential, commercial, and utility scales, and small natural gas- or diesel-fired units which operate only during the high demand days of the year, for example, hot summer afternoons in tropical and subtropical climates (depicted as up to 1000 hours in Figure 7.3). Medium-size natural gas- and coal-power plants satisfy the balance of daily loads (depicted as intermediate load in the same figure).

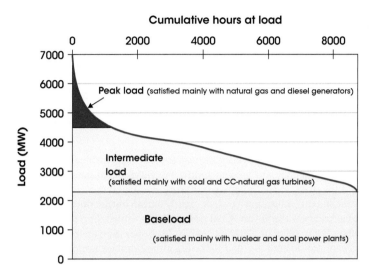

Figure 7.3 Example of a load duration curve.

7.2.1 Peaker plants

The market clearing price of electricity is set by the marginal price for the last MW to meet the load. During peak load hours, a large percentage of the generator fleet is dispatched, including "peaker plants," which are more expensive, less efficient, and more polluting than conventional generators.

These plants set a high electricity clearing price for all energy delivered in that time period, and it is therefore economically desirable to prevent their dispatch. Solar and wind generators have the potential to minimize the need for peakers, as peak demand is typically A/C-driven, and therefore effectively reduce the market clearing price while mitigating emissions.

Steam and gas turbines

Nuclear and coal power plants use steam turbines designed according to the Rankine cycle, thus a closed loop expansion and condensation of steam generated by the fission or combustion of fuel. Natural gas and diesel-fired turbines employ the Brayton cycle comprising alternative isentropic and isobaric process. There are two types of gas-fired power plants, viz. single cycle also called open-cycle gas turbine (OCGT) and combined-cycle gas turbine (CCGT) plants. OCGT plants consist of a single compressor/gas turbine that is connected to an electricity generator via a shaft. Air is compressed by the compressor, and its oxygen is burnt with natural gas in the combustion chamber of the gas turbine that drives both the compressor and the electricity generator. Almost two-thirds of the gross power output of the gas turbine is needed to compress air entailing high thermal losses, and the remaining one-third drives the electricity generator. CCGT use both a Brayron and a Rankine cycle as the later recovers rejected heat from the exhaust of the gas turbines to produce steam that drives a steam turbine and generates additional electric power. The Rankine engines have some flexibility within a range, by adjusting the flow of steam into the turbine, but the Brayton engines have even

---

(Continued)

> **(Continued)**
>
> greater flexibility as they have little thermal inertia and are capable to cycle and ramp quickly (10–30 minute startups, ~8%/min ramping).
>
> Typical thermal efficiency for utility-scale electrical generators operating at design capacity is around 33% for coal and oil-fired plants, between 35 and 42% for OCGT and 56–60% for CCGT. Their efficiencies decline when operating at partial-only capacity.

The power output of hydroelectric power plants, gas turbines, and CCGT can be effectively adjusted, and they are used to follow the variation of demand load throughout the day. Load following power plants run during the day and early evening when the demand is higher and are either shut down or greatly curtail output during the night, when the demand for electricity is the lowest. Coal power plants with sliding pressure operation of the steam generator can generate electricity at part load operation up to 75% of the nameplate capacity and can also be used for load following although their ramping rates are slower than those of gas turbines.

Intermediate and peak load power plants can also provide important voltage and frequency stabilization services in addition to supplying required loads.

Regulation reserves. Normally supplied by generators that have the ability to be dispatched up or down remotely. (Commonly referred to as automatic generation control or AGC.)

Contingency reserves. Can be supplied by generators that are online (spinning) and have generation capacity left to move up. Can also be supplied by "quick-start" generators that can be started and turned up within 15 minutes. The amount of power

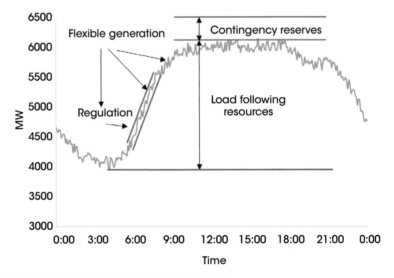

Figure 7.4 Power generators for load balancing and regulation.

that a generator can contribute is limited by its ramp rate and the room between its current dispatch level and its maximum capability. Contingency can also be supplied by demand response controlled by the system operator.

7.2.2 Baseload plants

Hydroelectric power plants can be efficient for both intermediate load following and baseload applications. Thermo-electric baseline generators (nuclear, large coal) use the Rankine cycle (discussed in text box above) and are built to operate at their maximum or near-maximum output 24 hours a day and are expensive to ramp or cycle as this incurs physical wear due to their high thermal inertia. CCGT power plants are also used for serving baseload and as the price of natural gas is being reduced they are displacing coal in baseline power plants.

A common critique of renewable power generation is that it is unsuitable for baseload supply; therefore, fossil power and nuclear power are needed. This critique is misleading. Baseload is a demand characteristic, not a supply technology characteristic. Nuclear or coal power plants are operated in baseload mode simply because they are not technically capable of operating in a more variable mode and must rely on high utilization to recover their high investment costs. In the future power system, the value of baseload will decrease. With higher shares of wind and solar electricity, supply and demand will be matched in a more flexible way. Variable renewable power generation can be combined with smart-grid technologies, demand response, energy storage, and more flexible generation technologies, including gas power plants and dispatchable renewable power supply options. A flexible, renewables-based power system is not only reliable but also economically efficient.

7.3 Connecting to the grid

7.3.1 Grid strength and fault levels

Over the past century, we have grown used to the idea of electricity grids as essentially "one-way" systems. Large AC generators in power plants (preferably out of sight, usually out of mind) feed electricity into the grid for delivery to homes, offices, and factories. Like water and gas supplies, the flow is one way. Who can imagine pumping water back into their kitchen taps, to be credited by the water company; or making their own gas, and pumping any excess back into the gas main? Yet, in recent years, it has been possible to buy a small wind turbine (WT), mount it on a pole in the garden, and pump electricity back into the cables that normally import it. In principle, an electricity grid can be supplied or tapped at any point.

Of course, a small domestic WT with a peak power output of, say, 2 kW is a very different proposition from a modern megawatt horizontal axis wind turbine (HAWT) rated at 4 MW – in fact, two thousand times different. Nobody suggests connecting a 4-MW HAWT to the domestic electricity supply, or a 2 kW machine directly to a large grid. But suppose a new wind farm is to be grid connected in a rural area. For a start, it is necessary to lay cables from the farm to the nearest suitable connection point (i.e., substation). But what is needed to make an efficient connection that is acceptable to the grid operator and nearby consumers? In this section, we deal with two important requirements:

- The wind farm, when operating in high winds at maximum output, must not overload local distribution cables and transformers.
- The wind farm must not cause unacceptable voltage fluctuations in the electricity supplied to local consumers.

To explore these issues, it is first necessary to appreciate the electrical layout of a typical grid, which is illustrated in Figure 7.5(a). On the left large power plant generators (we show just two, G_1 and G_2), generally with individual ratings well above 100 MW, feed three-phase electricity into a high-voltage (HV) transmission grid via step-up transformers (T_1 and T_2). Power is tapped at various points and fed into a medium-voltage (mv) distribution system via a step-down transformer (T_3) for delivery to local areas. And finally, it is transformed down again (T_4) to a voltage suitable for individual consumers (office, home, streetlights, etc.), who generally receive just one phase of the three-phase supply. Larger consumers such as factories are often connected to the mv system and supplied with three phases. The voltage levels in various parts of the system vary from country to country but are typically between 110 and 750 kV for HV transmission, 10 and 70 kV for mv distribution, and 120 or 240 V for final delivery to consumers. We have shown transformers in the figure but not the associated switchgear, fuses, and circuits that control power flow and prevent damage under fault conditions.

All this is conventional "one-way" delivery from large, centralized power plants to millions of individual consumers. There is no hint so far of renewable energy being fed into the grid, often by relatively small generators in far-flung parts of the network. To discuss this, we need

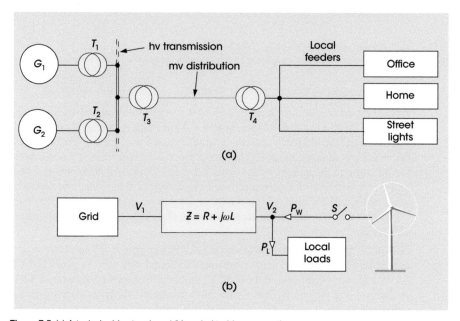

Figure 7.5 (a) A typical grid network and (b) a wind turbine connection.

some basic knowledge of AC circuits and associated terminology, and you may wish to refer back to the introduction given in Sections 3.4.1 and 3.4.2.1.

Part (b) of Figure 7.5 illustrates, in basic form, the electrical connection of a WT (or wind farm) to the mv distribution network. We assume that the "grid" on the left includes the large synchronous generators, transformers, and HV cables that make up a conventional high-power generation and transmission system, delivering a stable three-phase AC voltage (V_1) at a frequency of 50 or 60 Hz to the mv distribution network, represented by an impedance Z. On the right is a WT with a rated power P_W together with switchgear (S) that connects it to the nearest suitable point of the network. Also connected at this point are existing local loads that consume power (P_L). The voltage at the connection point is V_2.

We will start with the WT (or wind farm) disconnected. In this case, the local loads draw power from the grid in the normal manner. Ideally, the voltage V_2 they actually receive is very close to the nominal system voltage V_1. However, the current taken by the loads must flow through distribution cables and transformers, which have a combined impedance (Z) made up of resistance (R) and inductive reactance ($j\omega L$) in series. This causes a voltage drop equal to the product of current and impedance magnitudes and has the effect of reducing the voltage V_2 below V_1. Clearly, if the impedance is too large, or if too much current is being drawn, the voltage drop may become unacceptable. This is especially likely to happen if large loads are connected at the far end of a long distribution line.

You may be surprised at the representation of all supply cables and transformers by a single impedance Z. Typically, distribution cables are mainly resistive, causing real power losses; transformers are very efficient at transferring real power, but their windings are inherently inductive and need reactive power to set up and maintain their magnetic fields. So the combination of cables and transformers in a distribution network normally produces a complicated mix of resistance and reactance. Fortunately, AC circuit theory comes to the rescue, allowing us to lump all the individual elements together and represent them by a single impedance.

This idea has important implications for describing the quality of a grid connection. If Z is small (cables and transformers generously rated) in relation to the currents demanded by the loads, the voltage V_2 received by the loads remains very close to the nominal system voltage V_1 and the grid connection is said to be *strong*, or *stiff*. But if the impedance is comparatively large (low-rated cables and transformers), V_2 may deviate substantially from the system voltage, and the grid connection is said to be *weak*. Furthermore, the voltage will fluctuate substantially as individual loads are switched on and off – not a good recipe for customer satisfaction. Needless to say, strong grids are highly desirable from the customers' viewpoint and grid operators normally undertake to maintain supply voltages within, say, 10% of their nominal values. However, generous cables and transformers are expensive to install, especially in outlying areas, and a strong grid connection can easily become a weak one if more customers and loads are connected.

The usual way of quantifying grid strength at a particular point is in terms of its *fault level*. This is the product of the nominal system voltage and the current that would flow in the event of a short circuit. In Figure 7.5(b), the short-circuit current would be V_1/Z, so the fault level in volt–amperes (VA) is V_1^2/Z. For example, if the nominal voltage of a distribution system is 30 kV, and it has an impedance of 9 Ω at the connection point, the fault level is equal to:

$$\left(30 \times 10^3\right)^2/9 = 10^8\,\text{VA} = 100\,\text{MVA} \tag{7.1}$$

The fault level gives a good indication of the maximum load that can be connected without causing an unacceptable voltage drop. By definition, the short-circuit current is the maximum that could possibly flow, and it reduces the voltage V_2 to zero. But when a normal load is connected, the acceptable voltage drop is far smaller – say 5%. This in turn implies that the load VA should not exceed a few percent of the fault level. For example, loads totaling up to 5 MVA could probably be connected satisfactorily to a point where the fault level is 100 MVA.

We now come to an important question: what happens when the WT is run up to speed, connected to the grid, and starts generating? We should first note that large WTs often generate at 690 V and include their own step-up transformers; alternatively, in a wind farm, the power from all the turbines may be collected and transformed in the farm's substation. Whatever the details, we are assuming in Figure 7.5(b) that the turbine power P_W is connected to the grid at the same point as some local loads. Power cannot be stored, so the power flows from the turbine, into the loads, and through the impedance Z must always be in balance. Furthermore, *apparent power*, equal to the product of AC voltage and current and measured in VA, comprises *real power* measured in watts (W) and *reactive power* measured in reactive volt–amperes (VAR) – and both types of power must remain in balance.

All this leads to a complicated power balance at the grid connection point, making it difficult to predict exactly what will happen to the voltage V_2 as the turbine's power output varies and different loads are switched on and off. Generally, we may expect that whenever $P_W > P_L$ the voltage V_2 will tend to rise above V_1 as the turbine "pushes" real power into the grid through impedance Z; but when wind generation falls below local load demand, V_2 will fall below V_1. Detailed analysis is certainly possible[2] but beyond the scope of this chapter. However, it is clear that a strong grid with low impedance Z is highly desirable because it "ties" the voltage at the connection point more firmly to the nominal system voltage V_1, preventing unacceptable fluctuations as the wind strength and load demand vary. Large turbines attached to weak grids are a recipe for trouble!

Fault levels are good indicators of a grid's ability to accept power from a turbine or wind farm. For example, if a new wind farm is to be built with 10 large HAWTs and a rated output of 25 MW, it would be wise to seek a grid connection with a fault level above, say, 250 MVA. If this is not available, a network upgrade may be necessary. A key issue for grid operators as penetration levels of wind power continue to grow is the provision of grid capacity with adequate fault ride-through levels at all points of connection.

Weak grids caused major problems in California during the early days of the modern wind energy renaissance, sometimes referred to as the "Californian wind rush." A famous case centered on the Tehachapi Pass in the mountains north of Los Angeles where a huge number of turbines, relatively small by modern standards, were connected to an existing grid network.[3] Unfortunately, the 66 kV grid had been designed to serve a population of less than 20,000 people spread over a wide area, with a planned load of just 80 MW. But by 1986, the power output of Tehachapi turbines was peaking at 125 MW, and the system became overwhelmed. If ever there was an example of a weak grid, this was surely it. At the time there was very little understanding of the problem, which was exacerbated by the use of induction generators needing large amounts of reactive power that simply could not be transported over the cable network. Such mistakes, it seems safe to say, would not be repeated today. Actually, the Tehachapi problem was far greater than the connection of individual turbines or wind farms; it affected the operation and stability of the grid as a whole, and the lessons painfully learned were essentially those of *grid integration* – a theme we will develop a little later.

7.4   Electrical quality

In Section 7.3 we focused on the voltage fluctuations that affect an electrical grid, particularly when the grid is weak at the point of connection. The problem is important because electrical equipment, from industrial motors right down to consumer electronics, is designed to work at specific voltages and grid operators aim to keep voltage levels reasonably steady. However, there is more to "electrical quality" than relatively slow changes in grid voltage over time scales of minutes or hours, typically caused by turbines as the average wind speed varies. The *rapidity* of fluctuations is also very important.

A turbine or wind farm in a weak grid may produce rapid voltage changes that make lights flicker and annoy nearby consumers. Such *voltage flicker* is often caused by turbulent wind conditions, or turbine connection and disconnection. Other potential causes are disturbances to the wind flow as each blade passes the tower, referred to as *tower shadow*, and cyclic variations in rotor torque caused by wind shear. In a large three-bladed turbine, these effects occur at three times rotation speed, referred to as the *blade-passing frequency* – say between 15 and 50 times a minute. (We should distinguish voltage flicker from the flicker caused when sunlight is interrupted by blades as they sweep the sky.)

It is also important to realize that the amount of flicker caused by a large HAWT depends on its aerodynamic and electrical design. "Fixed speed" turbines based on induction generators are particularly susceptible, whereas modern variable-speed machines tend to iron out flicker effects by absorbing power surges as changes in rotor kinetic energy. The use of blade pitching to limit output power at high wind speeds may produce flicker if the pitch mechanism fails to respond to sudden gusts. Fortunately, the latest designs incorporating electronic power converters offer increased scope for rapid and flexible control of the voltage and power delivered to the grid.

Physiologists tell us that the human eye is most sensitive to brightness fluctuations at frequencies of around 8–10 Hz, but flicker comes in many guises and is highly subjective. Occasional connection and disconnection of turbines may be acceptable, whereas repeated on–off cycles as the wind hovers around cut-in speed can try the patience. Random flicker caused by wind turbulence is perceived as qualitatively different from regular, periodic, flicker caused by tower shadow. All are highly dependent on the strength of the grid at the point of connection and make any annoyance hard to predict or quantify.

Another important aspect of a grid's electrical quality concerns waveform distortion. AC voltages at all points on a grid are ideally sinusoidal in form, and so are the currents drawn by *linear* loads, as assumed in our introduction to AC circuits in Section 3.4.1. But *nonlinear* loads take non-sinusoidal currents from the supply. A familiar example is a dimmer switch controlling the brightness of an electric light, which works by switching the current rapidly on and off during each cycle of the voltage waveform. Dimmer lights are hardly likely to upset the operation of a grid, but large nonlinear loads are a different matter. By demanding non-sinusoidal currents, they tend to distort the voltage waveforms appearing on a grid – especially where it is weak – and in serious cases may produce overheating or failure of other equipment, faulty operation of protective devices, nuisance tripping of sensitive loads, and interference with communications circuits.[3] Electric utilities normally impose strict limits on the amount of voltage distortion that can be introduced into their networks by large nonlinear loads.

The other side of the coin concerns generation. Large synchronous generators that feed conventional electricity grids are expertly designed to produce sinusoidal voltage waveforms with minimal distortion, but WTs can be less than ideal. "Fixed speed," stall-controlled, machines cause little distortion because their induction generators are essentially linear machines. However, the power converters in today's variable-speed HAWTs incorporate electronic switching circuits that, by their very nature, can only approximate smooth sinusoidal waveforms. Here is a considerable challenge for designers, who must ensure that distortion caused by power converters remains within acceptable limits.

Waveform distortion is normally assessed in terms of *harmonics*. A pure sinusoidal voltage contains a single frequency, referred to as the *fundamental*. But if the waveform becomes distorted additional frequencies, known as harmonics, are introduced at integer multiples of the fundamental. For example, a distorted 50 Hz waveform may contain harmonic components at 100, 150, 200 Hz, and so on. The underlying concept is that of *Fourier analysis*, named after French mathematician and physicist Jean Baptiste Baron de Fourier (1768–1830), who showed that any periodic waveform, regardless of its shape, may be formed by adding together a set of sinusoids with appropriate amplitudes and phases. Switching waveforms tend to be rich in harmonics.

The amount of distortion in a waveform is usually measured in terms of its *total harmonic distortion (THD)*, which is the ratio of the energy in all harmonic frequencies to the energy in the fundamental, expressed as a percentage. THD in the high-power transmission section of a large electrical grid is normally very small – say below 1% – but it tends to grow as the electricity is transformed down and distributed to local areas and, finally, to individual consumers. A large WT incorporating electronic power converters tends to raise the THD on the distribution network, especially if the grid connection is relatively weak.

Two key roles of power converters in variable-speed WTs have already been shown in Figures 3.39 and 3.46: Conversion of AC to DC (rectification) and from DC to AC (inversion). The DC link between rectifier and inverter effectively isolates the frequency of the generator from that of the grid, allowing a turbine to run at variable speed. *Voltage-source inverters*, which use electronic switches to create an AC waveform from a source of DC voltage, are widely used in turbines, and we can illustrate their principles with some typical waveforms.

Figure 7.6(a) shows one cycle (period) of a pure sinusoidal voltage, the ideal output waveform for an inverter feeding an electricity grid. Part (b) shows a staircase-like approximation consisting of six discrete voltage steps – not a very accurate representation, but one that can be obtained with a simple switching circuit known as a *6-pulse inverter*. Three-phase versions can be made using just 12 electronic switches turned on and off at appropriate points in each cycle.[2] The high THD of this waveform may be reduced by increasing the number of steps in the staircase, at the expense of more complicated circuitry. When the number of steps is sufficient to reduce the THD to less than a few percent, manufacturers often describe an inverter output as "true sine wave" with a purity similar to grid electricity. True sine wave inverters are widely used to run sensitive low-power electronic equipment such as computers from a DC battery supply.

However, when it comes to the high-power inverters used in large variable-speed HAWTs with megawatt ratings, an alternative approach known as *pulse code modulation (PCM)* is widely used. It is illustrated in part (c) of the figure. The AC voltage waveform is synthesized as a series of constant-height pulses with widths determined by the value of the sinusoid at various points in the cycle. The pulses are widest when the sinusoid is at its peak positive value, and vice-versa. For clarity, we have shown just 10 pulses (and 20 switching operations)

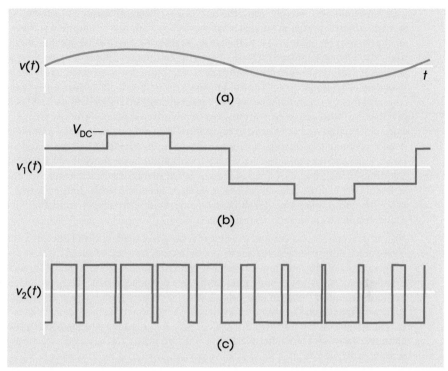

Figure 7.6 (a) A sinusoidal voltage waveform and its approximation by (b) a staircase waveform, and (c) a PCM waveform.

per cycle of the sinusoid, but in practice there are many more (switching rates are commonly in the kilohertz range). This has the beneficial effect of greatly increasing the frequencies of waveform harmonics, moving them far above the fundamental and reducing their nuisance value or, alternatively, making them easier to suppress by *filtering* (see later).

The PCM waveform in part (c) is not obviously a useful approximation to the sinusoid in part (a). But imagine that there are actually 100 pulses per cycle, rather than the 10 shown, and that we draw a smooth curve through the switching waveform to give a *running average*. It is not hard to see that this average will be high where the sinusoid is high, and vice-versa; in other words, by performing a running average we may extract the underlying fluctuations of the sine wave. In electronic terms, this is achieved by *low-pass filtering*. That is, the low frequencies in the waveform are passed, or retained, but the high-frequency ones are suppressed. In this case, we need to preserve the fundamental and suppress the harmonics. Filter theory is beyond the scope of this book, so we just note that it relies upon the frequency-dependent properties of inductors and capacitors to discriminate in favor of some frequencies at the expense of others.[2, 3] PCM techniques, together with filtering, can form the basis of high-power inverters with very low THD levels.

We have concentrated on the effects a turbine can have on the quality of grid electricity, and for anyone interested in the design and operation of WTs, these are clearly important issues; yet there is another side to the story – the effects that fluctuations of grid voltage and frequency, and grid outages, can have on the operation and safety of turbines. Supply and demand of active power on the grid must always be in balance, or the grid frequency will deviate far from its set point (60 Hz in the USA, 50 Hz in Europe), causing connected appliances to shut down or get damaged. The frequency of the system would vary as load and generation change. If there is more generation than demand, frequency goes up; if there is less generation than demand, frequency goes down. During a severe overload caused by tripping or failure of generators or transmission lines, the power system frequency will decline because of an imbalance of load versus generation. Loss of an interconnection, while exporting power, will cause system frequency to rise. Also, temporary frequency changes are an unavoidable consequence of changing demand. AGC is used to maintain scheduled frequency and interchange power flows and the presence of many generators and a large, distributed load allows for easy frequency management. Control systems in power plants detect changes in the network-wide frequency and adjust mechanical power input to generators back to their target frequency. Modern wind power plants (WPP) can also respond to frequency regulation by curtailing or increasing power instantaneously (see Section 7.7).

Not surprisingly, these also tend to be more serious in a weak grid and affect the older type of fixed-speed, stall-regulated, turbines more than modern variable-speed machines with power converters. An interesting case is India, where large numbers of stall-regulated machines were installed over many years in comparatively weak grid networks.[3] The experience gained is no doubt standing India in good stead as its wind energy industry mushrooms. Another example comes from Germany, where the concentration of turbines connected to mv distribution systems in the northwest of the country (see Figure 3.53) has highlighted concerns about two-way interaction between grids and turbines.[3]

We have covered a number of relatively local issues that arise when turbines and wind farms are connected to an electricity grid. It is now time to consider the effects that wind energy has on grid systems as a whole – and how they cope with increasing amounts of renewable generation.

7.5 Large-scale wind power

A large modern electricity grid connects hundreds of generators to thousands of kilometers of transmission and distribution lines and finally to millions of consumers. Often taken for granted by the public, electricity grids are among the most important strategic and economic assets of any modern society. From a technical point of view, they are complex engineering systems with operational features that can only be addressed at the system's level.

An important factor is grid strength at the point of WT connection. Local voltage fluctuations, flicker, and waveform distortion are certainly important to local consumers, but they hardly ruffle the feathers of a large grid. What matters to system planners and operators is the reliable and economic supply of electricity over a regional, national, or even international network.

Grid integration of large amounts of wind power has received a great deal of academic and practical attention over the past 30 years,[3] and much has been learned from countries at the forefront of developments including Denmark, Germany, Spain, and the USA. It is only fair to add that plenty of misinformation has been peddled by vested interests that oppose the growth of renewable electricity. We will examine the main issues in the following pages, but first it is helpful to list previous topics that relate to the operation of large electricity grids:

- The power produced by WTs and conventional power plants in relation to the annual electricity demands of households (Section 1.4).
- The mushrooming growth of wind energy capacity, national and global, and the development of offshore wind (Section 1.5).
- The wild nature of the wind and the importance of wind forecasting (Chapter 2).
- The nature of AC electricity and the characteristics of electrical generators (Section 3.4).

Conventional grids are powered by synchronous generators driven by steam, gas, and hydroelectric turbines that are firmly "tied" to the grid and must rotate at speeds dictated by the grid frequency. Individual synchronous generators are able to control their terminal voltage and power factor by varying field excitation. Operators keep the system frequency and voltage within narrow limits by matching total generation of real and reactive power to total demand, requesting changes in output from various power plants in a highly skilled balancing act.

Wind power tends to disturb the complex, but well-established, control of conventional synchronous generation in several ways:

- Large HAWTs use a variety of generating systems, synchronous and asynchronous, with and without power converters.
- The rated power of a small-to-medium wind farm (say 5–50 MW) is much smaller than that of a single, large, conventional synchronous generator (say 100–750 MW).
- The wind is uncontrollable.

None of these differences is of much concern to grid operators as long as the penetration of wind power remains small. But as it rises, and as the installed capacity of individual wind farms, especially offshore, begins to rival that of conventional power plants, the picture changes and successful grid integration moves to the top of the agenda. It is time to consider some large-scale system issues that affect the operation of a grid as a whole.

7.6   Intermittency and variability

Turbines stop running when the wind dies. People watching a turbine start and stop in response to varying wind speeds may be tempted to conclude that intermittency makes wind an unreliable source of energy.

In any event, the intermittency of a wind farm is less than that of its individual turbines. Different wind conditions across the site mean that some machines may be generating while others are stationary. When all are working, some generate more power than others and individual outputs vary in response to wind turbulence. Since the fluctuations are at least partly unsynchronized (uncorrelated) between different turbines the net effect is reduced variability in the output of the farm as a whole. This important effect, which grows with the capacity and extent of a wind farm, is referred to as *power smoothing*.[2]

Moving up the power scale, the operators of large electricity grids are certainly not concerned with the intermittency of individual turbines, or even of wind farms – unless they are very large. What matters is the *total* wind generation at a given time, gathered across the network from widely dispersed wind farms experiencing different wind conditions. It is therefore rather misleading to describe wind energy as intermittent at the power system level. It is certainly variable, but the power smoothing effect is greatly enhanced by a good geographical spread of wind farms because, for example, wind conditions in Scotland hardly ever match those in the south of England. In future, there will be trading of wind power between nations including those bordering the North Sea, leading to even more "geographical smoothing." But the big picture is rarely understood, or appreciated, by casual observers of stationary turbines.

Wind farms are not the only generators to display variability. No generating plant operates continuously or is 100% dependable at times of peak demand. In fact, the public might be quite surprised to learn how often a conventional plant is shut down in fault conditions or for maintenance – and how long repair can take, especially in the case of nuclear power where safety is such an issue. A large grid system does not notice the shutdown of a few megawatt HAWTs, but it must certainly respond rapidly to the unexpected loss of gigawatts from a coal-fired or nuclear plant.

However, we must admit that wind power's variability, as opposed to "on–off" intermittency, is significant for system operators. The wind is wild and does not blow to order; wind energy cannot be stored but must be used when available. Figure 7.7 shows a typical record

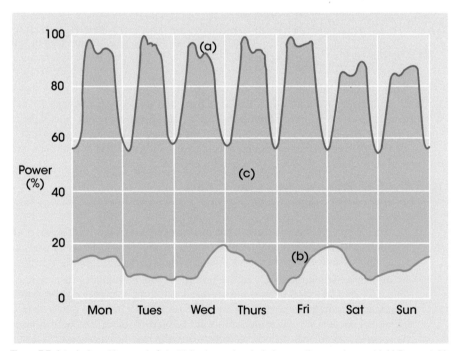

Figure 7.7 A typical weekly record of electricity demand and wind generation on a large grid. (a) Demand; (b) wind energy generation; (c) net load.

of electricity demand and wind generation in a large grid over a 7-day period, expressed as percentage of peak demand. The demand curve (a) displays a fairly regular daily cycle with a morning peak as the population rises and starts work, a flattening off in the middle of the day, and a smaller peak in the early evening. Loads are smaller at the weekend because industry and commerce are largely inactive. Wind generation, shown by curve (b), has peaks and troughs dictated by the weather rather than human activity. In this example wind supplies up to about 20% of peak demand but not, of course, in synchronism with it. The shaded area between the red and blue curves, labeled (c), represents the power that must be supplied on a minute-by-minute basis by all non-wind generation such as fossil-fuel, hydro, nuclear and other renewables, which we call net load herein.

A grid's ability to accommodate continually changing patterns of wind generation is sometimes questioned. But curve (a) emphasizes a crucial point: grid operators are already used to high levels of variability, not in supply, but *in demand*. They must cope with large daily swings, often up to 40 or 50% of peak demand, and schedule generation capacity accordingly. Power plants are continually being brought in and out of service, adjusting their outputs, and matching demand on a minute-by-minute basis. Actually, consumer demand can be even more fickle than suggested by the regular peaks and troughs in the figure. Famous – we might say notorious – examples are the sudden surges that occur at half-time in a World Cup football match, or during breaks in popular TV dramas, when people watching television rush to their electric kettles to brew quick cups of tea or coffee. The effect, known as *TV pickup*, is well known to the UK's national grid operators who have learned to cope with gigawatt fluctuations within a few minutes, balancing the sudden "ramping" of demand with generation by fast-response power plant. The speed of such changes rivals anything the wind can do!

We see that variability is a major and inherent feature of large electricity grids. Demand fluctuates on a wide range of time scales, from minutes to months and even years. Faults occasionally shut down large power plants or transmission lines. The injection of wind power certainly adds to the overall variability of the system, but not to the extent that many people assume. The example in Figure 7.7 shows wind contributing about 14% of demand over a whole week. It is interesting to note that as of 2023 Denmark generates about 53% of its annual electricity from the wind exceeding their 2050 goal, Portugal 26%, Spain 23%, and Germany 17%, and none of these penetration levels presents serious technical or operational difficulties.

Successful adaptation to wind's variability is made much easier by accurate forecasting. The science of wind forecasting is less developed than that of electrical load forecasting, and it is easier to predict wind speeds an hour or two ahead rather than days ahead. Errors for single wind farm sites tend to be larger than for regions.[4] But great advances have been made over the past 20 years, not least because of the value of accurate forecasts to owners and operators of wind farms and electricity grids. Fortunately, system operators are hardly concerned with the output of individual wind farms, unless very large, and can schedule alternative power plants based upon forecasts of wind generation over the whole network.

The central message is that intermittency is not a serious problem for electricity grids unless caused by the sudden, unscheduled, shut down of large power plants.[6] Wind energy simply contributes to the highly variable picture already produced by fluctuations in demand. That there are no insuperable technical problems, at least up to penetration levels of around 20%, seems proven by the experience of pioneering countries that are already integrating substantial amounts of wind energy into their grid networks.

7.7   Grid-friendly wind power plants

"Grid-friendly" WPP help to stabilize the grid by incorporating voltage regulation, active power controls, ramp-rate controls, fault-ride though, and frequency response.[7] A plant-level control system which controls many individual inverters to affect plant output at the grid connection point, is a key enabler of such features. The power plant controller (PPC) monitors system-level measurements and determines the desired operating conditions of various plant devices to meet the specified targets. It manages the inverters ensuring that they are producing the real and reactive power necessary to meet the desired settings at the point of interconnection (POI). For example, when the plant operator sends an active power curtailment command, the controller calculates and distributes active power curtailment to individual inverters. In general, the inverters can be throttled back only to a certain specified level of active power without causing the DC voltage to rise beyond its operating range. Therefore, the plant controller dynamically stops and starts inverters as needed to manage the specified active power output limit.

These advanced plant features enable WPP to behave more like conventional generators and actively contribute to grid reliability and stability, providing significant value to utilities and grid operators. They also use the active power management function to ensure that the plant output does not exceed the allowed ramp rates, to the extent possible. These capabilities were demonstrated in a series of tests conducted by the California Independent System Operator (CAISO), Avangrid Renewables, National Renewable Energy Laboratory, and General Electric (GE) during 2019 at Avangrid Renewables' 131 MW Tule Wind Farm, located in CAISO's balancing authority east of San Diego. The plant comprises 57 WTs, each of 2.3 MW rated capacity. The test results, much like those from a similar tests on inverter-controlled photovoltaic power plants (Gevorgian + O'Neal), showed that

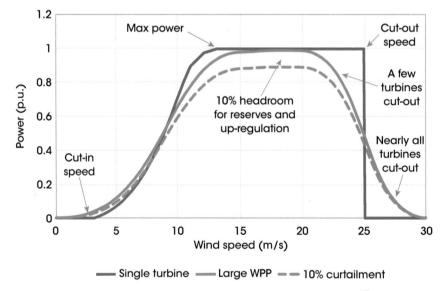

Figure 7.8 Power profile of wind turbine providing ancillary services (Source: Lutan *et al.* [15]).

a commercial WPP with inverter-based smart control can provide balancing or regulation up and down, voltage regulation control, active power control through ramping capability, and frequency response.[5, 15] A key component of the Tule WPP is the PPC. It is designed to regulate real and reactive power output from the WPP so that it behaves as a single large generator. Although the plant comprises individual inverters, with each inverter performing its own energy production based on local wind speed, the PPC coordinated the power output to provide typical large power plant services, such as APC and voltage regulation, through reactive power regulation. These tests proved that solar and WPP can provide critical grid services as long as they are equipped with PCC and they maintain excess capacity by normally operating below their maximum capabilities. In these tests, the plants were operating at 10% below their maximum capacities, allowing for 10% headroom for reserves and upregulation (Figure 7.8).

7.8   Capacity credit and backup generation

From an environmental point of view, the most obvious benefit of wind energy is that it is "carbon free." A unit of electricity generated by a WT obviates the need to generate it by other means, including the burning of coal, gas, or oil. For this reason, wind energy has often been described as a "fuel saver" that reduces carbon dioxide emissions. When we say that Denmark currently generates about 20% of its annual electricity from the wind, it is clear that the Danes' need for other "fuels," local or imported, is reduced accordingly.

So far, so good. But how about the ability of WTs to displace other types of generators? This is a key question because all types of power plants are expensive to build and maintain. For over 30 years, skeptics have repeated the claim that, since the output of WTs is intermittent, wind is an inherently unreliable energy source that cannot reduce the need for conventional generating capacity. They maintain that it is simply a fuel saver, without *capacity credit*, and that there will always be a need for *backup generation*.

All this relates to our discussion of intermittency and variability in Section 7.6. If you watch an individual turbine, or even a wind farm, you may be tempted to conclude that the wind is highly unreliable. But grid operators are concerned with total generation by all wind farms on a grid and regard wind power as one more variable input to the system. In their eyes, all types of generation and consumption are, to some extent, "unreliable."

Back in 2006, the UK's Energy Research Centre (UKERC) produced a detailed report[5] on these issues, based on a review of more than 200 international studies. None of the studies suggested that renewable energy generation at levels up to 20% of demand would compromise the reliability of the British electricity system over the following 20 years, although it might lead to some increase in costs and would affect the way the system is operated. We use some of the report's main findings in the following discussion.

Wind generation affects the performance and reliability of a large grid system in two rather distinct ways: short-term management and long-term planning.

Short-term management requires grid operators to *balance* electricity generation and consumption continuously, scheduling power plants over time scales from minutes to hours in order to keep the system frequency and voltage within statutory limits. Operators must be able to call on *balancing reserves* that can boost generation at short notice. The introduction of substantial amounts of wind power means that the output of other plant will need to be adjusted more frequently, and the UKERC report found that extra backup generation amounting to 5–10% of installed wind capacity may be needed to cope with peaks in demand that are uncorrelated with peaks in wind generation.

An important feature of balancing reserves is their speed of response, or *ramp rate*. Clearly, a sudden increase in electricity consumption (we mentioned the "World Cup problem" in Section 7.6, or a sudden loss of output from other generators due to fault conditions, cannot be offset by reserves that take hours to come online. Fast-response generators with high ramp rates such as hydroelectric and gas turbines can work up to full output within minutes and are said to be *dispatchable*. But the outputs from nuclear and, to a lesser extent, coal-fired power plants take far longer to adjust, and any continuous "cycling" tends to increase wear and tear on turbines and generators. For these reasons, they are normally used to supply the system's minimum demand level, or *baseload*, typically around 50% of the peak (see Figure 7.7).

In addition, the ramp rates associated with a particular type of reserve capacity are affected by whether or not it is kept *spinning*. Synchronous generators that are "up-to-speed" and locked to grid frequency do not necessarily have to generate power. They can just spin at synchronous speed, consuming minimal fuel, and react quickly when power is demanded. Conversely, *non-spinning* reserves must be brought up to speed before synchronizing with the grid. In the case of a steam turbine, this must be done gently to avoid severe thermal stresses as it approaches working temperature.

Grid operators normally have to accept wind power whenever it is generated because it is a question of "use it or lose it." This means, of course, that the output from other generators must be adjusted accordingly. The ramp rates of reserve capacity must be able to cope with the fastest rates of change in wind generation – caused, for example, when a storm sweeps rapidly across the North Sea. However, this does not mean that dedicated reserve capacity needs be provided to support individual wind farms; indeed, it would be very uneconomic to do so. The variability of wind, and the backup generation needed to support it, is only significant at the system level.

To summarize the short-term management issue, successful balancing of supply and demand in the presence of wind generation does require some extra backup plant that can match the ramp rates expected of variable wind. But the balancing act is no different in principle to that already required by other types of generation, conventional and renewable, and by the highly variable nature of consumer demand.

7.9   The variability challenge and solutions

The electric grid system and its market operations were designed to deal with variability of demand and supply on different timescales, mainly by dispatching controllable generators and, to a lesser degree, by using electricity storage systems. A high penetration of variable wind and solar electricity into the grid creates challenges for the grid operators who need to reliably satisfy load demands every hour of the day.

The following solutions are available for reducing or mitigating variability: (i) geographical diversity/transmission interconnections; (ii) forecasting; and (iii) energy storage. A combination of these solutions would in most cases provide the minimum cost solution as variability is reduced with geographical diversity and controlling it is easier and less expensive when we have accurate forecasting. Also demand management has a significant role in handling diurnal variability.

Below we discuss geographical diversity and high-voltage transmission lines; energy storage is covered elsewhere.[7, 8]

7.9.1 Geographical diversity/transmission interconnections

The wind speed at a single location can change drastically; however, the aggregate output of geographically dispersed WT systems is much smoother than that of a single point as the wind speed and direction vary with location.

7.9.2 Long-distance transmission lines

High voltage (HV) transmission lines
HV transmission lines can be AC of DC. The first are more common, but long-distance continuous interconnect is only feasible with DC. The high-voltage direct current (HVDC) lines have larger interconnection losses than high-voltage alternating current (HVAC), but much lower transmission losses.   While HVAC systems merely require transformers that step down the voltage to the next lower level HVDC systems rely on power electronics to convert between DC and AC networks. Transformers and power electronics are located in two converter stations at each end of the line. The line is bipolar at a voltage of ± 800 kV, so the load is carried by two cables in parallel. Each pole consists of six aluminum core steel reinforced (ACSR) conductors.

Geographical diversity can greatly reduce the fluctuations of both wind and solar energies. As discussed before, increasing the geographical area of wind, by either connecting an aggregate of dispersed WT systems or by constructing large wind farm plants, it decreases the cloud-induced fluctuations of their total output. Long-distance electricity travel over long distances or the long-distance interconnects will necessitate HVDC lines. The technology is well established. Currently, HVDC transmission lines with a capacity of 6.4 GW are operating in China utilizing ± 800 kV technology and a project utilizing 1.1 MV technology is in progress. With 800-kV HVDC lines, the transmission of electricity over 2000 km lines entails a 7% electromagnetic loss versus a 22% or higher loss via HVAC power lines of the same distance. Also, construction of HVDC power lines requires 37% less land area than constructing HVAC ones. For the USA, Fthenakis and collaborators Zweibel and Mason have defined the needs of HVDC from the SW for solar and the high plains for wind to the rest of the countries[3, 4]. MacDonald et al.[9] discussed the feasibility of reducing CO_2 emissions in the USA grid by 80% from those in 190, by moving away from a regionally divided electricity sector to a national system enabled by HVDC transmission. The biggest challenge regarding long-distance transmission is that would need approvals from a plethora of regional and national jurisdictions.

Crossing national boards may be more involving, but we have plenty of boundary-crossing transmission lines in Europe, US-Canada, and elsewhere. Developing large global electricity grids is not a far-fetched idea; it could follow the paradigm of global telecommunications networks, which were enabled by fiber-optics technology connecting continents with underwater cables; it is expensive but it is technologically feasible.

7.9.3 Grid flexibility

The flexibility of a power system that is its ability to vary its output to meet the demand depends on the mix of its generators. Its flexibility is constrained mainly by the baseload generators, usually nuclear and coal power plants. The operation of nuclear power plants is quite inflexible and that of coal power plants is flexible only within a narrow range.

We discussed earlier the need for adjusting power output for load following and for satisfying peak demand. In addition to these duties, in order to enhance the reliability of the grid, a number of generators are operated at partial load with their surplus megawatts available to ensure reliability under rapid changes on demand (frequency regulation) or an unplanned outage of a unit or a transmission line (contingency reserves). Frequency and contingency reserves operating at partial load are called spinning reserves. There are also units that provide grid stability (voltage, frequency, and reactive power). Overall, a number of generators operate at partial load account for the functions of load following, frequency regulation, and spinning reserves. Wind and solar penetration into a power system displaces generation from conventional units, and often their combination can increase such displacement. Figure 7.9 shows a result from simulations based on hourly wind, solar, and load data in NY state.[10] The figure shows the load, solar photovoltaics (PV), and wind outputs for a summer day where a peak of the load for that year occurred; it is shown that wind and PV together serve the load better than each of them separately. From a utility's perspective, the main advantages of this penetration are fuel economy, emissions reduction, and reduced need for increasing overall system capacity (Figure 7.10).

The limit of the penetration of renewable energy into an electricity grid depends on the mix of generators in the system. As described previously, there are two main types of generators: Inflexible baseload units and the more flexible cycling units. The former units are designed to operate at full output, and they often provide most of the winter demand or 35–40% of the annual peak capacity. The penetration of wind- and solar power cannot drive the net load below the limit imposed by the number and the type of baseload generators, and the amount and type of reserves. This limit depends on the ability of conventional baseload generators to reduce significantly their output, and on economic- and mechanical constraints.

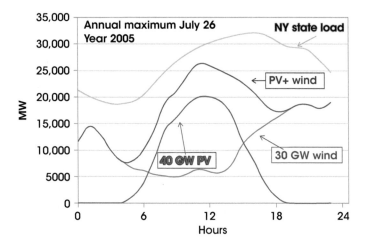

Figure 7.9 Synergy of PV and wind in New York State.

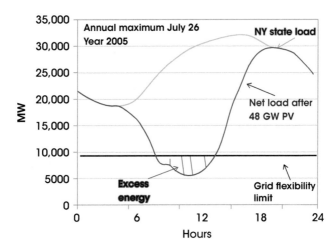

Figure 7.10 Effect of grid flexibility on PV energy delivery.

For example, coal plants can vary their output from full-to-half-capacity, but if this is done frequently, it would demand costly maintenance. (The flexibility limit that separates the flexible from the inflexible capacity and, hence, the flexibility of a system is defined as the percentage of the annual peak capacity that is flexible.)

High levels of wind- and solar-energy penetration may stress the system because of its flexibility limit; there will be hours throughout the year where the net load is brought below it. Then, the amount of energy below the flexibility limit cannot be absorbed and must be curtailed. This is more of a problem for incorporating wind power than solar power, since winds are stronger during the night when the load levels are low. The more flexible a power system is, the higher is the penetration achievable, and the less restraint on renewable electricity. The amount of energy to be cut back can be determined with a cost analysis; often it makes economic sense to curtail small amounts of energy as a trade-off in penetration.

The irregularity of renewable resources and the limit on the grid's flexibility both pose restrictions on the maximum penetration achievable in a system. There are some interesting findings on the maximum wind energy penetration that can be realized without storage. For example, in Texas during 2021, wind provided 24.8% of the annual demand, while a maximum of 66.5% penetration on March 22, 2021.[11]

7.10 Are 100% renewable energy grids feasible?

We should note that some countries already meet or come close to achieving this goal. Iceland, for example, supplies 100% of its electricity needs with either geothermal or hydropower. Hydropower is the major power resources in Norway (97%), Brazil (76%), and Canada (62%). However, most good sites for large hydropower resources have already been developed. So how do other areas achieve 100% renewable grids? For this, variable renewable energy (VRE), thus wind and solar, would have to be the major contributors and the grid. This has happened already, but only occasionally. For example, South Australia

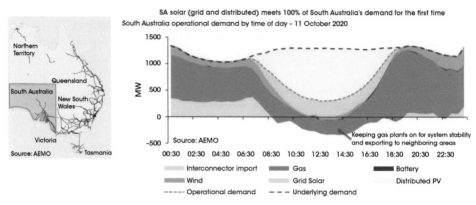

Figure 7.11 100% solar in Australia.

experienced times in October with 100% of the demand served with only solar; mostly distributed roof-top solar! (Figure 7.11).

Of course, as the solar resource is intermittent, such an aspirational goal can only be satisfied with a mixture of renewable energies.

It is possible that variable renewable generation technologies will continue to dominate the newly installed generation capacity over the next two decades, spurring a transition to 100% renewable-based power networks with a substantially altered landscape for the associated planning, management, stability, and control approaches.[12] More and more countries are setting net-zero emission targets, and in most countries, this means 100% RE supply. Sixty-one countries across the world have set 100% RE targets for at least their power sector. Denmark has set the target to reach 100% renewables across all energy sectors by 2050. In most transition pathways, solar energy and wind power increasingly emerge as the central pillars of a sustainable energy system combined with energy efficiency measures. Cost-optimization modeling and greater resource availability tend to lead to higher solar photovoltaic shares, while emphasis on energy supply diversification tends to point to higher wind power contributions. Recent research has focused on the challenges and opportunities regarding grid congestion, energy storage, sector coupling, electrification of transport and industry implying power-to-X and hydrogen-to-X, and the inclusion of natural and technical carbon dioxide removal (CDR) approaches.[12]

One of the main challenges pertaining to the integration of VRE resources into power networks is that they are integrated with a power electronic interface known as an inverter; they are therefore commonly referred to as inverter-based resources (IBRs). This stands in contrast to most conventional power plants where electricity is generated using synchronous generators. The task of the ISO is to unify the operation of synchronous machines and IBRs at all scales.[13]

7.11   Grid reliability

The second main impact of wind energy on a large grid network concerns medium to long-term system reliability. Planners aim for a safety margin in total grid capacity compared with anticipated peak loads, allowing for unexpected surges in demand and forced outages of large power plants. Unlike short-term system balancing, they must often look 15 or 20 years ahead, taking into account predicted growth in demand and the retirement of plants reaching the end of their working lives. The *system margin*, also called the *planning reserve*, is essentially concerned with strategic planning of generating capacity to ensure security of supply in the future – in popular parlance, "to prevent the lights going out."

No electricity system is 100% reliable because there is always a small chance of failures in power plants and transmission networks, especially when the demand is high. Even a substantial system margin cannot give an absolute guarantee of reliability. There is, very roughly, a 10% chance of a conventional power plant, fossil fuel or nuclear, being out of action at any particular moment, and in a worst-case scenario, the loss of a major plant can produce a domino effect on the whole system. For example, in 2003, there was a massive blackout in the northeastern USA caused by cascading power outages – and it had nothing to do with wind energy. In 2008, seven large UK power plants, including one nuclear station, unexpectedly shut down causing widespread power cuts described by one industry executive as a "gigantic coincidence" – which had nothing to do with wind energy – and the same nuclear plant was offline for more than 6 months in 2010. Given such uncertainties, planners must aim for a margin that represents a sensible compromise between reliability and cost, taking account of the existing and anticipated generation mix, including wind. The UKERC report discusses modeling this complex situation with a measure known as the *loss of load probability (LOLP)* – the probability that, at any time in the future, the system will be unable to satisfy demand.

The capacity credit for any type of generation is effectively a measure of the contribution it can make to system reliability and is normally expressed as a percentage of installed capacity. In the case of wind, this figure is predicted by the report to be approximately 20–30% in British conditions, assuming generation reaches 20% of total demand, and acknowledging the fact that peaks in wind power do not generally coincide with peaks in demand.

It is clear that wind capacity cannot displace conventional generation on a one-for-one basis, either for short-term system balancing or for longer-term system reliability. But taking an overall view, the international consensus among today's planners and system engineers is that large modern grid networks can accept up to 20% of renewable generation, including wind energy, without serious technical or operational problems, and that any difficulties arising tend to be managerial and organizational, requiring a shift of mind set away from the large centralized generation of the 20th century toward something more diverse and flexible in the 21st century.

This sounds optimistic and reassuring, yet there remains a lingering doubt about how much capacity credit should be given to wind generation, and what backup should be provided to cope with exceptional circumstances. As noted above, no grid system can ever be 100% reliable and wind energy simply adds its own element of uncertainty. The LOLP may be reduced to a very low level, but it can never reach zero – that is in the nature of engineering systems. So we will end with a story, an imagined scenario that might just conceivably occur in the not-too-distant future.

By the 2030s the countries of northwest Europe, linked by a North Sea Supergrid, have become used to trading offshore wind power according to where the wind blows strongest over the sea's vast stretches. It is rare for the winds to die simultaneously over the whole of the North Sea. But in the freezing winter of 2035, by which time the supergrid interconnects 120 GW of offshore wind capacity, a freak high pressure system settles over the North Sea for two weeks, virtually killing all winds. At the same time the hydroelectric reservoirs of Norway and Scotland have become unusually depleted; four large UK nuclear power plants are shut down for safety checks; and a trade dispute has severely reduced imported supplies of gas and oil from the Middle East. On 9th February vast numbers of Europeans decide to watch a momentous international football match on TV, and at half-time 10 million households switch on electric kettles to brew tea and coffee. The lights (and the TV's) go out – and do not come back on again for some considerable time.

Fact or fantasy; a miniscule risk to be ignored, or a significant one to be planned for? How often will extreme "low-wind" events occur, and how long will they last? These are big unanswered questions for wind energy, and only time will tell. We are back to the sort of statistics pioneered by Tippett and Gumbel (see Section 2.2.2) who considered the probabilities and return times of extremely unlikely events. In the meantime, we would surely be wise to understate the capacity credit of wind generation and diversify our energy supplies as much as possible, accepting that no sources can ever be completely reliable. And maybe, by 2035, all European consumers will be connected to electricity grids by very smart meters, utilities will vary the price of electricity hour by hour to control demand, and international football matches will have gone out of fashion.

7.12 The grid of the future

Technology advancements made it easier to deploy distributed variable renewable resources (DVRE). The integration of DVRE and distributed grids can increase efficiencies in the use of the existing grid as well as become part of the overall development strategy to balance supply and demand uncertainties and risks with a variety of different resources, assuring resilient, flexible, and safe power delivery to consumers. Furthermore, innovative changes to the regulatory climate will also affect paradigms of the electric power business.[14] The reregulation of electric power industries in the USA and elsewhere introduced wholesale electric markets. The advent of markets and environmental policies prompted significant changes in the fuel mix of generating stations that shifted from coal and nuclear generation to efficient natural-gas-fired combined cycle units. Recent developments include the advent of retail access offering more consumer choices and business opportunities as the internet facilitates more customer choices. Utility customers who own solar and wind systems are increasingly using the grid as a means to balance their own generation and demand and also as a backup supplier when their locally source generation is unavailable. More and more customers are becoming prosumers and expect to deliver excess generation back to the grid and be paid for it, without restrictions on their production. However, customers still expect the grid to be available to provide power when they need it. New applications of power electronics, such as smart inverters that can actively interact with the distribution system, facilitate a more efficient DVRE grid integration.[14]

Self-assessment questions

Q7.1   What is currently the installed cost for large wind turbines ($/MW) and what is the corresponding electricity cost ($/MWh) for production in regions with good wind resources?

Q7.2 How can variability of wind turbine output be smoothened?

Q7.3 What are expected synergies between solar and wind energy?

Q7.4 What are the general characteristics of "grid-friendly" wind farms?

Q7.5 When is HVDC preferable than HVAC and why?

Q7.6 How the grid flexibility defined?

Q7.7 What is the difference of peaker plants and base-load plants?

Q7.8 What three renewable energies are expected to be the major contributors to grids with 100% renewable energy in most countries? Comment if tis choice may not apply to your country.

Q7.9 How do you envision the grid of the future?

Q7.10 Why are offshore wind turbines expected to have a higher capacity ratio than onshore ones in the same region?

Problems

7.1 (a) Consider a wind farm consisting of 12-m diameter WTs with efficiencies of 40% on a grid given by the average spacing shown in the figure below. For a wind velocity of 8 m/s, what is the power output in kW per km²?

(b) Compare the result of part (a) with the 24-hour average output of a 1 km² photovoltaic array with an efficiency of 18% under average US irradiation of 1800 kWh/m²/yr.

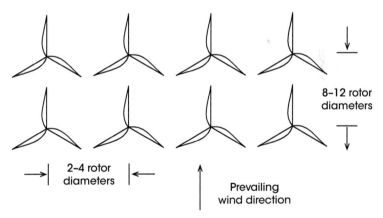

8–12 rotor diameters

2–4 rotor diameters

Prevailing wind direction

Optimal spacing of wind turbines in a wind farm. (Distances are shown as a function of the rotor diameter.)

7.2 A transformer rated at 120 V/600 V, 12 kVA has an equivalent circuit as shown in the figure. The low-voltage side is connected to a heater, rated at 6 kW, 120 V. The high-voltage side is connected to a 600 V, 60 Hz, single-phase power line. Find the actual power transferred, the magnitude of the measured voltage across the heater, and the efficiency (power out/power in) of the transformer.

7.3   A small WT generator (single phase) produces a 60 Hz voltage at 100 V rms. The output of the generator is connected to a diode bridge full-wave rectifier, which produces a fluctuating DC voltage. What is the average DC voltage? A silicon-controlled rectifier (SCR) is substituted for the diode rectifier. Under one condition the SCRs are turned on at 45° after the beginning of each half cycle. What is the average DC voltage in that case?

7.4   The parameters of an induction machine can be estimated from "no-load" and "blocked-rotor" tests. In both tests, voltage, current, and power are measured. Under the no-load test, the rated voltage is applied to the machine, and it is allowed to run at no load (i.e., with nothing connected to the shaft). Under these conditions, the slip is essentially equal to 0, and the loop with the rotor parameters may be ignored; the magnetizing reactance accounts for most of the impedance. In the blocked-rotor test, the rotor is prevented from turning and a reduced voltage is applied to the terminals of the machine. Thus, the slip is equal to 1.0 so the magnetizing reactance can be ignored, and the impedance of the leakage parameters can be determined. A third test can be used to estimate the windage and friction losses. In this test, the machine being tested is driven by a second machine, but it is not connected to the power system. The power of the second machine is measured, and from that value is subtracted the latter's no-load power. The difference is approximately equal to the test machine's windage and friction.

In a simplified version of the analysis, all of the leakage terms may be assumed to be on the same side of the mutual inductance, as shown as the above figure. In addition, the resistance in parallel with the magnetizing reactance is assumed to be infinite, and the stator and rotor resistances, R_S and R_R', are assumed to be equal to each other and the leakage inductances are also assumed to be equal to each other.

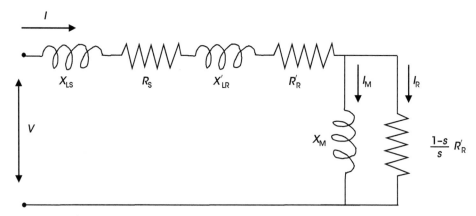

The following test data are available for a wye-connected, three-phase induction machine:

No-load test: $V_0 = 600$ V, $I_0 = 50$ A, $P_0 = 4000$ W

Blocked rotor test: $V_B = 75$ V, $I_B = 120$ A, $P_B = 6000$ W

Windage and friction losses equal 3200 W.

Find the parameters for the induction generator model.

7.5 A pump is used to draw water from a water reservoir 100 m deep in the ground. Assume water temperature of 15°C, density of 1 l/kg, and pump efficiency of 95% (5% in thermal losses).

(a) Determine the power required for a water flow of 1 m³/min through a pipe of a 12 cm diameter.

(b) Determine the exit velocity if the water exits through a 2 cm nozzle at the end of the pipe.

(c) Determine the power capacity of a WT that can power this turbine.

7.6 A compressor operating at steady state draws air at atmospheric pressure and 290 K with a velocity of 5 m/s and compresses it to 10 bar. At the exit with an area of 0.5 m² the temperature is 520 K and air velocity is 3 m/s. The rate of heat transfer from the compressor to its surroundings is 180 kJ/min.

(a) Assuming ideal gas law, calculate the power input to the compressor.

(b) repeat the calculation using a suitable compressibility factor and provide the reference.

Answers to questions

Q7.1 Capital cost of about $1 million/MW and leveled cost of electricity (LCOE) of $30–$50/MWh for on land WT and $50–$80 for offshore WT.

Q7.2 By interconnecting wind turbines at various locations (geographical diversity and transmission interconnections).

Q7.3 Wind may blow at night when there is zero solar and the capacity ratio of wind is higher in the winter when the capacity ratio of solar is lower.

Q7.4 They can provide voltage support, active power control, ramp-rate control, fault-ride through, and frequency response.

Q7.5 HVDC has larger interconnection losses than HVAC, but much lower transmission losses. That's why .HVDC is preferable for long distances transmission (e.g., >500 km) and for under the sea transmission.

Q7.6 The grid's ability to serve variable load. It depends on the mixture of generators - provide a more complete answer.

Q7.7 Base-load serving plants are nuclear plants, coal power plants, hydropower plants, and large combined-cycle natural gas plants that are scheduled to operate at all times. Peaker plants operate only at times of high demand and are typically single-cycle natural gas and diesel plants. Wind power plants can contribute to all demand categories.

Q7.8 Typically Solar, Wind and Hydro. Solar is the biggest in the USA; in other countries the biggest can be Wind or Hydro.

Q7.9 This is an open-end question.

Q7.10 Wind speeds are most often higher offshore as there is not resistance to wind from ground roughness (trees, buildings, other).

References

1. National Grid. serving the UK and the USA. www.nationalgrid.com (Accessed on May 4, 2024)
2. J.F. Manwell *et al*. *Wind Energy Explained: Theory, Design and Application*, 2nd edition. John Wiley & Sons Ltd: Chichester (2009).
3. T. Ackermann (ed.). *Wind Power in Power Systems*. John Wiley & Sons Ltd: Chichester (2005).
4. M. Milligan *et al*. Wind power myths debunked, IEEE Power and Energy Magazine, (November/December 2009).
5. V. Gevorgian and B. O'Neill, Advanced grid-friendly controls demonstration project for utility-scale PV power plants, NREL/TP-5D00-65368 (January 2016). http://www.nrel.gov/docs/fy16osti/65368.pdf
6. R. Gross *et al*. The costs and impacts of intermittency: an assessment of the evidence on the costs and impacts of intermittent generation on the British electricity network, UK Energy Research Centre (UKERC) (March 2006).
7. V.M. Fthenakis and P.A. Lynn. *Electricity from Sunlight: Photovoltaics Systems Integration and Sustainability*, 2nd edition. Wiley: Hoboken, NJ (2018). (ISBN 978-1-118-96380-7).
8. T. Nikolakakis and V. Fthenakis. The value of compressed air energy storage (CAES) for enhancing variable renewable energy (VRE) integration: a new unit commitment and economic dispatch model for Ireland. *Energy Technology*, 5(11), 2026–2038 (2017) https://onlinelibrary.wiley.com/doi/abs/10.1002/ente.201700151.
9. A. MacDonald *et al*. Future cost-competitive electricity systems and their impact on US CO_2 emissions. *Nature Climate Change*, 6, 526–531 (2016). T. Nikolakakis and V. Fthenakis. The optimum mix of electricity from wind- and solar-sources in conventional power systems: evaluating the case for New York State. *Energy Policy*, 39(11), 6972–6980 (2011) https://doi.org/10.1038/nclimate2921.
10. T. Nikolakakis and V. Fthenakis. The optimum mix of electricity from wind- and solar-sources in conventional power systems: evaluating the case for New York State. *Energy Policy*, 39(11), 6972–6980 (2011).
11. Electric Reliability Council of Texas (EPRI). https://www.ercot.com/files/docs/2021/12/30/ERCOT_Fact_Sheet.pdf (Accessed October 10, 2023).
12. C. Breyer *et al*. On the history and future of 100% renewable energy systems research. *IEEE Access*, 10, 78176–78218 (2022).
13. B. Kroposki *et al*. Achieving a 100% renewable grid – operating electric power systems with extremely high levels of variable renewable energy. *IEEE Power and Energy Magazine*, 15(2) (2017) http://ieeexplore.ieee.org/document/7866938/.
14. M.I. Henderson *et al*. Electric power grid modernization trends, challenges, and opportunities, IEEE (November 2017).
15. C. Loutan and V. Gevorgian, Avangrid Renewables Tule Wind Farm, Demonstration of Capability to Provide Essential Grid Services, California ISO, Avangrid Renewables, NREL, March 11, 2022, https://www.caiso.com/Documents/WindPowerPlantTestResults.pdf

8 Wind energy growth and sustainability (cost, resources, environment)

During 2000–2020, the wind energy contribution to electricity grids enjoyed an average growth of ~30% per year; this growth recently slowed down but still annual growth of about 10% is projected till midcentury. Is further growth sustainable and what is the maximum level it could reach in the foreseeable future? To answer this question, we need to reflect on what sustainable development is all about. Think of it as "development that meets the needs of the present without compromising the ability of future generations to meet their own needs." Wind turbines (WTs), as fuel-free energy sources, are inherently sustainable unless they are too expensive to produce, are manufactured using materials that are depletable, or are environmentally unsafe. Measurable aspects of sustainability include cost, resource availability, and environmental impact. The question of cost concerns the affordability of wind energy compared to other energy sources throughout the world. Environmental impacts include local, regional, and global effects, as well as the usage of land and water, which must be considered in a comparable context over a long, multigenerational horizon. Finally, the availability of material resources matters to current and future generations under the constraint of affordability. More concisely, wind energy must meet the need for generating abundant electricity at competitive costs while conserving resources for future generations and having environmental impacts much lower than those of conventional modes of power generation, preferably lower than those of alternative future energy options (Figure 8.1).

Onshore and Offshore Wind Energy: Evolution, Grid Integration, and Impact, Second Edition.
Vasilis Fthenakis, Subhamoy Bhattacharya, and Paul A. Lynn.
© 2025 John Wiley & Sons Ltd. Published 2025 by John Wiley & Sons Ltd.
Companion website: www.wiley.com/go/fthenakis/windenergy2e

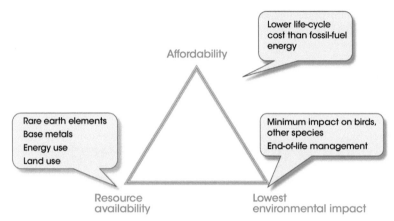

Figure 8.1 The three major pillars of wind large-growth sustainability (*Source*: Adapted from Fthenakis V.M. and Lynn P.A, Electricity from Sunlight: Photovoltaics Systems Integration and Sustainability, Wiley, 2nd edition, 2018.).

8.1   Affordability

8.1.1   Upfront capital and installation costs

In the previous chapters, we explored the technical aspects of large grid-connected wind turbines—the principles underpinning their design and the engineering needed to ensure safe and reliable operation. We also covered the technical criteria that determine the efficiency of turbines at capturing the wind and generating electricity, criteria that have a major impact on the viability of wind energy projects. It is now time to give a brief introduction to wind energy economics, and we start with a key question: how much do large horizontal axis wind turbines (HAWTs) cost to purchase and install, both onshore and offshore, and how are costs likely to change in the coming years?

The question may seem obvious, but the answer is far less so, for various reasons including:

- Large HAWTs are undergoing continuous development with improvements in design, operating efficiency, and manufacturing methods.
- Installation costs are site-dependent, and generally offshore installations are more expensive than onshore.
- Large wind farms tend to be much more cost-effective than individual turbines or small groups of turbines.
- Countries such as China and India have burgeoning wind energy industries and will increasingly look to export markets, where their products are likely to prove very competitive.

Before getting down to detail, it is helpful to consider a general feature of manufactured products in a growth industry—the tendency for prices to fall systematically as cumulative production rises, in accordance with the *learning curve* concept. Price reductions occur partly as a result of "learning-by-doing"—people get better at making things as they become more experienced, partly due to the introduction of new designs, techniques, and better utilization of materials, and partly due to the benefits of scale. The long experience of many industries has shown that prices tend to drop between 10 and 30% in real terms for every doubling of cumulative production. For mature technologies such as steel construction or electric motors, each doubling, given the very high current production levels, may require decades; but for relatively new technologies undergoing rapid expansion, doublings of cumulative production occur over much shorter time scales. As production levels soar, prices are driven down.

How does the modern wind energy industry fit this scenario? Certainly, it has been developing at a remarkable rate, with global installed capacity rising by around 30% per annum, corresponding to a doubling about every 3 years). In spite of occasional hiccups including the global economic crisis in 2008 and the inflation and supply chain challenges due to Covid and the war in Ukraine in recent years, the signs for continued growth are very positive. The slope of the learning curve—the rate at which prices are expected to fall—depends on the maturity of wind energy's contributing technologies including steel foundations and towers, rotor blades, gearboxes, generators, transmission cables, electronic power converters, and control systems. Clearly, some of these are more innovative than others. It also depends on changes to the basic product. The average size of grid-connected HAWTs continues to increase, and cumulative global production is bolstered not only by more turbines but also by bigger ones. It is cheaper to install and interconnect a small number of large turbines on a wind farm than a large number of small ones. The combined effects of "learning-by-doing" and increasing turbine sizes caused costs to fall by about 15% for every doubling of global cumulative capacity between 1982 and 2020.[1]

8.1.1.1 Onshore wind turbines

It is difficult to give actual cost figures. Manufacturers' turbine prices are often confidential and vary with the number of units ordered. Installation costs depend on many factors including wind farm size, the amount of site preparation and cabling, and the difficulty of grid connection. As a rough guide, average wind farm costs per installed megawatt are currently of the order of 1 million US dollars onshore and 2 million US dollars offshore—and, as discussed above, are expected to reduce steadily in real terms as time goes by.

Figure 8.2 shows typical cost breakdowns under four headings. Starting with the onshore case:

- *Turbines (t)*. The turbines typically account for about two-thirds (66%) of total installed costs.
- *Electrical (e)*. On-site cabling, substation equipment, and grid connection account for about 14%.
- *Civil (c)*. Civil engineering works including site access and preparation, turbine foundations, and buildings account for about 12%.
- *Other (o)*. The remaining 8% covers other items such as project management, legal costs, and insurance.

The percentages vary considerably between sites according to geography, soil conditions, the need for access roads, proximity to a suitable grid connection point, and so on. The figure should therefore be taken simply as an approximate guide.

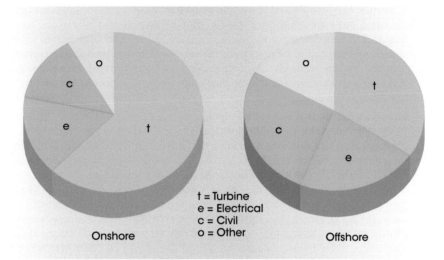

t = Turbine
e = Electrical
c = Civil
o = Other

Onshore

Offshore

Figure 8.2 Typical cost breakdowns for onshore and offshore wind (assuming that the cost of the offshore project is double that of the onshore project).

8.1.1.2 Offshore wind turbines

Offshore wind has really taken off in the new millennium and is currently undergoing rapid expansion. Its technology is generally similar to that of onshore wind, but there are important differences involving foundations, weather protection, and undersea cabling. Capital and installation costs per rated MW have often been quoted as almost double those of onshore wind, but it seems very likely that the learning curve for offshore wind will outpace that for onshore wind in the near future, bringing their costs closer together. And we must always bear in mind that offshore turbines generally work at higher capacity factors (CFs), producing more electricity per megawatt of installed capacity. Because the total project cost is much higher, the contribution of turbine costs is generally much lower—typically nearer one-third of the total. Electrical costs (including undersea cabling and substations) and civil costs (including foundations) are relatively high, and other costs must cover more complex planning and project management, the hire and scheduling of jack-up barges, cable laying and support vessels, and increased insurance charges. The detailed cost breakdown is again dependent on on-site conditions, including distance to shore.

Offshore wind farm development is a technically complex, lengthy, risky, and capital-intensive process, primarily because of the more demanding operations over the sea. In order to assess the economics of an offshore wind farm, it is essential to consider three phases: investment, operation, and decommissioning. The investment phase includes the development, design, and fabrication of all the components and their installation and is the most capital-intensive.

The investment phase can last up to 9 years, with most of this time spent in planning and obtaining consent—activities that can take up to 5 years on their own. After the investment, the wind farms are expected to produce electricity to the grid for at least 20–30 years. During this operation phase, regular maintenance is required to keep the downtime minimum and prolong the life span of the whole project. Details of operation and maintenance (O&M)

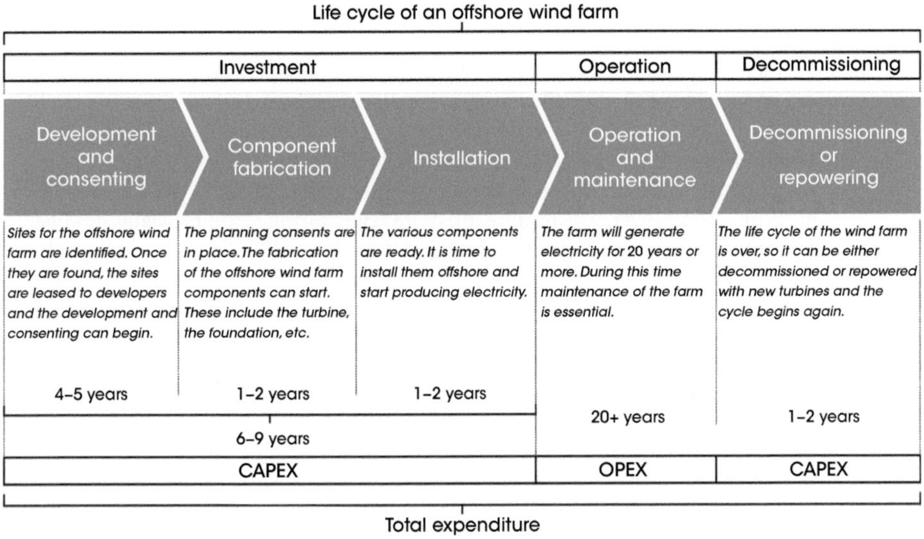

Life cycle of an offshore wind farm

Investment			Operation	Decommissioning
Development and consenting	Component fabrication	Installation	Operation and maintenance	Decommissioning or repowering
Sites for the offshore wind farm are identified. Once they are found, the sites are leased to developers and the development and consenting can begin.	The planning consents are in place. The fabrication of the offshore wind farm components can start. These include the turbine, the foundation, etc.	The various components are ready. It is time to install them offshore and start producing electricity.	The farm will generate electricity for 20 years or more. During this time maintenance of the farm is essential.	The life cycle of the wind farm is over, so it can be either decommissioned or repowered with new turbines and the cycle begins again.
4–5 years	1–2 years	1–2 years	20+ years	1–2 years
6–9 years				
CAPEX			OPEX	CAPEX

Total expenditure

Figure 8.3 Life cycle of an offshore wind farm in the early years of development.

statistics and protocols are discussed in Chapter 6. When the operation phase is over, there is the decommissioning phase where there can be two options: either the entire project is dismantled and disposed of, which marks the end of the farm's life cycle, or the farm is repowered with new turbines and a new life cycle begins. The life cycle of an offshore wind farm (along with all the different phases) is illustrated in Figure 8.3.

The total cost of the offshore wind farm can be divided into two fundamental cost categories: capital expenditure (CAPEX) and operational expenditure (OPEX). In general, CAPEX is defined as the initial one-time expenditure required to build an income-generating asset and achieve its commercial operation. On the other hand, OPEX can be defined as the ongoing cost, either once or recurring, associated with the operation of the asset. Both can further be divided into more detailed costs. For example, CAPEX can be divided into three main categories. These are the turbine costs (including all the wind turbine components), the construction costs (including foundation, electrical infrastructure, assembly, and installation costs), and the development costs (including the planning, insurance, construction and contingency financing, and decommissioning costs). Similarly, OPEX includes two main categories of costs, which are the operation and the maintenance costs. In a typical project, the contribution of these two costs toward the total cost is approximately 75% for CAPEX and 25% for OPEX.

8.1.1.2.1 Cost reduction

The onshore wind industry has been able to benefit from many years of cost reduction, unlike the offshore that has only just started. For the UK, the reduction of cost is remarkable: levelized cost of energy (LCOE) for offshore wind reduced from £160/MWh to £44.00/MWh. The main way to reduce the costs associated with offshore wind energy will be by utilizing efficiently the inherent advantages of offshore wind, but at the same time minimizing the cost disadvantages. In other words, the wind farms on the one hand must gain access to sites further from shore and consequently in deeper waters, where the wind is faster and more consistent, and on the other hand, they should achieve a significant cost reduction. Cost reduction is possible through the optimization of the key components and installation and maintenance activities of a wind farm. These key components and processes and some of the ways to increase their cost-effectiveness are described below:

(1) *Wind turbines.* These alone make up approximately 30–35% of the capital cost of the whole project. Therefore, their optimization is a key driver to the cost reduction of offshore wind power. Up-to-date, offshore wind turbines are typically modified versions of the largest onshore turbines, suitable for high-CFs. Since they are not designed specifically for the harsh conditions over the sea, reliability issues arise that often lead to significant downtime, in other words, lower power generation. If the turbines are located further from shore, the downtime may be even larger, as the maintenance operations are more challenging. Consequently, the main way to reduce these issues is to improve wind turbine technology. Technology is already being developed for improved generators with direct-drive, gearless nacelles, and high-power output (e.g., >12 MW) (Figure 8.4). There are also significant improvements that can reduce manufacturing and assembly costs, based on the optimization of the supply chain and the standardization of components.

(2) *Foundations of turbines.* Foundations make up approximately 20–25% of the capital cost of offshore wind and reflect the second most expensive part of the whole project, after the actual turbines. Currently, approximately 80% of the foundations used in wind farms are monopiles. However, this type of foundation cannot be used beyond 30-m water depths, and it is not considered economical for larger output wind turbines. As the offshore industry is trying to exploit the higher wind speeds far from shore, new types of foundations such as tripods and jacket structures are being considered. For water depths beyond 60 m, floating foundations are being developed. There are also significant opportunities to reduce foundation costs through economies of scale, reduced materials costs, and the development of new installation techniques.

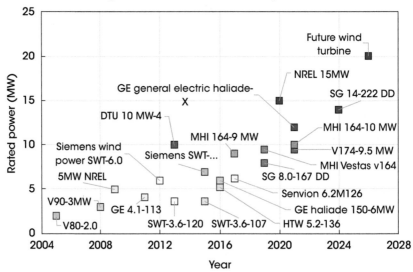

Figure 8.4 Turbine development during 2004–2022 and future projections.

(3) *Electrical connections.* There are two different types of connections: the high-voltage alternating current (HVAC) and the high-voltage direct current (HVDC). The main parameters to consider, when choosing connections, are the rated power and the distance to shore. Currently, most offshore wind farms use HVAC type of connections, and only 14% use HVDC. However, as the offshore farms move further from shore, the HVDC technology is likely to be a necessity because they have less transmission losses for longer distances. More details on HVDC connections are given in Sections 4.6 and 7.9.2.

(4) *Installation.* The offshore wind installation techniques have not yet been optimized for high volumes and speeds. For water depths up to 35 m, foundations and turbines are installed with the use of standard jack-up barges, but for deeper waters and larger turbines, special installation vessels may be needed. Therefore, more opportunities should emerge for wind turbines, foundations, and grid connections to be designed to optimize and reduce the cost of the installation process.

(5) *O&M.* The unexpected repairs arising from reliability issues of the turbines can significantly increase the OPEX cost. Therefore, unscheduled repairs should be minimized by increasing the reliability of wind turbines. In addition, the cost of maintenance can be reduced by the application of remote monitoring and with systems that allow access and repairs in adverse weather conditions. Furthermore, future repairs could be conducted from offshore accommodation facilities, in a similar manner to oil rigs. This could leverage the economies of scale from a collection of large offshore wind farms and significantly reduce travel times, downtime, and thus cost.

In Section 8.1.2, we consider the additional costs incurred over a wind farm's operational lifetime and the income generated by the production of electricity.

8.1.2 Operation, maintenance, and cash flow

The economic success of a wind farm depends upon recovering its costs and making a profit by selling electricity over its working lifetime. Capital and installation costs are "up front" and the fuel is free, but there are ongoing O&M charges spread over many years, and the whole project must be financed. In many ways, the production and selling of electricity from the wind are just like any other business.

One quick way of assessing viability is to estimate the *simple payback period (SPP)*, the number of years it takes to recoup the initial investment. This is found by dividing total costs by the average annual income expected from the electricity generated. Another approach is to estimate the *cost of energy (COE)* equal to total costs divided by the system's total expected energy production. However, SPP and COE calculations give only crude estimates of profitability because they fail to allow for the timescales over which costs are incurred and income is generated.

A more accurate approach known as *life-cycle costing*[2] is illustrated in Figure 8.5. This shows the *cash flows* that occur as costs arise and income is generated over the lifetime of the project, say 20 years. Negative cash flows (costs) are shown in orange; positive ones (income) are shown in blue. A major feature of wind energy, like other forms of renewable energy, is that the initial capital and installation costs (*A*) produce by far the largest negative cash flow. This is followed by many years of smaller negative flows to cover operation, maintenance, and,

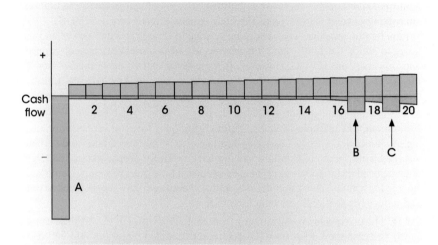

Figure 8.5 Positive and negative cash flows.

assuming the project is financed by a bank, interest payments on the loan. As the turbines and other equipment approach the end of their working lives, routine maintenance costs tend to rise and further cash injections (*B, C*) may be needed to repair or replace major worn-out components.

Offsetting the negative cash flows is the income generated, year on year, by selling electricity (shown in blue). The trend is shown upward on the assumption, generally held by energy experts, that electricity prices will continue to rise in real terms over the coming decades.

A key advantage of life-cycle analysis (LCA) over SPP and COE estimates is that it acknowledges the *time value of money*—a major consideration for a long-term project. In a nutshell, the cash flow expected in the future should not be given the same monetary value today. To relate this to personal experience, would you rather have 100 dollars today, or the expectation of 150 dollars in 10 years' time? Your answer will probably depend on predicting future interest rates (you could put the money in the bank), or the confidence you have about future payments, or you may prefer to purchase something for 100 dollars today. Life-cycle costing analysis takes the time value of money into account by referring all future cash flows to their equivalent in today's money using a *discount rate*. This is the rate above general inflation at which money could be invested elsewhere—say between 1 and 5%. In this way, the *present worth* of a complete long-term project can be estimated, and compared with alternatives, allowing for a more realistic investment decision to be made. A positive value of present worth is generally taken as a good indication of economic viability.

So far so good, provided we recognize that investment decisions based on LCA make many assumptions about costs, technical performance, system and component lifetimes, interest rates, and the future price of electricity. Figure 8.5 looks simple enough, but it is hard to put accurate numbers against the various items. In the case of wind energy, uncertainties arise under several headings:

■ *Initial costs.* Although the capital costs of turbines and equipment are based on firm prices quoted by manufacturers, installation costs are harder to predict. For example, unforeseen difficulties may arise with turbine foundations or cable laying, and progress may be delayed by bad weather, especially offshore.

- *O&M*. Annual costs for operating and maintaining large HAWTs typically amount to around 2% of initial capital costs. Wind farm costs include items such as insurance, taxes, and land rental; and maintenance includes periodic inspections, testing, and blade cleaning. Offshore maintenance is generally more expensive than onshore. As a wind farm ages, occasional replacement or repair of major components becomes more likely but is very hard to predict. Over the life of a wind farm, O&M costs have a major influence on profitability.

- *Technical performance and reliability*. Wind farm income is derived entirely from the electricity generated over its working lifetime. As we have seen in previous chapters, the wind regime at the chosen site and the technical performance of the turbines are key determinants. The cubic relationship between wind speed and power means that annual electricity production is affected by quite small changes in average wind speed. The "availability" of turbines—the percentage time they are able to generate electricity—should clearly be as high as possible, but occasional downtime for unscheduled maintenance and repair is inevitable.

- *Value of wind-generated electricity*. Wind energy's value to an electric utility is usually assessed in terms of "avoided costs," the savings in fuel costs (FCs) of alternative generation, and the reduced need for alternative capacity. As far as a wind farm operator is concerned, the "fuel" is free, and it makes sense to sell wind energy at almost any price rather than waste it. In today's increasingly deregulated energy markets, where electricity is traded as a commodity, the tension between these viewpoints is supposed to lead to contracts between "willing buyers" and "willing sellers" at mutually acceptable prices. However, the whole area of electricity pricing, present and future, and the income that wind farms can expect are heavily influenced by government policies toward renewable energy. For example, legislation in Germany requires grid operators to buy all renewable electricity produced within their supply areas at guaranteed minimum prices, giving considerable stability. Conversely, changes in policy, particularly of the stop–go variety, breed uncertainty, and loss of confidence. And finally, the economics of a wind farm depend greatly on how long it lasts. If it can continue generating for a few years longer than originally planned, with modest maintenance costs, the extra income becomes a pure bonus.

8.1.2.1 Levelized cost of electricity (LCOE)

So far so good, provided we recognize that the decision, even when based on careful life-cycle cost analysis, contains uncertainties about technical performance, system and component lifetimes, interest rates, and the future price of electricity. The investment on a commercial or industrial system carries less uncertainty as it is often linked with a power purchase agreement (PPA) with a utility that guarantees revenue which, over the course of the investment period, carries all initial and recurrent expenses. Wind as well as other renewable energy systems may have a high upfront cost, but they use free "fuel" during the many years they operate. On the other hand, the operating costs of coal and natural gas-burning power plants are high and uncertain as they depend on the price and availability of fuel. Thus, to compare the COE technologies with different operating conditions, we need to sum up the capital and operating costs and assign them to the amount of electricity produced over a period of time. For this purpose, the LCOE is defined as a constant unit cost, per kWh or MWh, of a payment stream over a period of time, which has the same present value as the total cost of constructing and operating a power generating plant over its life. It represents the constant level of revenue necessary each year to recover all expenses over the life of a power plant. In simple words, LCOE represents the total life-cycle cost divided by the revenue from the total lifetime electricity production. The costs include the initial capital investment and the lifetime operating expenses, less the end-of-life value discounted into the present value. The revenue from energy production depends on the location (wind resource), the performance ratio, and the life of the system. An important constituent in the LCOE is the cost of financing (debt service) of the relatively high equity investment needed upfront. The schematic in Figure 8.6 illustrates this concept. The cost of financing is linked to the real and/or perceived risk of the asset. Long-term PPA with utilities and government feed-in-tariff programs reduce the risk to investors. Deployment of wind power plants removes the uncertainty of fuel supply and prices.

Levelized Cost of Electricity (LCOE)

$$LCOE = CR + PTI + O\&M + FC$$

with

$$CR = \frac{OCC}{CF \times 8760} \times \frac{WACC(1 + WACC)^t}{(1 + WACC)^t - 1}$$

where

CR: capital recovery

OCC: overnight capital cost

O&M: annualized operating and maintenance expenses (variable and fixed)

PTI: annualized property taxes and insurance

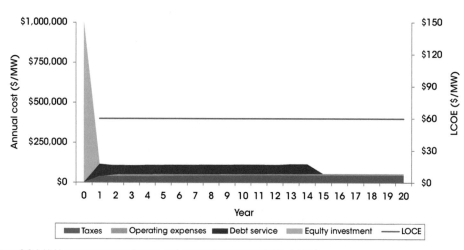

Figure 8.6 Initial investment, annual costs, and levelized cost of electricity (LCOE).

FC: annualized fuel costs

CF: capacity factor of the plant

WACC: weighted average cost of capital

t: economic life of the plant

We summarized the ideas behind conventional life-cycle cost analysis, with its positive and negative cash flows and levelized cost of electricity that averages all costs over a certain period. But what if the economics are affected by a government decision to offer capital grants to offset the initial purchase price, or suddenly to change or terminate grants that are presently available? And what if the price paid for renewable electricity is bolstered by special tariffs that may be altered or removed by a change of government? Over the years, there have been many such stop–go incidents in countries as wide apart as Australia, Spain, and the USA. One of the biggest threats to rational decision-making and steady growth in the wind energy market is uncertainty about government policy, and one of the biggest benefits is consistent long-term support.

You may be wondering why governments offer financial support to wind energy in the first place. There are two principal reasons. First, the products of a new high-tech industry tend to be very expensive at the start, before cumulative production gathers pace. If governments wish to pursue urgent policy objectives such as the reduction of carbon emissions, they may decide to stimulate market development with financial incentives. Second, Figure 8.6 makes clear that wind, like other renewable energy technologies including solar and wave, has its major costs "up front," with no fuel charges. This is quite different than conventional electricity generation based on fossil fuels. Projects with high initial costs that must be set against future income are commonplace for large corporations but tend to be far more problematic for small businesses, organizations, and individuals who find it hard to raise the initial capital.

Our short excursion into economics has taken no account of the environmental issues that, in many people's minds, are just as important as the business concerns of financiers and accountants. We now turn our attention to resource availability and environmental issues, which form some of the main justifications—and raise a few doubts—surrounding the modern renaissance of wind energy.

8.2   Resource availability

The main environmental credentials of wind energy are established beyond doubt: its important contribution to reducing carbon emissions, cleanliness in operation, lack of spent fuel or waste, and general public acceptability in terms of visual impact. We have already referred to such advantages at various points in this book. But there are further environmental considerations as wind energy accelerates into multi-gigawatt annual production—can Planet Earth provide the necessary quantities of raw materials, and are there any conflicts with the use of land or water for the hundreds of thousands of wind turbines that are needed for a significant contribution of wind energy in climate change mitigation scenarios?

This merits consideration of primary resources and of the potential to reuse/recycle them.

8.2.1 Raw material primary resources

The main wind turbine materials are steel, aluminum, composites, polymers, cast iron, and rare earth elements (REEs). The inland turbine foundation uses concrete and transmission lines use copper. A recent National Renewable Energy Laboratory (NREL) study lists the material breakdowns in modern wind turbines.[3] As shown in Figure 8.7, current land-based wind power plants require about 1200 metric tons (t) of material per megawatt, composed (by mass) of approximately 53% road aggregate, 34% concrete, 9% steel, 2% composites and polymers, 1% cast iron, and 1% other materials. Future land-based wind plants may contain a larger proportion of concrete due to bigger foundations required for larger and taller turbines. These changes could shift the material breakdown of the future land-based wind plants by mass to 46% concrete, 39% road aggregate, 10% steel, 3% composites and polymers, 1% cast iron, and 1% other materials. Offshore wind plants currently require about 300 metric tons of material per megawatt, composed (by mass) of 87% steel, 5% other metals (e.g., Al and Ni) and their alloys, 4% composites and polymers, 3% cast iron, and 1% other materials.

The same NREL study indicates that in a high deployment scenario, there are six types of materials that exceed 20% of the current US production, namely, carbon fiber, electrical steel,

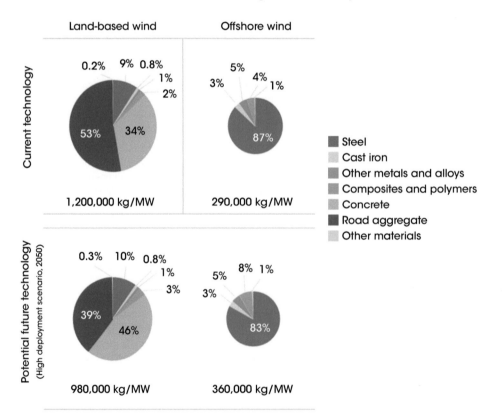

Figure 8.7 Material intensities of current and potential future wind energy technologies (*Source*: Eberle *et al.*[9]/NREL/Public Domain).

aluminum, cobalt, nickel, and REE (i.e., dysprosium, neodymium, and praseodymium). Their use is expected to increase because the entire renewable energy industry is experiencing high growth. The issue with aluminum and steel is more related to the energy used in their manufacture, whereas increasing the supply of composite materials used in blades and REE used in permanent magnets has additional environmental impacts.

8.2.1.1 Blades

Blades are made mainly of carbon fiber, fiberglass, and balsa wood, and the wind industry drives a significant portion of global demand for these materials. As reported by the Union of Concerned Scientists demand for balsa wood outstripped supply in 2020. The supply is highly concentrated in Ecuador and Peru, and its high demand in the wind industry may have increased Amazon rainforest logging.[1]

8.2.1.2 Rare earth elements (REE)

Let us now discuss the production of REE. REE are relatively plentiful in the entire Earth's crust, but they are contained in only trace concentrations in mineral ores, and to obtain rare earths at usable purity requires processing enormous amounts of raw ore, thus the name "rare" earths (Wikipedia, 2024. https://simple.wikipedia.org/wiki/Rare_earth_element). The list includes the Lanthanides [atomic numbers (57–71), Yttrium (21), and Scandium (39), Table 8.1]. The diverse nuclear, metallurgical, chemical, catalytic, electrical, magnetic, and optical properties of the REE have led to a variety of applications. Among this group, the key elements for clean energy technologies are neodymium, praseodymium, dysprosium, and terbium, used to manufacture neodymium–iron–boron (NdFeB) permanent magnets which are components in generators for wind turbines and in traction motors for electric vehicles. Most land-based turbines use an electromagnetic generator, in which copper coils rotate through a magnetic field to produce electricity. But another option, popular in offshore wind, is a permanent magnet generator, which contains a ring of brick-shaped rare Earth magnets that spin with the rotor to produce electricity. With such magnets, offshore wind turbines can be directly driven, avoiding the high maintenance gear boxes. Neodymium and praseodymium strengthen the magnet, while dysprosium and terbium make it resistant to demagnetization at high temperatures. A large wind turbine uses about 1.5 metric tons of these REEs. The World Bank predicts that the demand for neodymium for energy technologies in 2050 will be 37% of total 2018 neodymium production. Global demand for neodymium for wind turbines is estimated to increase 48% by 2050. The economic development of REE mineral deposits presents challenges as they are in low concentrations in ores of multicomponent minerals.

Table 8.1 Rare earth elements (symbols, atomic numbers, names).

La (57)	Lanthanum	Eu (63)	Europium	Tm (69)	Thulium
Ce (58)	Cerium	Gd (64)	Gadolinium	Yb (70)	Ytterbium
Pr (59)	Praseodymium	Tb (65)	Terbium	Lu (71)	Lutetium
Nd (60)	Neodymium	Dy (66)	Dysprosium		
Pm (61)	Promethium	Ho (67)	Holmium	Sc (21)	Scandium
Sm (62)	Samarium	Er (68)	Erbium	Y (39)	Yttrium

---

[1] Union of Concerned Scientists (2022). https://blog.ucsusa.org/charlie-hoffs/what-happens-to-wind-turbine-blades-at-the-end-of-their-life-cycle/

The USA is currently the second largest producer of REE-containing minerals after China, with US production[2] totaling 16% of global production compared to China's 58%. However, most ore extracted from US sources is exported for processing, and US consumption of REE relies heavily on imports; China provides 80% of US REE imports. Although recycling REE technologies exist, secondary production has not yet matured, and all commercial-scale US consumption of REE comes from primary sources.

There are many opportunities for innovation to change material requirements for wind energy. In general, such opportunities involve material substitution, weight reduction, lifetime extension, component reuse, and material recycling.

8.2.2 Wind turbine decommissioning

An obvious question is what is the fate of a wind turbine at the end of its productive life? The first step will be to decommission the turbine, and this may be more challenging in offshore than in onshore projects, so in this section, we present the decommissioning steps involved in decommissioning one of the early projects, the Lely Wind Farm (Netherlands) where the monopile foundations were also removed. Typically, the piles are cut below the seabed, at an appropriate distance; often this distance is around 3 m below the natural seabed. Figure 8.8 shows the photos of step-by-step decommissions, and they are self-explanatory.[4]

The Lely Wind Farm (four 2-MW turbines in the Ijsselmeer, located approximately 800 m offshore) built in 1992 generated electricity for 24 years and was fully decommissioned in 2017. Crane barges and tugs were used to dismantle the turbines and tower sections. The monopiles were extracted using a vibratory hammer.

In this context, it is necessary to provide an overview of the current legislation on offshore installations: For the UK, decommissioning of offshore structures is regulated by UK law and by the OSPAR Decision 98/3 on the Disposal of Disused Offshore Installations; an expansion of the law to cover offshore wind turbines is under discussion (https://oap.ospar .org/en/ospar-assessments/quality-status-reports/qsr-2023/other-assessments/renewable-energy/).

8.2.3 Wind turbine recycling

The good news is the steel, iron, aluminum, copper, and concrete can be completely recycled. Recycling the blades and the REE-containing magnets is a greater challenge, as they are made of different materials that need multistep separation and purification processes in order to be recovered.

8.2.3.1 Blades

In 2021 in the USA, 8000 blades were retired. By 2050, the world may be getting 15–18% of its energy from wind but may also have 43 million metric tons of retired blades to address. Currently, most retired wind turbine blades end up in landfills. Although landfilling of blades

---

[2]United States Geological Survey (USGS). https://pubs.usgs.gov/periodicals/mcs2023/mcs2023-rare-earths.pdf

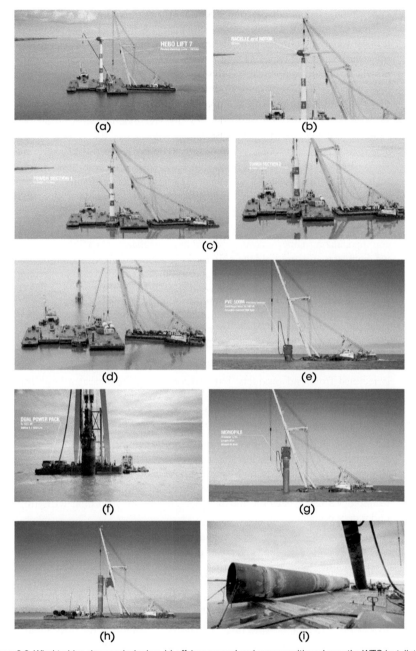

Figure 8.8 Wind turbine decommissioning: (a) offshore vessel and crane positioned near the WTG installation; (b) lift of nacelle and rotor; (c) lift of tower section 1; (d) lift of tower section 2; (e) superstructure on the vessel; (f) vibratory hammer used to remove the monopile from the ground; (g) power pack used along with the vibratory hammer; (h) lift of the monopile on the vessel; (i) monopile on the vessel (*Source*: Bhattacharya (2019)).

does not pose a threat of soil or groundwater contamination and would represent only a tiny fraction of global solid waste streams, the life cycle of wind turbines would be more circular if there were more options to reuse or recycle the blades. Innovative companies are employing repurposing and recycling technologies to help avoid that fate. Currently, these involve mostly downcycling applications like shredding down fiberglass blades and turning them into cement or textiles or railroad ties and plastic pellets; upcycling processes that will separate blade fibers from epoxy and use the epoxy in new blades are also under development. In parallel, companies have also started to develop new, more recyclable blade technologies with new types of blade epoxy resins.

The major western original equipment manufacturers (OEMs) are members of the DecomBlades consortium announced in 2021, with the aim of developing recyclable wind turbine blades, and as of 2023, they have started limited production of prototype recyclable wind turbines. In parallel, important industry–university collaborations have been formed to further pursue blade recycling.

A UK project between Aker Offshore Wind and the University of Strathclyde focuses on commercializing a fiber-recovery method used to separate glass fibers from resin components in composites and reprocess them. In the USA, the Carbon Rivers project, funded by the US Department of Energy (DOE) in collaboration with the University of Tennessee, Knoxville, is scaling up a recovery process for wind turbine blade recycling that has the capability to recycle fiberglass from 50,000 metric tons of blades annually.

8.2.3.2 Rare earth elements (REE)

Due to the high cost of separation and purification, currently only a small amount of REE is recycled from end products, but recycling could help meet a large fraction of future demand. Adding recycled rare earths as a new source to the supply chain is expected to reduce environmental contamination and energy costs associated with their primary mining and separations.

Recycling of REE is currently accomplished through a combination of processes, including hydrometallurgy (acid leaching to extract rare earth oxides and salts), pyrometallurgy (roasting and melting at high temperatures), or electrochemistry (using electricity to separate materials based on their differing abilities to gain or lose electrons). In addition, technologies using bacteria and microorganisms for absorbing REE from products are under development.[3]

In summary, REE are crucial materials in wind turbine and automobile permanent magnets as well as in many consumer products. These elements currently have a significant environmental burden as consumer products are discarded into the trash. Recycling rare-earth-containing products would provide a steady, domestic source of rare earths to manufacturers while also reducing waste and the same applies to recycling composite materials used in blades.

In general, recycling of wind turbine materials strengthens the three pillars of sustainability for large wind turbine market growth; it provides an important secondary resource, it

---

[3]Union of Concerned Scientists, Just and sustainable solutions for the mining and recycling of rare earth elements in wind turbines (December 12, 2022). https://blog.ucsusa.org/charlie-hoffs/just-and-sustainable-solutions-for-the-mining-and-recycling-of-rare-earth-elements-in-wind-turbines/

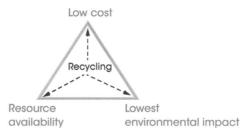

Figure 8.9 Recycling strengthens the three pillars of WT sustainability.

relieves the pressure for price increases that result from increasing demand and eliminates environmental risks from disposal (Figure 8.9).

Research on wind recycling is getting a new boost in the USA as the Bipartisan Infrastructure Investment and Jobs Act includes $40 million for such research. Consequently, on January 18, 2024, the US DOE announced funding of 20 projects of the first phase of a two-phases Wind Turbine Materials Recycling Prize which was launched in July 2023 as part of the American-Made Challenges Program.[4]

8.3 Environmental and social aspects

In Section 8.1, we considered the costs incurred, and the income generated, over the lifetime of a wind farm—the classic approach to assessing a project's economic viability. However, this form of LCA ignores the *environmental* costs and benefits of wind generation that have such important implications for global energy policy and society as a whole. It is now time to paint with a broader brush and attempt a balanced account of the effects, negative as well as positive, that wind energy has on the environment.

It is very important to realize that all methods of energy production—whether based on coal, gas, oil, nuclear, or renewable sources—have environmental impacts that are ignored by the traditional notion of "cost." A narrow economic view of industrial processes assesses everything in terms of money, ignoring other factors that common sense tells us should be taken into account in any sensible appraisal of value. For example, the "cost" of generating electricity in nuclear power plants has traditionally been computed without taking any account of accident or health risks or the need to store dangerous waste products safely for thousands of years; in the case of coal-fired plants, without acknowledging their large and unwelcome carbon dioxide emissions and health and safety effects to miners; and, in the case of wind farms, without placing any value on landscape.

There are two main reasons for this apparent short-sightedness. First, aspects such as health, safety, environmental protection, and the beauty of landscape cannot easily be quantified and assessed within a traditional accounting framework. We all know they are precious and, in many cases, at least as important to us as money, but appropriate tools and

---

[4]US-DOE Wind Energy Technology Program. https://www.energy.gov/eere/wind/wind-turbine-sustainability+

methodologies for including them are only now being developed and accepted. It is surely vital to do this because so many of our current problems and dissatisfactions are bound up with the tendency of conventional accounting "to know the price of everything and the value of nothing."

The second reason relates to the important notion of *external costs* of energy generation. These costs, which are mostly environmental or societal in nature, have traditionally been treated as outside the energy economy and to be borne by society as a whole, either in monetary terms by taxation or in environmental terms by a reduction in the "quality of life." They contrast with the *internal costs* of running a business—for plant and buildings, fuel, materials, staff wages, and so on—that are paid directly by a company and affect its profits. If Planet Earth is treated as an infinite "source" of raw materials and an infinite "sink" for all pollution and waste products, it is rather easy to ignore external costs. For example, it seems doubtful whether the 19th-century pioneers of steam locomotion ever worried about burning huge quantities of coal or the 20th-century designers of supersonic civil airliners about fuel efficiency and supersonic bangs. One of the positive changes now taking place is the growing worldview that external costs should be worked into the equation—not just the local or national one, but increasingly the global one. In other words, external costs should be *internalized* and laid at the door of the responsible industry or company. In a nutshell, "the polluter should pay."

External and internal costs associated with industrial production are shown in Table 8.2. The external ones, representing charges or burdens on society as a whole, are split into environmental and societal categories, although there is quite a lot of overlap between them. You can probably think of some extra ones. Internal costs, borne directly by the organization or company itself, cover a wide range of goods and services.

Table 8.2 External and internal costs.

External: Environmental				
CO_2	Emissions & waste	Resource depletion	Accidents	Species loss
Internal				
Buildings	Plant & machinery	Office systems	Transport	Fuel
Materials	Wages	Pensions	Advertising	Insurance
Human health	Noise	Visual intrusion	Land use	Security
External: Societal				

The distinction between external and internal costs is somewhat clouded by the fact that many items bought in by a company, for example, fuel and materials, have themselves involved substantial external costs during manufacture and transport.

In the case of electricity generation, a proper analysis of the environmental burdens should take account of all contributing processes "from cradle to grave," whether conducted on- or off-site. Needless to say, such an *environmental* LCA is a challenging task. One of the special difficulties facing renewable generation, including wind energy, is that so many of its advantage stems from the *avoidance* of external costs. They are therefore hidden by conventional accounting methods. Renewables tend to produce very low carbon dioxide emissions, cause little pollution, make little noise, create few hazards to life or property, and have wide public support. Yet politicians and economists rarely mention the reduction or avoidance of external costs as an advantage of wind energy. Fortunately, energy experts and advisers to governments are taking increasing notice of environmental LCA and assessing the risks and benefits of competing technologies on a more even footing.

Monetization of the external costs of energy life cycles, including those of renewable energies, is well-documented in reports by the National Research Council, the Harvard School of Public Health, and others. These studies show the greatest health and environmental effects in the life cycle of coal power generation are those from toxic air pollutants during combustion, followed by the impacts of mining and greenhouse gas (GHG) emissions. More specifically, the Harvard study estimated that the total external costs of coal-fired electricity during extraction, transport, processing, and combustion are \$345 billion or 18 cents/kWh of electricity produced. These estimates are based on health costs, health insurance, and damage costs that are not included in the electricity costs but are paid by society at large. If these were fully accounted for, then the price of electricity from coal in the USA would have been much higher than the LCOE from wind turbines. Carbon dioxide capture with carbon sequestration (CCS) is advocated widely as enabling the continuation of coal burning for power generation, but CCS, while reducing GHG emissions, will lead to increases in toxic emissions and health, safety, and environmental (EHS) impacts in mining regions, as coal consumption per unit of electricity output would increase. The same applies to the proliferation of natural gas from gas-shale resources, as hydraulic fracturing increases the impacts of extracting gas on both the air and water pathways, and conversion to liquefied natural gas (LNG) for exportation further increases the upstream safety risks.

Where does all this leave wind energy in its remarkable and seemingly unstoppable progress toward 1000 GW of global installed capacity? We will start with the positive environmental impacts.

To many governments around the world, and their citizens, the major environmental benefit of wind power is the avoidance of the emissions caused by fossil-fuel power plants—principally carbon dioxide, but also other polluting gases—and reduced risk of climate change. Figures are widely quoted for individual turbines, wind farms, and nations. For example, a well-sited 2-MW onshore HAWT in the UK is typically quoted as saving about 1800 tons of CO_2 per annum. The British Wind Energy Association estimated that the UK's total installed wind capacity of 5300 MW gave annual savings of around 5.9 million tons of CO_2 plus 140,000 tons of sulfur oxides and 41,000 tons of nitrogen oxides. However, such calculations must depend on the energy mix of the country concerned and the types of alternative generation that are being displaced. France, which produces around 75% of its electricity in nuclear plants, and Norway, which approaches 100% hydroelectric generation, are clearly very different from the UK. This is not to doubt the global benefits of wind energy but simply to point out that claimed emission reductions should be interpreted in a national context.

Further environmental benefits of wind energy include:

Fast energy payback. It takes energy to produce energy. All types of electricity generation consume energy in the production of plants and buildings, and most subsequently waste a lot of energy due to the low efficiency of the thermodynamic cycle. Wind farms incur "imbedded" energy during turbine manufacture and installation, but typically all this is paid back within 3–5 months by the electricity generated. This is detailed in Section 8.6.

Energy diversity and security. Wind farms add to a nation's energy diversity, reducing dependence on other forms of generation including those relying on imported fuel from areas of the world that are politically unstable. Wind farms are less vulnerable to catastrophic accidents or terrorist attacks than large, centralized power plants, including nuclear. Security of supply is best guaranteed by a broad energy mix and should be considered an industrial and societal benefit rather than a strictly environmental one.

Having outlined the main environmental benefits of wind energy, we are ready to address some less comfortable issues—the potential negative impacts that, unless carefully handled, can lead to a loss of public support.

8.4 Land use

Ground-mount wind farms occupy significant amounts of land. However, the space between wind turbines can be used for other applications. Data from wind farms in the USA show land use in the range of 5–8 acres/MW. This is the land that wind turbines occupy to receive the wind energy and thus the free "wind fuel" plus the large open areas needed to prevent wake effects from one turbine affecting the performance of other turbines, and the access roads and facilities. Now is this a lot of land? To answer this question, we would need to make comparisons with the land used by the coal, natural gas, and nuclear life cycles.

Fossil-fuel-based generation such as coal has a seemingly low land footprint, as power plants take up a relatively small surface area for their large power output. However, the picture changes when life-cycle land use is assessed, accounting for the direct (mining and fuel processing, plant footprint) and indirect (land usage for materials and building infrastructure needed to operate the mines) land transformed. The life-cycle land usage for different sources of electricity can be compared on a "surface area per energy unit" basis, although in that case, it is important to define a finite time scope as the land used for wind and solar could virtually generate electricity forever (in the case of this study: 25 years, the wind turbine system lifetime). Land usage for wind turbine (WT) and wind would perhaps better be described with a "surface area per power unit" as their power source for that surface area will exist for virtually infinite time. In contrast, the energy source from surface coal mines is exhausted after the fuel is extracted.

Historical data show that wind farms in the USA often use less land during their life cycle than coal during its life cycle.[5]

For the coal fuel cycle, the direct land transformation is primarily related to the coal extraction, electricity generation, and waste disposal stages, while the indirect land use refers to the upstream land use associated with energy and materials inputs during the fuel cycle. Land use statistics during coal mining vary with factors including heating value, seam thickness, and mining methods. Surface mining in the Western USA tends to disturb less area (per unit coal mined) than in other areas due to the thick seams, 2–9 m. Central states where seam thickness is only 0.5–0.7 m transform the largest area for the same amount of coal mined. On the other hand, underground mining transforms land mostly through indirect pathways. Wood usage for supporting underground coal mines accounts for the majority of indirect land transformation. Currently, in the USA, about 70% of coal is mined from the surface.

For operating a coal power plant, land is required for facilities including powerhouse, switchyard, stacks, precipitators, walkways, coal storage, and cooling towers. The size of a coal power plant highly varies; a typical 1000-MW capacity plant requires between 330 and 1000 acres which translates into 6–18 m^2/GWh of transformed land based on a CF of 85%. Another study based on a 500-MW power plant located in the Eastern USA estimated 32 m^2/GWh of land transformation. Also, coal-fired power plants generate a significant amount of ash and sludge during operation. Disposing of the solid wastes accounts for 2–11 m^2/GWh, 50% each for ash and sludge, in the US condition.

The natural gas life cycle would use less land, but when it is extracted from the dilute formations on shale rocks then the land occupation is expected to be as large as that of coal.

In addition to the use of the land, we should consider its transformation and possible damage. WT installations do not damage the land (Figure 8.10). Once the installation is completed, WT operation will not disturb the land, in contrast to coal mining which often

Figure 8.10 Environmentally friendly use of land by wind turbines (*Source*: Altitudedrone/Adobe Stock).

---

[5]V.M. Fthenakis and H.C. Kim. Land use and electricity generation: a life cycle analysis. *Renewable and Sustainable Energy Reviews*, 13, 1465–1474 (2009).

pollutes the land. Also, fossil or nuclear fuel cycles need to transform certain amounts of land continuously in proportion to the amount of fuel extracted. Restoring land to its original form and productivity takes a long time and often is infeasible. Accounting for secondary effects including water contamination, changes in the forest ecosystem, and accident-related land contamination would make the WT cycle even better than other fuel cycles. For example, water contamination from coal and uranium mining and from piles of uranium mill tailings would disturb adjacent lands. Additionally, land transformed by accidental conditions, especially for the nuclear fuel cycle, could change the figures dramatically. The Chernobyl accident contaminated 80 million acres of land with radioactive materials, irreversibly disturbing 1.1 million acres of farmland and forest in Belarus alone.

8.5   Water use

Wind turbines have a major advantage against any thermoelectric power generation (e.g., coal, natural gas, nuclear, biomass, and concentrated solar power) as it does not need any water for power generation. Electricity generation via conventional pathways accounts for a major part of water demand as water cooling is employed in most power plants. Overall, it was estimated that WT deployment in the U.S.-SW would displace 1700–5600 m^3/GWh of water demand when it displaces grid electricity.[6]

8.6   Life-cycle analysis

In Sections 8.2 and 8.4, we considered WT requirements for raw materials and land—two environmental issues that surface before WT production even begins. Further important environmental questions arise during a WT system's lifetime, which starts with extraction and purification of raw materials; proceeds through manufacture, installation, and many years of operation; and ends with recycling or disposal of waste products. The whole sequence is referred to as a *life cycle*, and it is important to appreciate its environmental consequences. Note that this form of LCA is not the same as the classical economic life-cycle costing version previously introduced, which deals with cash flows and financial decisions. We are now moving on to something much broader, with important implications for global energy policy and society.

In this brief introduction, we will consider LCA under two main headings:

- *Environmental impact.* What impacts (e.g., GHG emissions) are incurred or avoided?
- *Energy balance.* How does the amount of electrical energy generated over a system's lifetime compare with the energy expended in making, installing, and using it?

We now move on to the much-discussed topic of *energy balance*. Clearly, it takes energy to produce energy. But, how does the total amount of electrical energy generated by a WT module or system over its lifetime compare with the input energy used to manufacture, install, and use it? Closely related to the energy balance is the *energy payback time (EPBT)*, the

---

[6]V.M. Fthenakis and H.C. Kim. Life-cycle of water in U.S. electricity generation. *Renewable and Sustainable Energy Reviews*, 14, 2039–2048 (2010).

number of years it takes for the input energy to be paid back by the system. We naturally expect WT to have favorable energy balances and payback times, especially in view of its claims to be clean and green.

Two initial points are worth making. First, energy payback is not the same as economic payback. The latter is concerned with repaying a system's capital and maintenance costs (including the COE consumed) by a long-term flow of income and is essentially a financial matter; energy payback is much more about the environment. Second, the environmental benefits of a short payback time depend on the present energy mix of the country, or countries, concerned. If the required input energy is largely derived from coal-burning power plants, it is more damaging than if it comes from, say, hydroelectricity.

Major energy inputs to a WT system occur during the following activities:

- extraction, refining, and purification of materials.
- manufacture of wind turbines, and balance of system (BOS) components.
- transport and installation.

Interestingly, some of the most significant energy inputs are for components such as aluminum for WT components and concrete foundations for support structures in large WT plants.

The other side of the energy balance—the total electrical energy generated by a system over its lifetime—depends on the following factors:

- efficiency of WT,
- the level of wind resources at the location of deployment,
- alignment of the WT array and shading (if any), and
- the life of the system.

Now, let us discuss the LCA in a more structured way.

The energy balance is most favorable for systems that are efficiently produced in state-of-the-art factories and installed at optimal sites in sunshine countries. Things get even better if systems last for longer than their projected or guaranteed lifetimes—but of course this is hard to predict. Some life-cycle studies carried out in the early years of the new millennium painted a rather gloomy picture of WT's environmental and health impacts, due largely to the fossil-fuel energy used during WT manufacture. However, up-to-date peer-reviewed studies that take proper account of advances in WT engineering have corrected this picture. The EPBTs of current WT power plants are between 0.5 and 1.8 years, depending on the wind resource in the place of installation and the type of technology used.[7,8]

To better understand this result, we would need to discuss the elements of the LCA. LCA is a comprehensive framework for quantifying the environmental impacts caused by material

[7]Luiz Felipe Souza Fonseca Monica Carvalho, Greenhouse gas and energy payback times for a wind turbine installed in the Brazilian Northeast, Front. Sustain., Sec. Quantitative Sustainability Assessment Volume 3 - 2022 | https://doi.org/10.3389/frsus.2022.1060130

[8]Nasssar E.C.M et al, Carbon footprint and energy life cycle assessment of wind energy industry in Libya, Energy Conversion and Management Volume 300, 15 January 2024, 117846, https://doi.org/10.1016/j.enconman.2023.117846

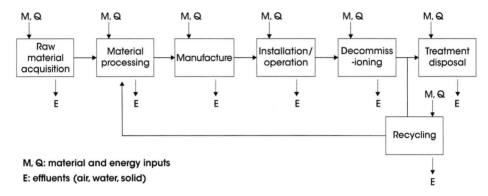

M, Q: material and energy inputs
E: effluents (air, water, solid)

Figure 8.11 The life-cycle stages of wind turbines.

and energy flows in each and all the stages of the "life cycle" of a product or an activity. It describes all the life stages, from "cradle to grave," thus from raw materials extraction to end of life. The cycle typically starts from the mining of materials from the ground and continues with the processing and purification of the materials, manufacturing of the compounds and chemicals used in processing, manufacturing of the product, transport, installation if applicable, use, maintenance, and eventual decommissioning and disposal and/or recycling. To the extent that materials are reused or recycled at the end of their first life into new products, then the framework is extended from "cradle to cradle." This life cycle for WT is shown in Figure 8.11.

The life-cycle **cumulative energy demand (CED)** of a WT system is the total of the (renewable and nonrenewable) primary energy (PE) harvested from the geo-biosphere in order to supply the direct energy (e.g., fuels, electricity) and material (e.g., Si, metals, glass) inputs used in all its life-cycle stages (excluding the wind energy directly harvested by the system during its operation). Thus,

$$\text{CED} \left[\text{MJ}_{\text{PE-eq}}\right] = E_{\text{mat}} + E_{\text{manuf}} + E_{\text{trans}} + E_{\text{inst}} + E_{\text{EOL}} \tag{8.1}$$

where

E_{mat} [MJ$_{\text{PE-eq}}$]: PE demand to produce materials comprising of WT system

E_{manuf} [MJ$_{\text{PE-eq}}$]: PE demand to manufacture WT system

E_{trans} [MJ$_{\text{PE-eq}}$]: PE demand to transport materials used during the life cycle

E_{inst} [MJ$_{\text{PE-eq}}$]: PE demand to install the system

E_{EOL} [MJ$_{\text{PE-eq}}$]: PE demand for end-of-life management

The CED of a WT system may be regarded as the energy investment that is required in order to be able to obtain an energy return in the form of WT electricity.

EPBT is defined as the period required for a renewable energy system to generate the same amount of energy (*in terms of equivalent PE*) that was used to produce (and manage at end of life) the system itself.

$$\text{EPBT} \left[\text{years}\right] = \text{CED} / \left(\left(E_{\text{agen}}/\eta_{\text{G}}\right) - E_{\text{O\&M}}\right) \tag{8.0}$$

E_{agen} is then converted into its equivalent PE based on the efficiency of electricity conversion at the demand side using the grid mix where the WT plant is being installed. Note that

E_{agen} is measured (and calculated) in units of kWh, and we first have to convert it to MJ (1 kWh = 3.6 MJ) and then use nG to convert MJ of electricity to MJ of PE (MJ_{PE-eq}). Thus, calculating the PE equivalent of the annual electricity generation (E_{agen}/η_G) requires knowing the life-cycle energy conversion efficiency (η_G) of the country-specific energy mixture used to generate electricity and produce materials. The average η_G for the USA and Europe are, respectively, approximately 0.30 and 0.31.

Energy return on (energy) investment (EROI) is defined as the dimensionless ratio of the energy generated over the course of its operating life over the energy it consumed (i.e., the CED of the system). The electricity generated by WT needs to be converted to PE so that it can be directly compared with CED. Thus, EROI is calculated as:

$$\text{EROI} \left[MJ_{PE-eq}/MJ_{PE-eq} \right] = T \times \left[(E_{agen}/\eta_G) - E_{O\&M} \right] /\text{CED} = T/\text{EPBT}$$

where T is the period of the system operation; both T and EPBT are expressed in years.

EROI and EPBT provide complementary information. EROI looks at the overall energy performance of the WT system over its entire lifetime, whereas EPBT measures the point in time (t) after which the system is able to provide a net energy return.

GHG emissions and global warming potential (GWP): The overall GWP due to the emission of a number of GHGs along the various stages of the WT life cycle is typically estimated using an integrated time horizon of 100 years (GWP_{100}), whereby the following CO_2-equivalent factors are used: 1 kg CH_4 = 32 kg CO_2-eq, 1 kg N_2O = 296 kg CO_2-eq, and 1 kg chlorofluorocarbons = 4600–10,600 kg CO_2-eq. Electricity and fuel use during the production of the WT materials and components are the main sources of the GHG emissions for WT cycles, and specifically the technologies and processes for generating the upstream electricity play an important role in determining the total GWP of WTs, since the higher the mixture of fossil fuels is in the grid, the higher are the GHG (and toxic) emissions.

An average between inland and offshore WT is reported in the Wind Vision 2015 DOE study[9] to be 13 g CO_{2eq}/kWh of which the majority (12 g/kWh) is attributed to upstream material and component manufacturing. The LCA results strongly depend on the available electricity mix. The EPBT of WT is between 6 months and a year and the Energy Return to Energy Investment is 20–30 assuming a 20-year operation life.

8.7   Landscape and amenity

The interaction of large wind turbines with natural landscapes is probably the most contentious issue surrounding the development of wind farms, especially onshore. Visual intrusion is the most-cited reason for opposition to wind farms, and there are small but highly vociferous groups in many countries that object to what they see as "industrial-scale" machines imposed upon the beauty of the natural world. The whole issue is highly subjective and depends on the number, size, and arrangement of turbines as well as the geography and scale of the landscape (Figures 8.12 and 8.13). While most readers of this book presumably

---

[9]Wind Vision: a new era for wind power in the United States, US-DOE (2015). https://www.energy.gov/eere/wind/articles/wind-vision-new-era-wind-power-united-states

Figure 8.12 Beauty in the eye of the beholder (*Source*: With permission of French Wind Energy Association).

Figure 8.13 Figures in a windswept landscape (*Source*: With permission of French Wind Energy Association).

approach wind energy with a positive attitude—or at least an open mind—it is important to recognize genuine concerns and, where possible, to counter mischievous half-truths put about by vested interests.

The first point to make is that there are no easy energy options for the 21st century. Over recent years, it has become increasingly clear that fossil-fuel combustion is threatening a global environmental crisis and that, even if this were not so, there is a severe problem of resource depletion. China, India, and many other populous nations are presently industrializing on a huge scale, and it seems unimaginable that Planet Earth can survive unscathed if we continue on our present course. Nuclear power is often championed as an alternative to renewable sources, but its enthusiasts rarely mention the unsolved problems of radioactive waste disposal, the huge costs of decommissioning, and the dangers of nuclear accidents. While nobody should pretend that renewable energy is the answer to all our problems, or that it avoids all environmental concerns, it is surely wise to develop wind and other renewable technologies as an important part of the energy mix. Seen in this light, the visual intrusion of wind turbines in sensitive landscapes becomes just one of the many environmental issues raised by electricity generation.

In any case, there is no widespread hostility to wind farms. For example, survey after survey of public attitudes in the UK has shown high levels of support—typically around 80% of those questioned—and the support grows stronger among people living close to new farms when they actually experience them in operation and realize that the scare stories of pressure groups are exaggerated. In Scotland, where onshore wind farms sited on high uplands and hill ridges have become increasingly common in recent years, public support is consistently high among people living close to the turbines. By contrast, opposition, although noisy, comes from a very small percentage of the population. Of course, objections to the visual intrusion of large wind turbines are greatly influenced by other attitudes. If you are skeptical about the threat of climate change, you are likely to see any attempts to alleviate it as pointless; if you believe the stories about wind energy being inefficient, you may well join a pressure group to oppose it. And there is inevitably a strong element of the "nimby" (not in my backyard) syndrome in local objections to new wind farms, whereas those of us who take a wider view see them as evidence, not of a casual indifference toward the natural world but rather of our increasing desire to live in harmony with it. As so often, beauty is in the eye of the beholder.

At times like these, it is helpful to take a historical perspective. In the 19th century, there was huge opposition by English landowners to the building of railways and vocal attempts by vested interests to prevent their spread across the countryside; yet in time they became so accepted and appreciated as an industrial and social good that when thousands of miles of branch railway lines were condemned as uneconomic in the 1960s, there was great public misgiving. Today, there is plenty of nostalgia about the passing of the "age of steam." It is interesting that a project proposed by the UK government in 2011 to build a new high-speed rail link between London and Birmingham, which was vehemently opposed by many people living close to the selected route, was finally constructed.

Another good example is provided by the many thousands of UK electricity transmission towers (pylons) that have marched unopposed across the countryside since the 1950s. True, there are complaints when a new high-voltage transmission line is proposed, but once permitted and built, the noise quickly dies down. It is a truism that people, especially in rural areas, tend to dislike change, but in most cases, they adapt to it well enough in the longer term, and we may be reasonably confident that our grandchildren will accept wind farms as a normal part of the rural scene, just as our grandparents accepted the railways.

However, it is important to avoid complacency. The general acceptability of wind turbines depends markedly on scale—their individual size and their total numbers. Germany has proved its commitment to wind energy in recent years with well over 20,000 large onshore wind turbines but when, following the 2011 Fukushima nuclear disaster in Japan, Chancellor Merkel announced the premature closure of seven nuclear power plants and a great expansion of renewables, there were strong rumblings of discontent in regions that already have their "fair share" of wind farms. Even in a population that consistently vote for renewable energy by margins of 80% or more, and steadfastly opposes nuclear expansion, there may come a time when "enough is enough." It must be added that, on past experience, the German government seems unlikely to abandon its plans; but in the case of wind turbines, they will probably move some of them out to sea.

The current boom in offshore wind, especially in the North Sea, undoubtedly softens the problem of visual intrusion. Offshore turbines are typically at least 10 km away from human habitation and, by definition, down at sea level (Figure 8.14). The most powerful HAWTs with rotor diameters larger than the *London Eye*, the famous Ferris wheel that offers tourists such spectacular views over the capital city, are generally only acceptable offshore. This is not to say that their visual presence is negligible; indeed, those who love gazing out to sea at distant, uninterrupted horizons may still have cause for regret. But as we have already said, energy always comes at a price over and above its monetary cost.

We have called this section *Landscape and amenity* because large wind turbines affect people in more ways than visibility. Two other concerns that are often raised are *turbine noise* and *flicker*.

Figure 8.14 Distant horizons (*Source*: With permission of Orsted).

The noise problem tends to be exaggerated. The latest HAWTs produce very little mechanical noise from gearboxes and generators, and in most cases, the aerodynamic noise associated with rotation is unnoticed by people unless they live within a few hundred meters. The swishing of turbine blades is often masked by similar noise produced when the wind—especially strong or turbulent wind—meets trees and buildings. Nevertheless, there are two significant sources of aerodynamic noise[2] that can cause irritation or sleep disturbance:

Wind shadow. As each HAWT blade passes the tower, it enters a disturbed flow that produces an "impulsive" noise at a blade-passing frequency (see Section 3.3.3).

Airfoil noise. Complex interactions between airflow and airfoil sections, especially near blade tips and trailing edges, generate noise over a wide frequency range. In general, its severity increases with the tip speed ratio (see Section 3.2.3).

We have already mentioned the problem of flicker in Section 5.2.2. It takes two forms: supply voltage variations caused by local turbines in relatively weak grids produce flickering of lights in homes and commercial premises and sunlight flicker caused by blades interrupts the sun's rays as they sweep the sky. The first is mainly a nighttime concern, the second a daytime one—but only in sunny weather! Both have strictly local impacts, and although we should not dismiss them, they normally affect very few people.

Environmental impacts of wind farms on landscapes and amenities undoubtedly exist and cause annoyance to some. This is true of all forms of electricity generation and distribution. Developers are well aware of the problems, and in most countries, there are stringent hurdles to overcome before a wind farm is given permission by the planning authorities. The present surge in offshore wind in densely populated countries such as the UK, Denmark, Germany, and the Netherlands promises to reduce opposition that might otherwise be directed at new developments onshore.

8.8 Birds and bats

There have been many reports over the years of birds being killed by wind turbines, and this remains a significant environmental concern as wind energy continues its international progress. We will give the first word on this issue to the *Royal Society for the Protection of Birds (RSPB)*,[5] whose more than one million UK members and 200 nature reserves covering 130,000 hectares make it one of the largest and most influential environmental organizations in Europe. In 2009, the RSPB's head of Climate Change Policy noted that:

The need for renewable energy could not be more urgent. Left unchecked, climate change threatens many species with extinction. Yet that sense of urgency is not translating into action on the ground to harness the abundant wind energy around us … if we get it right, the UK can produce huge amounts of clean energy without time-consuming conflicts and harm to our wildlife. Get it wrong and people may reject wind power. That would be disastrous.

Such brave and unambiguous words encapsulate the attitude of most environmental organizations toward renewable energy in general and wind energy in particular. Translating a global view into local action and believing that birds are far more threatened in the long term by climate change than by regrettable, but occasional, encounters with wind turbines, the RSPB comes out clearly in favor of wind farms provided they are sited away from sensitive areas frequented by rare or endangered species.

The modern history of wind turbines and bird strikes goes back to the Californian "wind rush" of the 1980s, when protected species including golden eagles were being killed by turbines and high-voltage transmission lines in the Altamont Pass.[2] Subsequently, there have been reports from around the world of occasional, but sometimes serious, strikes as unsuspecting birds lose battles with rotating blades and power lines, and there are additional concerns about habitat. We now have a clear idea of the main risks:

Migration routes and corridors. Bird populations are not evenly distributed. Flight paths and patterns are often localized and concentrated, especially in flocks migrating, foraging, roosting, or flying in restricted geographical areas such as mountain passes.

Electrocution. There may be significant extra risks from electrocution by overhead power distribution lines.

Disturbance of habitat. The presence of wind turbines can disturb or displace breeding and foraging birds. Rather than engage with wind turbines, birds may move away from them, effectively losing preferred habitats. The installation of wind turbines and associated infrastructure including roads may cause significant habitat damage.

Considerable evidence of mortality has been gathered over the years. In general, bird deaths caused by wind turbines and farms are a minute proportion of the deaths due to all human activities. Expressed in terms of fatalities per unit of electricity generated, wind energy fares far better than fossil-fuel generation.[2] And we may feel that, compared with the millions of songbirds shot every year for "sport" as they migrate through southern Europe, and the tens of millions of birds killed annually by domestic cats in the UK, wind turbines have rather little to answer for!

However, as with the visual intrusion of turbines, we must not become complacent—especially where the bird species under threat are scarce or endangered. The RSPB's admirably balanced policy toward wind energy leads it to oppose wind farms in some locations, and we may use one of its campaigns as an example.

The Scottish branch of the RSPB has objected to less than 10% of all the wind farm proposals it has monitored in recent years, proving that it does not engage in opposition for opposition's sake. But Scotland has vast areas of wild country, and much of it is a paradise for rare birds. Several applications have been made to build wind farms on the Isle of Lewis, the most northerly of the Western Isles. To put this in context, Figure 8.15 is a map of Scotland with the most important areas for bird protection and conservation marked in green. Based on RSPB data which were redrawn with lower resolution to show just areas of high sensitivity, it is immediately apparent that the West Highlands and Western Isles are especially significant, and that the Isle of Lewis ranks among the most sensitive of all. This may surprise people who have visited Lewis and know its bleak windswept landscapes—not obviously a bird sanctuary, at least not for delicate species. But in fact, there are major concerns for a variety of species including golden eagles, red-throated and black-throated divers, and merlins that depend on the Lewis peatlands, an area designated for special protection.

In view of the RSPB's 2009 statement supporting wind energy, quoted above, it may seem surprising that it had previously opposed a major wind farm application for Lewis with considerable vehemence, on the grounds of scale (181 large HAWTs), overhead cables and pylons,

Figure 8.15 Areas of high sensitivity for rare and endangered wild birds in Scotland (*Source*: Adapted from RSPB).

eight electrical substations, and associated roads. Its opposition focused on the potential destruction of high peatland, known as blanket bog, which is especially important for rare birds and is found in only a few areas in the world. After much campaigning by environmental interests, the Scottish government turned down the proposal in 2008, and the RSPB declared it "an extremely commendable decision that is absolutely right for Scotland." In fact, the Isle of Lewis is now getting its wind farms, but the developers are showing far more sensitivity toward the problem of scale, the needs of wildlife and conservation, and the importance of public opinion. As we noted earlier, conventional economics tends to neglect everything except money, but the public is increasingly aware of environmental costs. Those of us who support the growth of wind energy for *global* environmental reasons must accept that *locally* the boot is sometimes on the other foot.

Wind turbines can be a threat to bats as well as birds. Back in 2005, the *Washington Post* reported that a researcher at the University of Maryland had found hundreds of bat carcasses at a wind farm on Backbone Mountain, reinforcing concerns as the US wind industry expanded away from its traditional areas in the West and Great Plains into more eastern states. Bat fatalities are worrying because many species are in decline worldwide, and although bats lack the mass appeal of birds, they are considered remarkable by biologists and scientists for the sonar echolocation they use to hunt night-flying insects, including many agricultural pests. Research in North America[6] has increased understanding of the interactions between bats and wind turbines and suggested ways of reducing fatalities. Key points include the following:

- Bats on well-defined migration routes are especially vulnerable.
- There is strong evidence that deaths are due not only to collisions with turbines but also to critical lung injuries caused by sudden changes in air pressure in the vicinity of turbine blades.
- Bats are mainly active in calm conditions. Fatalities may be greatly reduced at times when bats are on the wing by increasing the cut-in wind speed of turbines, preventing rotation during periods of low wind speed. The cost in terms of annual energy production is very modest—typically below 1%.

European bats are also vulnerable to wind turbines, and significant fatalities have been reported in Germany, France, and Portugal. The first part of a comprehensive study by the Bat Conservation Trust in association with the University of Bristol, aimed at understanding and quantifying bat deaths caused by onshore wind farms in the UK, was published in 2009.[7]

However, in our days, artificial intelligence (AI) systems are developed and used for alarming birds so that they avoid collisions on wind turbines . For example, the IdentiFlight bird detection system blends artificial intelligence with the high-precision optical technology to detect eagles and other protected avian species. In an operating wind farm, IdentiFlight contributes to eagle conservation by helping protect them from collisions with rotating wind turbine blades. Automatic detection and species determination occur within seconds for birds flying within a one kilometer hemisphere around an IdentiFlight tower. The IdentiFlight system has completed real-world testing and validation in pilot programs at wind farms with elevated eagle activity and is now commercially deployed at projects around the world. For example. The "Top of theWorld" wind farm in Wyoming, USA, employs 48 IdentFlight towers covering the 110 wind turbines, 17 000 acres site of the farm. (Source: https://www.identiflight.com/installations-2)

8.9 Farming

Finally, we should say a few words about farming and the impact that wind turbines have on farm animals. The antiwind lobby sometimes claims that farmland is "used up" by wind turbines, implying that turbines make a site unsuitable for agriculture. This, of course, is

Figure 8.16 Undisturbed pastures (*Source*: With permission of French Wind Energy Association).

nonsense. The actual amount of land required by turbine foundations and associated infrastructure including roads is a tiny fraction—typically under 1%—of the total site area of a wind farm, and crops may be grown right up to the turbine foundations. Indeed, from the farmer's point of view, wind turbines are often an economic blessing, offering a valuable extra source of income. The situation is almost as clear with farm animals including sheep and cattle, which generally ignore wind turbines and often graze right up to the towers (Figure 8.16). There have been occasional reports that sheep are disturbed by turbine noise, but other studies claim that they enjoy the shade provided by turbine towers! All in all, today's clear consensus is that wind farms are compatible with farming in its many forms.

8.10   Seabirds, fish, and marine conservation

Offshore wind farms seem to present few environmental risks. Although they consume more materials and energy per rated megawatt during manufacture and installation than onshore farms, they offer societal benefits including reduced visual intrusion and negligible nuisance from turbine noise. Public attitudes are generally positive.

In 2006, the Danish Energy Authority published a report on the Horns Rev 1 and Nysted Offshore Wind Farms (see Section 4.4.1) which found no evidence of significant negative impacts on birds, fish stocks, marine mammals, or undersea ecology. In 2009, the UK government's Department of Energy and Climate Change published a comprehensive environmental study suggesting that up to 7000 large HAWTs could be installed off the nation's coastlines without adverse effects on the marine environment. And in 2011 the Scottish branch of the RSPB, which campaigns strongly against certain onshore wind farm proposals, welcomed a strategic approach by the Scottish government to environmental planning of offshore wind capacity.[5] These are just a few of the positive indications that offshore wind is generally seen as having minor, and manageable, negative impacts on the natural world.

Within this generally optimistic scenario, how do the risks to seabirds and birds on migration compare with those posed by onshore wind farms? Fortunately, the available evidence is reassuring:

- Birds generally show much more successful "avoidance responses" to offshore wind farms, with reduced risk of collision.
- Although there is some displacement of sea birds from former feeding areas, the overall effect on bird populations is negligible.

It seems that, compared to the dangers to wild birds caused by many human activities and industries, offshore wind is benign. However, the situation varies from site to site, and new wind farm applicants are invariably required to submit *environmental impact assessments (EIAs)* covering the threat to birds. For example, during the initial planning of the 1-GW London Array (see Chapter 4, Section 4.4.2), the EIA identified a previously unknown and internationally important population of rare red-throated divers wintering on part of the site. The RSPB objected to the proposal, but following discussions it was agreed to go ahead with a "stage 1" comprising 175 turbines (630 MW) installed in the least sensitive part of the site, requiring the developer to monitor its effects on the birds over a 3-year-period before "stage 2" (370 MW) could be sanctioned. As noted by the RSPB:

> The case shows how early and continuing consultation and negotiation by a developer, both willing to listen to concerns and act on them, can lead to a positive outcome for birds and renewable energy.

Apart from birds, the most obvious environmental issues raised by offshore wind energy relate to marine conservation and fishing.[8] Early indications are positive and suggest that offshore wind farms may actually benefit the marine environment (Figure 8.17). For example, the gravel or boulders that are often used to prevent scouring of the seabed around monopile foundations may act as valuable "artificial reefs," encouraging colonization by a wide variety of marine organisms and increasing biodiversity. Another possibility is to conserve fish stocks by restricting, or even prohibiting, fishing within wind farms. At a time when individual farms with areas in excess of 100 km^2 are being planned and built, this is clearly an important option. In the UK, active discussions take place between the government, the wind energy industry, marine conservationists, and fishing interests to decide whether new offshore wind farms can be colocated with marine conservation zones.

As so often, the attitudes of fishermen tend to reflect the conflict between earning a living now and preserving fish stocks for the future. They may give a general welcome to the prospect of wind farms protecting the marine environment and helping conserve fish stocks, but if prevented from entering traditional fishing grounds they tend to object that their livelihood is under threat. As far as wind farm operators are concerned, it is clearly important to prevent large vessels from going too close to turbine towers, especially in rough weather, and if fishing is allowed at all, it is unlikely to include trawling with the attendant risk of snagging cables on, or just below, the seabed. Operators are also concerned about legal and management aspects: if fishing is restricted or prohibited within wind farms, who

Figure 8.17 Offshore wind: high hopes for environmental benefits (*Source*: With permission of Orsted).

will monitor it and take responsibility for enforcement? These and other questions are very much up for discussion as the offshore wind industry gathers pace, and the next decade will surely be crucial for developing satisfactory agreements between the interested parties.

Self-assessment questions

Q8.1   The EPBT of wind turbines in upstate New York, where the average wind speeds are 10 mph, is 1.2 years. What would be the EPBT of the same wind turbine in a northern Texas site where the average wind speeds are 20 mph?

Q8.2   What is the EROI of the same turbine in the two regions?

Q8.3   How does the land use of wind farms compare with that of coal life cycles and photovoltaic life cycles in the USA?

Q8.4   What critical elements are used in wind turbine manufacturing?

Q8.5   How can wind turbines affect avian lives and how such effects can be avoided?

Q8.6   What base metals are used in wind turbine manufacturing?

Q8.7   What are the risks associated with offshore wind turbines?

Q8.8   What are the main cost and performance differences between onshore and offshore wind turbines?

Q8.9 What types of costs are referred to as "external costs" in electricity life cycles and why?

Q8.10 What parts of wind turbines can be recycled and how?

Problems

8.1 Estimate the LCOE of a 10-MW wind turbine from the following data:

Capital Cost: $1 million/MW; O&M cost: $3/MWh; life expectancy: 25 years; Interest rate: 5%

8.2 Determine the raw material requirements for a wind farm comprising twenty 2-MW wind turbines based on the material breakdown in the table below:

Condensed Bill of Materials for Wind Turbines Used in Analysis

OEM							
Turbine make	Micon	Nordex	Micon	Vestas	Vestas	Vestas	Vestas
Turbine model	NM52	N-62	NM72	V82 1.65	V90 2.0	V100 2.0	V110 2.0
Nameplate capacity	0.9 MW	1.3 MW	1.5 MW	1.65 MW	2.0 MW	2.0 MW	2.0 MW
Hub height	60.7 m*	69 m	80 m	78 m	80 m	80 m	80 m
Rotor diameter	52.2 m	62 m	72 m	82 m	90 m	100 m	110 m
Mass (kg per kW)							
Steel	111.2	104.5	110.1	96.3	82.2	83.9	92.2
Fiberglass/resin/plastic	18.8	23.8	20.9	18.2	16	14.1	14.2
Iron/cast iron	7.2	17.3	8.7	17.8	20.5	13.3	13.3
Copper	1.6	1.5	1.2	1.8	0.9	0.6	0.7
Aluminum	N/A	N/A	N/A	1.9	2.1	1.7	1.9
Total	139.9	148.2	141.7	138.9	124	115	124
% of Total Turbine Mass							
Steel	79%	71%	78%	69%	66%	73%	74%
Fiberglass/resin/plastic	13%	16%	15%	13%	13%	12%	11%
Iron/cast iron	5%	12%	6%	13%	17%	12%	11%
Copper	1%	1%	1%	1%	1%	1%	1%
Aluminum	N/A	N/A	N/A	1%	2%	1%	2%
Total	99.20%	99.40%	99.40%	97.80%	98.00%	98.70%	98.50%

Sources: Liberman (2003); Vestas (2006); Garrett and Ronde (2011); Razdan and Garrett (2015a, 2015b).

8.3 List the external costs of wind power and discuss how these can be estimated.

8.4 Compare the EPBT and EROI of land-based and offshore wind turbines assuming the same turbine type and size. Include transportation in your evaluation.

8.5 Compare the last stage, that is, decommissioning/recycling, of the wind turbine life cycle with the last stages of conventional fossil-fuel and nuclear life cycles.

8.6 Summarize the differences in the potential impacts of wind turbine installations on land and in the sea.

8.7 What can be noise effects on land-based and offshore wind turbine installation and operation?

8.8 A homeowner installs a wind turbine with a rotor diameter of 3 m to supplement electricity from the public utility. The cost of the turbine, the associated electronics, and the energy storage system (batteries) is $15,000. If the turbine has an efficiency of 40% and the energy is utilized and/or stored at an efficiency of nearly 100%, what is the payback period for the investment? Assume that maintenance costs are minimal, electricity from the public utility costs $0.10 per kWh, and the wind velocity is constant at 10 m/s.

Answers to questions

Q8.1 It will be proportionally lower, thus 0.5 years.

Q8.2 For a life expectancy of 20 years, EROI would be X and Y respectively.

Q8.3 Coal life cycles use more land than either wind or PV because of coal surface mining. Wind turbines use more land than PV because of the required spacing between turbines, unless this space is used for other purposes.

Q8.4 REEs used in permanent magnets, such as neodymium, praseodymium, dysprosium, and terbium.

Q8.5 Moving blades can kill birds; this risk can be minimized by avoiding installations in bird mitigation passages.

Q8.6 Mostly carbon fiber, steel, copper, and aluminum.

Q8.7 They are more difficult to repair under bad weather conditions.

Q8.8 Offshore wind turbines cost more to install, but they can produce more energy as winds are stronger and more persistent offshore than on land.

Q8.9 Health, environmental, and social costs are not accounted for in the price of electricity but are paid by the general public.

Q8.10 Actually, all parts of the wind turbine parts can be recycled.

References

1. M. Bolinger *et al.* Levelized cost-based learning analysis of utility-scale wind and solar in the United States. *iScience*, 25(6), 104378 (2022).
2. J.F. Manwell *et al.* *Wind Energy Explained: Theory, Design and Application*, 2nd edition. John Wiley & Sons Ltd: Chichester (2009).

3. A. Eberle *et al*. Materials used in U.S. wind energy technologies: quantities and availability for two future scenarios. NREL/TP-6A20-81483. National Renewable Energy Laboratory: Golden, CO (2023). https://www.nrel.gov/docs/fy23osti/81483.pdf.
4. S. Bhattacharya. *Design of Foundations for Offshore Wind Turbines*, Section 3.3, Wiley (2019).
5. Royal Society for the Protection of Birds (RSPB). www.rspb.org.uk (accessed May 4, 2024).
6. Bats and Wind Energy Cooperative (BWEC). www.batsandwind.org.
7. The Bat Conservation Trust. www.bats.org.uk (accessed on May 2, 2014).
8. J.C. Wilson. Offshore wind farms: their impacts, and potential habitat gains as artificial reefs, in particular for fish, University of Hull (2007). https://tethys.pnnl.gov/sites/default/files/publications/Their_Impacts_and_Potential_Habitat_Gains_as_Artificial_Reefs.pdf.

Index

Note: *Italicized* and **bold** page numbers refer to figures and tables, respectively

Onshore and Offshore Wind Energy: Evolution, Grid Integration, and Impact, Second Edition.
Vasilis Fthenakis, Subhamoy Bhattacharya, and Paul A. Lynn.
© 2025 John Wiley & Sons Ltd. Published 2025 by John Wiley & Sons Ltd.
Companion website: www.wiley.com/go/fthenakis/windenergy2e